AF388909

Welding, Joining and Coating of Metallic Materials

Welding, Joining and Coating of Metallic Materials

Editors

Michael Zinigrad
Konstantin Borodianskiy

MDPI • Basel • Beijing • Wuhan • Barcelona • Belgrade • Manchester • Tokyo • Cluj • Tianjin

Editors
Michael Zinigrad
Ariel University
Israel

Konstantin Borodianskiy
Ariel University
Israel

Editorial Office
MDPI
St. Alban-Anlage 66
4052 Basel, Switzerland

This is a reprint of articles from the Special Issue published online in the open access journal *Materials* (ISSN 1996-1944) (available at: https://www.mdpi.com/journal/materials/special_issues/Welding_Metallic_Materials).

For citation purposes, cite each article independently as indicated on the article page online and as indicated below:

LastName, A.A.; LastName, B.B.; LastName, C.C. Article Title. *Journal Name* **Year**, *Article Number*, Page Range.

ISBN 978-3-03936-726-9 (Hbk)
ISBN 978-3-03936-727-6 (PDF)

Contents

About the Editors

Michael Zinigrad is the Rector and the Head of Materials Research Center at the Ariel University (AU), Israel. Prof. Zinigrad earned his M.Sc. in 1968 in Metallurgical Engineering at the Dnepropetrovsk Metallurgical Institute, his Ph.D. in 1972 in the metallurgy of non-ferrous, noble and rare metals at the former USSR Academy of Science (currently Russian Academy of Sciences), Institute of Electro-chemistry in Sverdlovsk/Yekaterinburg, and his D.Sc. in 1982 in physical chemistry at the former USSR Academy of Science, Institute of Metallurgy in Sverdlovsk/Yekaterinburg. After his repatriation to Israel, he took part in establishing the Faculty of Natural Sciences at AU and served as its first dean for 13 years. He also founded the Materials Research Center and currently serves as its head. Prof. Zinigrad has over 40 years of experience in materials science and engineering, and he conducts theoretical and experimental investigations into high-temperature physico-chemical processes. The results of his research are being applied to the automotive, aircraft and metallurgy industries, among others. Prof. Zinigrad has published over 200 scientific publications (including monographs, scientific papers and patents). He has supervised 19 Ph.D. students, 38 M.Sc. students and five post-docs.

Konstantin Borodianskiy is a senior lecturer at the Ariel University (AU), Israel. Dr. Borodianskiy earned his PhD in 2011 in Chemistry of the Materials at the Bar-Ilan University, Israel. After the completion of his postdoctoral studies at the University of Windsor, Canada he established the Metallurgy and Applied Nanoscience research lab at AU. His scientific interests focus on the metallurgy of non-ferrous alloys, the improvement of metallic properties by nanomaterials and the development of advanced coatings and surfaces. He published 18 peer-reviewed papers and two chapters in books. Dr. Borodianskiy is a member of several international conference organizing committees in the field of material sciences, a guest editor of several peer-reviewed journals and an active reviewer of some high-quality scientific journals. He currently (June 2020) supervises one Ph.D. and five M.Sc. students.

Preface to "Welding, Joining and Coating of Metallic Materials"

This book is a collection of several reviews and original research papers that cover recent developments in the field of welding, joining and coating of metallic materials. These processes are highly applicable to industry, due to the materials having desirable properties and performance. Therefore, welding, joining and coating are also being studied in detail by scientists worldwide.

The following materials are the focus of this book: titanium, aluminum and magnesium alloys, as well as various types of steel, intermetallics and shape memory alloys. Particular emphasis is placed on the microstructural investigations during the welding, brazing and coating processes. However, in order to solve complex issues, works on modeling and simulation are also presented in the book.

Different metallurgy-related processes are covered, such as spot welding using metallic interlayers, the welding of dissimilar metals, the study of joints applicable as biomaterials. The advances in coating technologies covered in the book are related to the micro arc oxidation (MAO) method, the plasma vapor deposition (PVD) approach and the chemical vapor deposition (CVD) technique.

We would like to take this opportunity to thank all of the authors who have contributed to this book.

Michael Zinigrad, Konstantin Borodianskiy
Editors

 *materials*

Editorial

Welding, Joining, and Coating of Metallic Materials

Michael Zinigrad and Konstantin Borodianskiy *

Department of Chemical Engineering, Ariel University, Ariel 40700, Israel; zinigrad@ariel.ac.il
* Correspondence: konstantinb@ariel.ac.il; Tel.: +972-3-9143085

Received: 5 June 2020; Accepted: 8 June 2020; Published: 10 June 2020

Abstract: Welding, joining, and coating of metallic materials are among the most applicable fabrication processes in modern metallurgy. Welding or joining is the manufacture of a metal one-body workpiece from several pieces. Coating is the process of production of metallic substrate with required properties of the surface. A long list of specific techniques is studied during schooling and applied in industry; several include resistant spot, laser or friction welding, micro arc oxidation (MAO), chemical vapor deposition (CVD), and physical vapor deposition (PVD), among others. This Special Issue presents 21 recent developments in the field of welding, joining, and coating of various metallic materials namely, Ti and Mg alloys, different types of steel, intermetallics, and shape memory alloys.

Keywords: welding; solidification; coating; metals; materials properties; build-up

---

Metals and alloys fabrication, known as metallurgy, has been known since ancient times as the art of making tools and devices for practical applications. This is a scientific field of research that focuses on the study of metals' properties and production from the Stone Age through the Bronze Age and the Iron Age to the modern age of today. Modern metallurgy focuses not only on products for daily use but also on the fabrication of novel materials for aerospace, automotive, marine, nuclear, electric, electronic, and other industries. Investigation of the metallic structure is of the highest scientific interest since it has a primary effect on the properties and performance of the developed product. These properties are divided into three classes: (1) mechanical properties as strength, toughness, hardness, and ductility; (2) physical properties as thermal and electrical conductivity; and (3) chemical properties as corrosion resistance.

One of the main tasks of modern metallurgy is a joining process that involves the assembling of two or more pieces into one. Joining may be conducted by welding, brazing or soldering techniques. Recent scientific works in this field focus on understanding the physical processes, structural evolution, and the correlation between the created structure and final properties of the metal or alloy. In order to solve complex tasks, knowledge of the interdisciplinary basics in chemistry, physics, mathematics, and engineering is required. In addition to those, the advanced scientific topic of nanoscience is also involved in recent works in order to create material with the highest properties as possible.

An additional task of modern metallurgy is surface engineering, mostly applied in order to improve corrosion and wear resistance. Recent advances in the research of functional coatings and surface engineering focus on the environmentally friendly techniques of fabrication associated with performance improvement and cost-effectiveness. A wide variety of advanced properties may be achieved using different coating technologies, such as bio-inert surfaces, antireflective or antifriction layers, and corrosion resistive coatings, among others.

The current Special Issue contains 21 scientific works that cover recent developments and investigations related to the welding, joining, and coating of metallic materials. These works cover state-of-the-art issues in welding processes, such as resistance spot welding, laser welding, and friction stir welding. The published works are mostly focused on the microstructural study of metallurgy as well as the properties and performance investigation of the created joints. Furthermore, several works on the modeling and calculation of the joints are also presented in this Special Issue.

Microstructural studies and investigations on the properties of different types of steel using welding processes are reported [1–3]. Zhang et al. investigate the spot welding process of complex parts conducted on NiTi shape memory alloy using a copper interlayer [4]. These joints may be applicable as biomaterials in the medical industry. A complex process of dissimilar welding is presented by Silvayeh et al. [5], who show numerical calculations of the intermetallic layer thickness of aluminum to steel welding. Statistical analysis of dissimilar welding of aluminum to copper and investigation of the joint microstructure and properties are illustrated by Yang and Jiang [6]. In addition to welding technology, research work on the transient liquid phase bonding approach is published in this Special Issue. AlHazaa and coauthors report the successful bonding between Ti-6Al-4V and Mg AZ31 alloys using a zinc interlayer [7].

As mentioned above, coating technology is also covered by this Special Issue. One of the most promising coating methods in recent years is plasma electrolytic oxidation (PEO) or the micro arc oxidation (MAO) approach, as it is also known. A novel approach of MAO treatment is reported by Sobolev and coauthors who demonstrate oxidation of Ti-based alloy in molten salt for possible biomedical applications [8]. Additionally, other authors also illustrate the influence of MAO parameters on the formation of the oxide coating on Ti-6Al-4V alloy [9]. Other coating approaches are presented by Ali et al. who show stainless steel thin film formation on copper substrates by the physical vapor deposition (PVD) method [10]. Meanwhile, Zenkin et al. implement the chemical vapor deposition (CVD) process on diamond–iron interactions [11]. Chang and Amrutwar show improvements on tool steel mechanical properties using the plasma nitriding method [12].

Finally, we can point out the metallic materials of greatest interest in modern welding, joining, and coating technology whose studies are presented in the current Special Issue. These materials are titanium and magnesium alloys, various types of steel, intermetallics, and shape memory alloys. All of them are attractive to modern science and engineering due to their ability to achieve a combination of advanced properties that open new horizons of application.

Funding: This research received no external funding.

Conflicts of Interest: The authors declare no conflicts of interest.

References

1. Chen, Y.; Wang, H.; Cai, H.; Li, J.; Chen, Y. Role of Reversed Austenite Behavior in Determining Microstructure and Toughness of Advanced Medium Mn Steel by Welding Thermal Cycle. *Materials* **2018**, *11*, 2127. [CrossRef] [PubMed]
2. Zhang, X.; Yao, F.; Ren, Z.; Yu, H. Effect of Welding Current on Weld Formation, Microstructure, and Mechanical Properties in Resistance Spot Welding of CR590T/340Y Galvanized Dual Phase Steel. *Materials* **2018**, *11*, 2310. [CrossRef] [PubMed]
3. Cui, L.; Peng, Z.; Yuan, X.; He, D.; Chen, L. EBSD Investigation of the Microtexture of Weld Metal and Base Metal in Laser Welded Al–Li Alloys. *Materials* **2018**, *11*, 2357. [CrossRef] [PubMed]
4. Zhang, W.; Ao, S.; Oliveira, J.; Zeng, Z.; Huang, Y.; Luo, Z. Microstructural Characterization and Mechanical Behavior of NiTi Shape Memory Alloys Ultrasonic Joints Using Cu Interlayer. *Materials* **2018**, *11*, 1830. [CrossRef] [PubMed]
5. Silvayeh, Z.; Götzinger, B.; Karner, W.; Hartmann, M.; Sommitsch, C. Calculation of the Intermetallic Layer Thickness in Cold Metal Transfer Welding of Aluminum to Steel. *Materials* **2018**, *12*, 35. [CrossRef] [PubMed]
6. Yang, C.-W.; Jiang, S.-J. Weibull Statistical Analysis of Strength Fluctuation for Failure Prediction and Structural Durability of Friction Stir Welded Al–Cu Dissimilar Joints Correlated to Metallurgical Bonded Characteristics. *Materials* **2019**, *12*, 205. [CrossRef] [PubMed]
7. AlHazaa, A.; Alhoweml, I.; Ali Shar, M.; Hezam, M.; Abdo, H.-S.; AlBrithen, H. Transient Liquid Phase of Ti-6Al-4V and Mg-AZ31 Alloys Using Zn Coatings. *Materials* **2019**, *12*, 769. [CrossRef] [PubMed]
8. Sobolev, A.; Wolicki, I.; Kossenko, A.; Zinigrad, M.; Borodianskiy, K. Coating Formation on Ti-6Al-4V Alloy by Micro Arc Oxidation in Molten Salt. *Materials* **2018**, *11*, 1611. [CrossRef] [PubMed]

9.   Sobolev, A.; Kossenko, A.; Borodianskiy, K. Study of the Effect of Current Pulse Frequency on Ti-6Al-4V Alloy Coating Formation by Micro Arc Oxidation. *Materials* **2019**, *12*, 3983. [CrossRef] [PubMed]

10. Ali, N.; Teixeira, J.; Addali, A.; Saeed, M.; Al-Zubi, F.; Sedaghat, A.; Bahzad, H. Deposition of Stainless Steel Thin Films: An Electron Beam Physical Vapour Deposition Approach. *Materials* **2019**, *12*, 571. [CrossRef] [PubMed]

11. Zenkin, S.; Gaydaychuk, A.; Okhotnikov, V.; Linnik, S. CVD Diamond Interaction with Fe at Elevated Temperatures. *Materials* **2018**, *11*, 2505. [CrossRef] [PubMed]

12. Chang, Y.-Y.; Amrutwar, S. Effect of Plasma Nitriding Pretreatment on the Mechanical Properties of AlCrSiN-Coated Tool Steels. *Materials* **2019**, *12*, 795. [CrossRef] [PubMed]

Article

Role of Reversed Austenite Behavior in Determining Microstructure and Toughness of Advanced Medium Mn Steel by Welding Thermal Cycle

Yunxia Chen [1], Honghong Wang [2,*], Huan Cai [2], Junhui Li [2] and Yongqing Chen [2]

[1]   School of Materials Science and Engineering, Shanghai Dianji University, Shanghai 201306, China; chenyx@sdju.edu.cn

[2]   The State Key Laboratory of Refractories and Metallurgy, Wuhan University of Science and Technology, Wuhan 430081, China; caihuan7963@163.com (H.C.); 15671628834@163.com (J.L.); chenyqwork@163.com (Y.C.)

*   Correspondence: wanghonghong@wust.edu.cn

Received: 24 September 2018; Accepted: 24 October 2018; Published: 29 October 2018

Abstract: Reversed austenite transformation behavior plays a significant role in determining the microstructure and mechanical properties of heat affected zones of steels, involving the nucleation and growth of reversed austenite. Confocal Laser Scanning Microscope (CLSM) was used in the present work to in situ observe the reversed austenite transformation by simulating welding thermal cycles for advance 5Mn steels. No thermal inertia was found on cooling process after temperature reached the peak temperature of 1320 °C. Therefore, too large grain was not generated in coarse-grained heat-affected zone (CGHAZ). The pre-existing film retained austenite in base metal and acted as additional favorable nucleation sites for reversed austenite during the thermal cycle. A much great nucleation number led to the finer grain in the fine-grained heat-affected zone (FGHAZ). The continuous cooling transformation for CGHAZ and FGHAZ revealed that the martensite was the main transformed product. Martensite transformation temperature (Tm) was higher in FGHAZ than in CGHAZ. Martensite transformation rate was higher in FGHAZ than in CGHAZ, which is due to the different grain size and assumed atom (Mn and C) segregation. Consequently, the softer martensite was measured in CGHAZ than in FGHAZ. Although 10~11% austenite retained in FGHAZ, the possible Transformation Induced Plasticity (TRIP) effect at −60 °C test temperature may lower the impact toughness to some degree. Therefore, the mean absorbed energy of 31, 39 and 42 J in CGHAZ and 56, 45 and 36 J in FGHAZ were exhibited at the same welding heat input. The more stable retained austenite was speculated to improve impact toughness in heat-affected zone (HAZ). For these 5Mn steels, reversed austenite plays a significant role in affecting impact toughness of heat-affected zones more than grain size.

Keywords: reverse transformation; grain growth of reversed austenite; welding thermal cycle; advanced medium Mn steels; impact toughness

---

1. Introduction

Grain size plays a significant role in determining the strength and toughness of materials. During the welding process, austenite tends to grow thermally and the austenite grain size provides the initial condition for the subsequent phase transformation during cooling and then affects the final microstructure and resulting mechanical properties [1,2]. It is reported that increasing the austenite grain size shifts the continuous cooling transformation diagram to longer reaction time and increases the possibility of martensite formation. Martensite formation may lower the toughness [3–5]. Large austenite grain size is of particular concern during welding where the HAZ experiences rapid thermal cycles with high peak temperature which give rise to austenite grain growth, especially in the region adjacent to the fusion zone (coarse-grained heat-affected zone, CGHAZ). Therefore, the austenite growth behavior has aroused much interest in the past several decades [6–8].

Alloying elements are of importance in affecting the austenite grain growth. Microalloying elements, like Nb, V and Ti, suppress the austenite growth as (Nb,V,Ti)(C,N) precipitates pin the austenite grain boundaries before complete dissolution at 1200 °C on heating [9–12]. Some elements, like B, easily segregate at grain boundaries. Thus, they reduce the austenite growth rate and result in finer grain at room temperature in CGHAZ. Ni is an austenite former element. When enriched in retained austenite in the original microstructure, it may enhance the reversed austenite nucleation site and lead to the final finer grain [13,14]. Other main elements, e.g., Mn should play a role in influencing the austenite grain growth due to either austenite former element or segregation to the grain boundaries. However, there is a lack of information about the Mn effect on austenite growth during welding.

Mn has been utilized in advanced cryogenic steels to replace Ni recently. Like Ni, Mn can enlarge the γ phase and prompt reversed austenite. Additionally, Mn enriches the reversed austenite and enhances its stability, thus improving the cryogenic toughness. In this present work, 5% Mn cryogenic steel was experimented. The following issues were investigated based on reversed austenite transformation at a high temperature of the welding thermal cycle:

(1)   The reversed austenite nucleation characterization on heating.
(2)   The reversed austenite growth kinetic.
(3)   The effect of different austenite grain size on continuous cooling transformation temperature, and thus martensite transformation and reversed austenite transformation.
(4)   The role of microstructure in influencing the cryogenic toughness of heat-affected zone.

2. Experimental

The material studied is an advanced medium Mn steel. Its chemical composition is shown in Table 1 and its mechanical properties are shown in Table 2. The microstructure consists of main martensite and 10~15 vol.% retained austenite. The high yield strength of 650 MPa, ultimate strength of 770 MPa, and superior cryogenic toughness of around 200 J at −60 °C were obtained by optimized intercitical annealing. The transformation temperature was measured as Ac1 = 626 °C and Ac3 = 790 °C on a slow heating rate of 3 °C/min by Gleeble 3800 (Dynamic Systems Inc., Poestenkill, NY, USA) [15].

Table 1. Chemical composition of investigated 5Mn steel (wt.%).

C	Si	Mn	S	P	Ni	Cr + Cu + Mo	Al$_t$
0.04~0.06	0.2~0.25	4.9~5.2	≤0.0012	≤0.009	0.27~0.30	0.2~0.5	≤0.023

Table 2. Mechanical properties of investigated 5Mn steel.

R$_{el}$/MPa	R$_m$/MPa	A/%	KV$_2$ (−60 °C)/J
645~650	770~775	25~27	200~205

A He–Ne CLSM (VL2000DX-SVF17SP, Lasertech Yokohama, Chiba, Japan) with infrared image furnace and laser scanning confocal microscopy was employed to in situ observe the reverse transformation of austenite on heating, and martensite transformation on cooling during the simulated welding thermal cycle. The CLSM samples, with approximate 4 mm in diameter and 6 mm in height, were carefully machine polished, and then set into an alumina crucible with 0.5 mm in thickness. The sample chamber was evacuated and then filled with argon to prevent the sample from being oxidized during heating. Using focused infrared light heating mode, the specimens were heated to the peak temperature of 1320 °C at the rate of 5 °C·s^{-1}, and then cooled down to 200 °C at the rate of 5 °C·s^{-1}. The live pictures were taken every 15 s from 240 mm × 240 mm surface areas.

Gleeble 3800 was used to detect the expansion variation during the simulated welding thermal cycle. The sample was 6 mm in diameter and 70 mm in length. The peak temperature was 1320 and 850 °C to simulate the coarse-grained heat affected zone (CGHAZ) and fine-grained heat affected zone (FGHAZ) respectively. The different cooling rates of 1~60 °C·s^{-1} were set to draw the simulated heat-affected zone continuous cooling transformation diagram (SHCCT). The transformation temperature and transformation kinetic were detected based on the thermal expansion curves.

The mean grain size was estimated on the CLSM sample using a circular-intercept method from image analysis of at least 500 grains. Intercept lengths were determined and then converted into nominal grain diameters using standard tables. The electron backscattering diffraction (EBSD) was applied to analyze the misorientation of the microstructure.

The Metallographic specimens of simulated HAZs were polished and etched with 2% nital before conventional light microscopy. Transmission electron microscopy (TEM, JEM-2100UHR STEM/EDS, JEOL Ltd., Tokyo, Japan) studies were carried out on thin foils. Thin foils were prepared by cutting thin wafers from the simulated HAZ samples, and grinding to 0.1 mm in thickness. Three millimeter discs were punched from the wafers, and were electropolished using a solution of 5% perchloric acid/95% acetic acid. Foils were examined by conventional transmission electron microscope operated at 120 kV using standard bright field and dark field imaging techniques.

The volume fraction of reversed austenite was determined using XRD (XPert PRO MPD, PANalytical B. V., Almelo, Holland) with Cu Kα radiation at a scanning speed of 1°·min^{-1} and step size of 0.02°. The specimens of simulated HAZ were first mechanically polished and then electropolished using an electrolyte consisting of 12% perchloric acid and 88% absolute ethyl alcohol at room temperature. The integrated intensities of (200)γ, (220)γ, (311)γ, (110)α, (200)α, (211)α, (220)α diffraction peaks were used to determine the austenite volume fraction.

Hardness was measured on simulated HAZ samples with a load of 200 g, using a MICROMET5101 Vickers hardness tester (Matsuzawa Co., Ltd., Tokyo, Japan). Standard Charpy v-notch (CVN) impact tests were performed on simulated samples with a different cooling rate at −60 °C, using a NI500A imapct tester (NCS Co., Ltd., Beijing, China) on specimens of dimensions 10 × 10 × 55 mm^3.

3. Results and Discussion

3.1. Nucleation and Growth of Reversed Austenite

Figure 1a shows that reversed austenite started to nucleate when the temperature reached around 640 °C. Due to the limitation of magnification, only nucleation at the grain boundaries was observed clearly; the nucleation inside the grain cannot be presented. Theoretically, the pre-existing retained austenite were the preferable sites for austenite nucleation and growth [16]. In this investigated steel, the 5~10% retained austenite played a beneficial role in improving nucleation number. Therefore, after completion of reversed austenite transformation, the grain size was small. As seen in Figure 1b, the grain size was around 3~10 μm at the temperature of 850 °C.

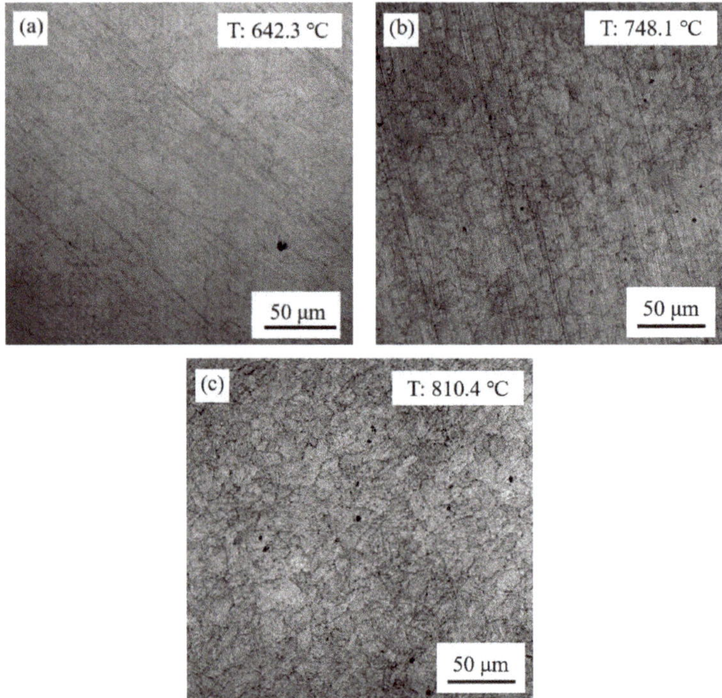

Figure 1. Reversed austenite at different temperatures on heating during thermal cycle: (**a**) 642.3 °C, (**b**) 748.1 °C, (**c**) 810.4 °C.

After completion of reversed austenite transformation on heating, austenite started to grow, as shown in Figure 2. The austenite grain size at different temperatures was measured on the CLSM samples using the circular-intercept method. The austenite growth rate was plotted in Figure 3. It reveals that the grain size growth rate was small below the temperature of 1100 °C on continuous heating. There was a sharp increase in austenite growth from 1100 °C up to 1250 °C. When the peak temperature of 1308 °C was attained, the maximum grain size of ~52 μm was observed and almost no growth was found during the following continuous cooling until it reached austenite decomposition temperature.

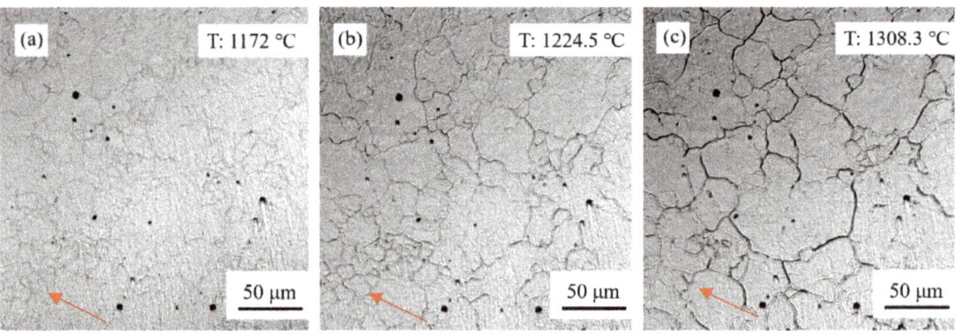

Figure 2. Austenite growth at different temperatures on heating during thermal cycle: (**a**) 1172 °C, (**b**) 1224.5 °C, (**c**) 1308.3 °C.

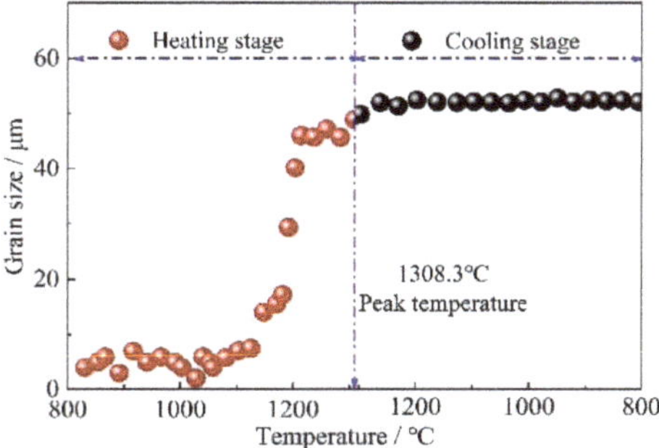

Figure 3. Austenite grain size VS temperature in the heating and cooling stage.

The reversed austenite generally nucleated at grain boundaries, as seen in Figure 1. Especially, the pre-existing retained austenite took a positive role in improving austenite nucleation sites. The filmy retained austenite grew directly in one dimension, which implies a larger nucleation number in the present investigated steel. The large initial nucleation number caused the slow grain growth. On the other hand, the grain grew in the boundary migration way, which is controlled by the relatively short distance of carbon diffusion. Therefore, the grain growth rate was small below 1100 °C.

During the temperature increases, the element diffusion coefficient became lager. The large diffusion of both carbon and Mn changed the method of grain growth from grain boundary migration to grain annexation. The sharp increase of grain growth rate was present during continuous heating from 1100 °C up to 1250 °C.

In low-alloyed steel, the austenite grain continuously grows up during following cooling process above peak temperature [2], which is referred to as "hot-inertia" [17]. Interestingly, in this present steel, there was almost no growth of austenite measured, as shown in Figure 3. It is reported that Mn atoms are likely to segregate toward grain boundaries during continuous cooling [18]. Mn segregation at the grain boundary was assumed to pin the grain boundary migration and hinder the growth of austenite grain. Further study is needed to figure out the Mn segregation behavior and its effect on grain growth.

3.2. Effect of Prior Austenite Grain Size on the Continuous Cooling Transformation Temperature and Transformation Rate

Figure 4 shows the simulated heat-affected zone continuous cooling transformation diagram. Figure 4a is a SHCCT diagram for coarse-grained HAZ with the peak temperature of 1320 °C, while Figure 4b is for fine-grained HAZ with the peak temperature of 850 °C. Both diagrams indicate that martensite was the main phase. However, the martensite's start transformation temperature in FGHAZ was higher than that in CGHAZ by 17~31 °C at the same cooling rate, while the finish transformation temperature in FGHAZ was higher than that in CGHAZ by 31~58 °C, with an exception at the $t_{8/5}$ of 7.5 s. Moreover, the transformation temperature range in FGHAZ was smaller than that in CGHAZ, as shown in Table 3.

Table 3. Continuous cooling transformation temperature at different $t_{8/5}$.

$t_{8/5}$/s	5	7.5	20	30	60	Peak Temperature
Start transformation temperature/°C	371	380	383	388	394	T_p = 1320 °C
	397	403	414	405	423	T_p = 850 °C
Finish transformation temperature/°C	220	235	212	213	200	T_p = 1320 °C
	251	214	256	261	258	T_p = 850 °C
Transformation temperature range/°C	151	145	171	175	194	T_p = 1320 °C
	146	189	158	144	165	T_p = 850 °C

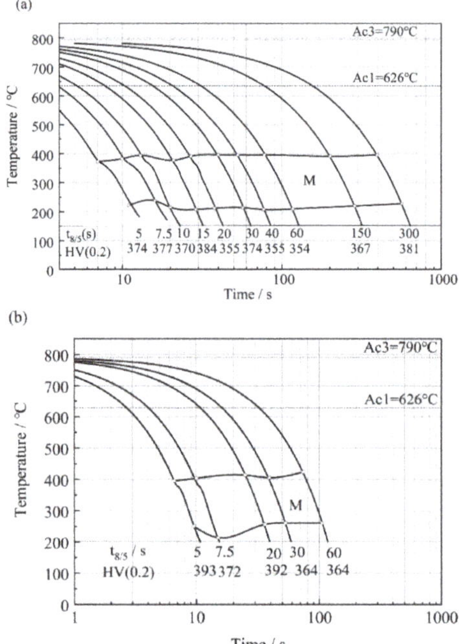

Figure 4. Simulated heat-affected zone continuous cooling transformation diagram for peak temperature of (**a**) 1320 °C; (**b**) 850 °C.

Mn has a definite segregation tendency due to its differential diameter of 5×10^{-12} m with the Fe element. During the welding thermal cycle, Mn may segregate towards gain boundaries by the nonequilibrium segregation mechanism. The higher the peak temperature is, the higher the segregation concentration of Mn at grain boundaries is. For FGHAZ, the small grain size allows Mn and C to diffuse short distance to grain boundary. Furthermore, the great number of grain boundaries can enhance more Mn and C segregates at the grain boundaries. Therefore, the concentration of Mn and C should be lower in FGHAZ than in CGHAZ. Consequentially, the martensite start temperature was higher in FGHAZ than in CGHAZ.

The in situ observation of martensite transformation by SLCM (Figure 5) shows that lath martensite nucleated inside austenite grain and then grew quickly along the length direction. The austenite was divided by the first generated longer lathes. Then the lath widened. In the fine austenite grain, the martensite grew in the limited space, so short and thin martensite lath was found in FGHAZ. On the other hand, long and thick martensite lath was found in CGHAZ.

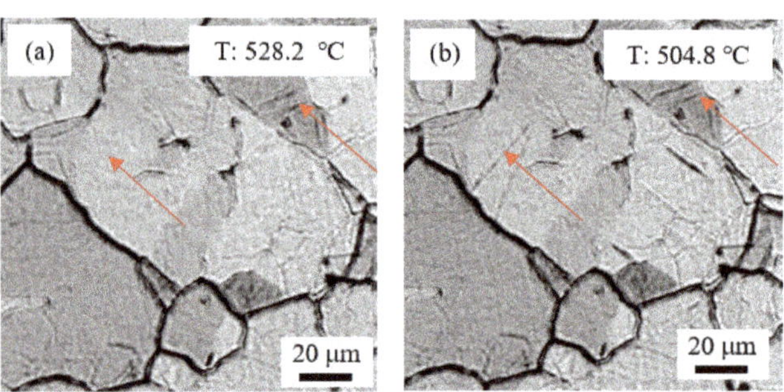

Figure 5. *Cont.*

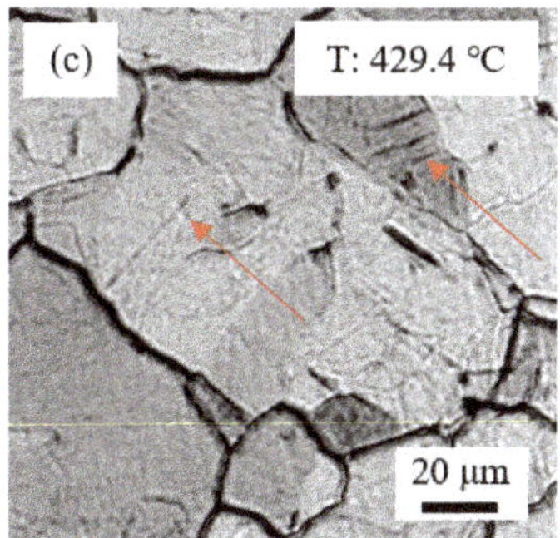

Figure 5. Martensite transformation in the cooling stage in CGHAZ at different temperature of (**a**) 528.2 °C, (**b**) 504.8 °C, (**c**) 429.4 °C.

Figure 6 reveals the martensite transformation rate vs martensite transformed volume fraction. At the peak temperature of 1320 °C, the transformation rate from austenite to martensite was measured to be a maximum of 0.60, while at the peak temperature of 850 °C, the transformation rate from austenite to martensite was measured to be a maximum of 0.98. The transformation rate is closely related to the transformation start temperature and transformation temperature range. The higher start transformation temperature often leads to a relatively high transformation rate. Therefore, because the martensite transformation temperature was higher at the peak temperature of 850 °C than that at 1320 °C, as well as the grain size was smaller at the peak temperature of 850 °C than that at 1320 °C, FGHAZ expressed the relatively higher transformation rate.

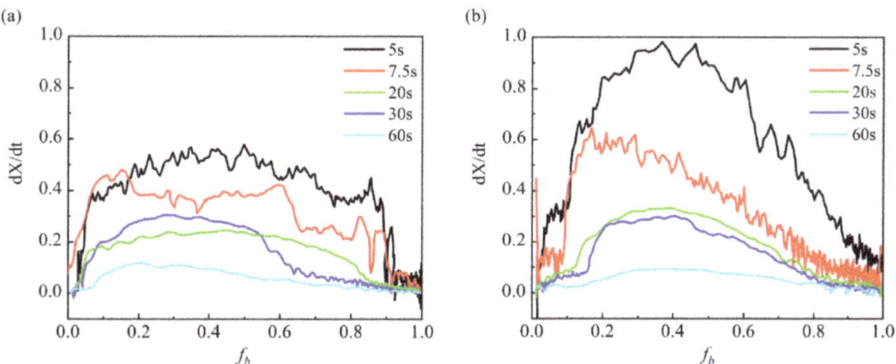

Figure 6. Martensite transformation rate VS martensite transformed volume fraction with different cooling rate at peak temperatures of: (**a**) 1320 °C, (**b**) 850 °C.

3.3. Martensite Transformation and Retained Austenite

Figure 7 shows the crystallography of CGHAZ and FGHAZ by EBSD. It is clear that CGHAZ had the larger grain (Figure 7a) compared with FGHAZ (Figure 7c). The EBSD superimposed figure in Figure 7b–d of band contrast and grain boundaries indicated that the number of boundaries with the misorientation >15° in FGHAZ was greater than in CGHAZ, which means that the effective grain size in FGHAZ was smaller than that in CGHAZ. The comparison is presented in Figure 7e. This agreed well with the in situ observation of martensite transformation.

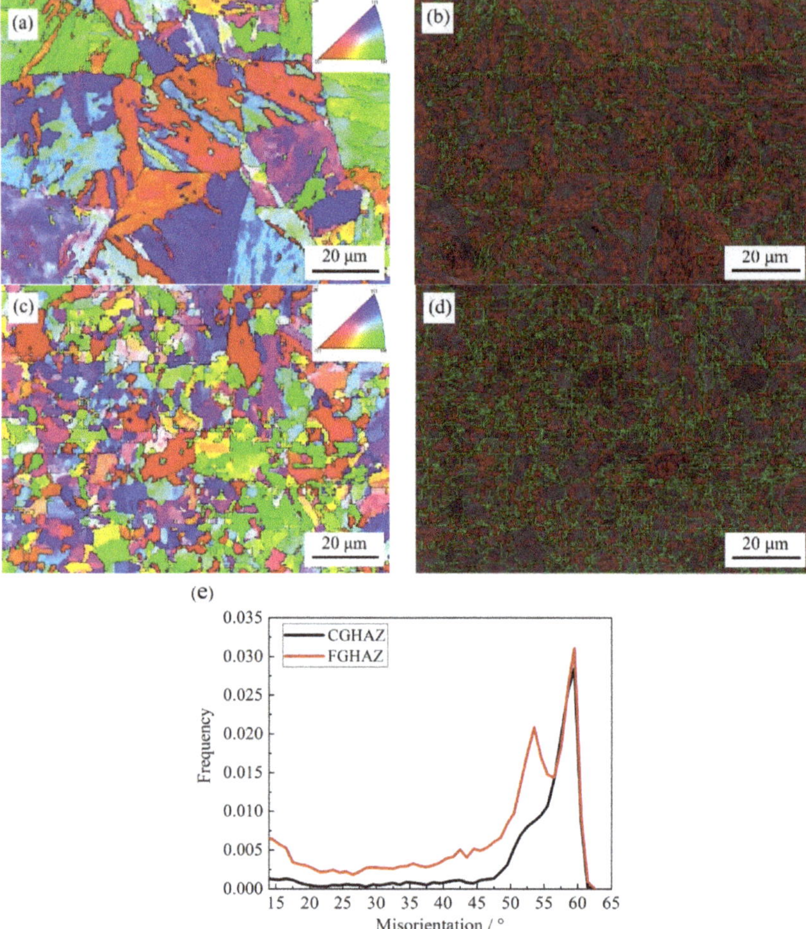

Figure 7. Crystallographic analysis in sample Z direction of simulated (**a**) CGHAZ, (**c**) FGHAZ, EBSD superimposed figure of band contrast and grain boundaries (misorientation < 15° in red line, misorientation > 15° in green line) of simulated (**b**) CGHAZ, (**d**) FGHAZ with $t_{8/5}$ of 25 s. (**e**) Curves of misorientation angle-relative frequency of simulated HAZs.

Apart from martensite transformation, some austenite was retained in FGHAZ because of the fast continuous cooling transformation (shown in Figure 4), as well as the shorter transformation temperature range. The volume fraction of retained austenite in FGHAZ was measured to be 9~11% by XRD, while only 0.5~1.2% retained austenite in CGHAZ (Figure 8a,b).

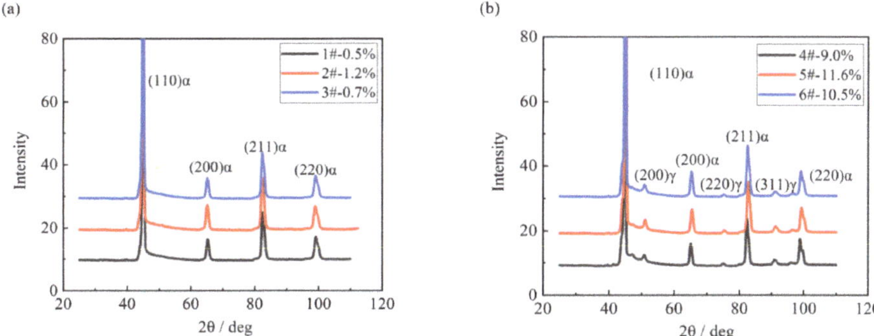

Figure 8. Comparison of volume fraction of retained austenite by XRD in simulated (**a**) CGHAZ and (**b**) FGHAZ.

The retained austenite morphology in FGHAZ by TEM is presented in Figure 9. The retained austenite was found along the martensite lath, as shown in Figure 9a. In the bright field micrograph (Figure 9b), the retained austenite was found to be film-like, and thickness was around ~150 nm. The retained austenite is demonstrated by the field micrograph in Figure 9c and the diffraction pattern in Figure 9d. By using EDS line-scan, the concentration of Mn in retained austenite was measured and there was ~1.8 enrichment factor of Mn found in retained austenite, which is shown in Figure 10.

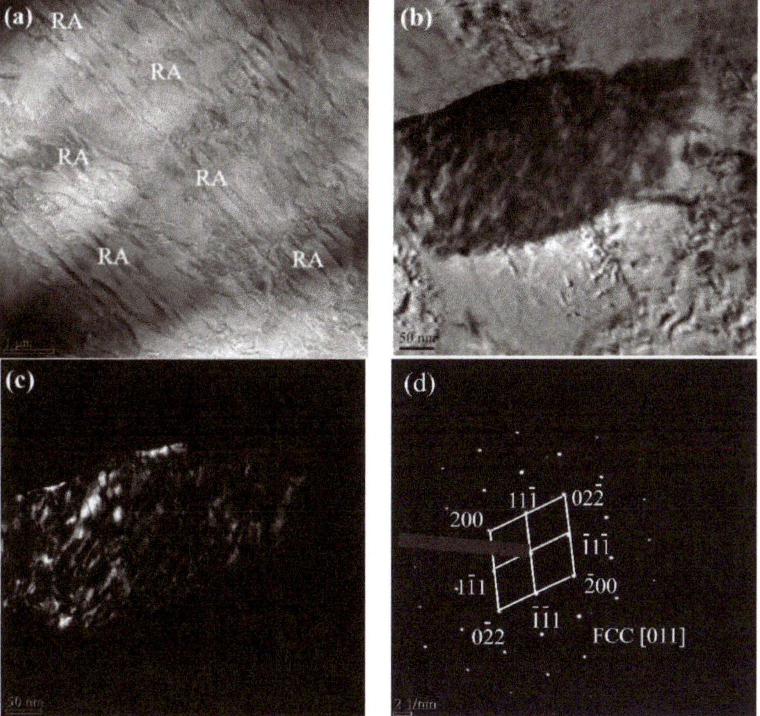

Figure 9. TEM showing (**a**) RA morphology, (**b**) Bright field micrograph and (**c**) Dark field micrograph (**d**) diffraction pattern showing austenite in simulated FGHAZ at t$_{8/5}$ of 10 s.

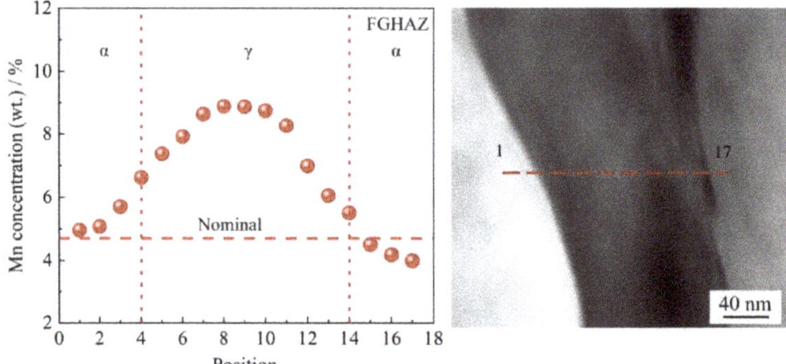

Figure 10. TEM micrograph of retained austenite and EDS line-scan along the red line in retained austenite of simulated FGHAZ at $t_{8/5}$ of 10 s.

The enrichment of Mn of 9% in retained austenite lowers the martensite transformation temperature. It is possible for retained austenite to transform into martensite in an impact toughness test temperature of $-60\,^{\circ}$C due to the lower stability of retained austenite.

3.4. Hardness

The SHCCT diagram in Figure 4 shows that the hardness of FGHAZ was around 363~393 HV (0.2) and around 354~384 HV (0.2) of CGHAZ. Both at the peak temperature of 1320 $^{\circ}$C and 850 $^{\circ}$C, the continuous cooling transformed product was martensite. Figure 4 also reveals that at the same $t_{8/5}$ range of 5~60 s, each average hardness value of FGHAZ was larger than that of CGHAZ by around 10 HV. The harder martensite in FGHAZ was attributed to the higher transformation rate, as well as fine grain.

3.5. Impact Toughness and the Microstructure Effect

The impact toughness of simulated CGHAZ and FGHAZ was present in Figure 11. At $-60\,^{\circ}$C, the mean absorbed energy of simulated CGHAZ was 42, 39 and 31 J with $t_{8/5}$ of 10 s, 20 s and 30 s, while it was 36, 45 and 56 J in the FGHAZ at the identical cooling rate. The impact toughness seemed to be independent of the cooling time $t_{8/5}$ at the different peak temperatures. In addition, the FGHAZ did not show more superior impact toughness than CGHAZ, which differentiates from conventional high strength low alloy (HSLA) steel. Even much finer effective grain and more 10~11% retained austenite, the impact toughness was not much better in FGHAZ than in CGHAZ. The harder martensite was the main factor in impact toughness. Moreover, the retained austenite may transform into martensite at $-60\,^{\circ}$C during the impact test, which is referred to as the transformation-induced plasticity (TRIP) effect [19]. This is due to the relatively lower enrichment factor of 1.8 in retained austenite, which means lower stability of retained austenite. The martensite due to the TRIP effect led to the decrease in impact toughness.

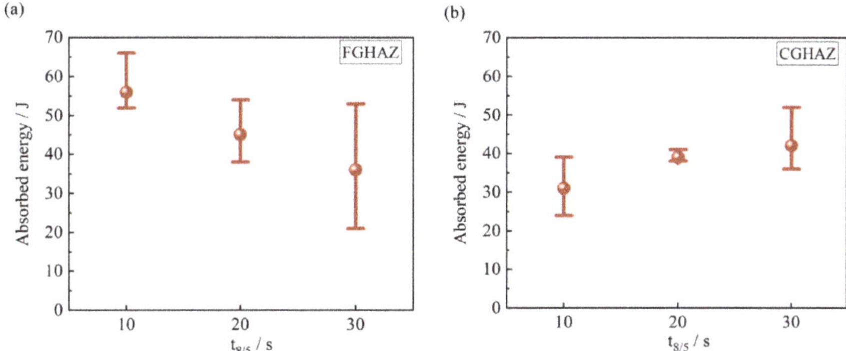

Figure 11. Absorbed energy of simulated (**a**) FGHAZ and (**b**) CGHAZ at the variation of $t_{8/5}$.

4. Conclusions

(1) The pre-existing retained austenite in the investigated 5Mn steel enhanced the reversed austenite nucleation number during heating of the welding thermal cycle and led to the relatively small grain size in simulated CGHAZ and FGHAZ.

(2) Austenite growth dominantly took place in the heating process and there was almost no austenite growth in the cooling process above peak temperature during the simulated welding thermal cycle. Therefore, the relatively small size was found in CGHAZ.

(3) The higher transformation rate was measured to be a maximum of 0.98 in the simulated FGHAZ and a maximum of 0.60 in the simulated CGHAZ. Thus, a 10~11% austenite was retained in FGHAZ.

(4) Compared to CGHAZ with the larger grain size, the impact toughness in fine grain FGHAZ was not significantly improved. This was due to the transformation of relatively harder martensite and the TRIP effect during the impact test in simulated FGHAZ. Therefore, how to retain austenite will be the future work of how to improve impact toughness of HAZ.

Author Contributions: Conceptualization, writing—original draft preparation, writing—review and editing, Y.C. (Yunxia Chen) and H.W.; supervision, investigation and funding acquisition, Y.C. (Yunxia Chen); methodology, resources and project administration, H.W.; software and formal analysis, H.C.; visualization and data curation, H.C., J.L. and Y.C. (Yongqing Chen).

Funding: This research was funded by the National Natural Science Foundation of China, grant number 51809161 and Shanghai Municipal Natural Science Foundation, grant number 18ZR1416000.

Acknowledgments: The authors gratefully acknowledge the administrative and technical support of Shanghai Dianji University and Wuhan University of Science and Technology.

Conflicts of Interest: The authors declare no conflict of interest.

References

1. Guthrie, R.I.L.; Jonas, J.J. Properties and selection: Irons, steels and high performance alloys. In *ASM Handbook*, 10th ed.; ASM International: Materials Park, OH, USA, 1990; Volume 1, pp. 115–116.
2. Militzer, M.; Hawbolt, E.B.; Meadowcroft, T.R.; Giumelli, A. Austenite grain growth kinetics in Al-killed plain carbon steel. *Metall. Mater. Trans. A* **1996**, *27*, 3399–3409. [CrossRef]
3. Easterling, K. *Introduction to the Physical Metallurgy of Welding*; Elsevier Science & Technology: Oxford, UK, 1992; pp. 117–125. ISBN 978-0750603942.
4. Zhang, L.P.; Davis, C.L.; Strangwood, M. Effect of TiN particles and microstructure on fracture toughness in simulated heat affected zones of a structural steel. *Metall. Mater. Trans. A* **1999**, *30A*, 2089–2096. [CrossRef]
5. Eroglu, M.; Aksoy, M. Effect of initial grain size on microstructure and toughness of intercritical heat-affected zone of a low carbon steel. *Mater. Sci. Eng.* **2000**, *286*, 289–297. [CrossRef]

6. Enomoto, M.; Li, S.; Yang, Z.N.; Zhang, C.; Yang, Z.G. Partition and non-partition transition of austenite growth from a ferrite and cementite mixture in hypo- and hypereutectoid Fe-C-Mn alloys. *Calphad* **2018**, *61*, 116–125. [CrossRef]

7. Annan, K.A.; Siyasiya, C.W.; Stumpf, W.E. Austenite grain growth kinetics after isothermal deformation in microalloyed steels with varying Nb concentrations. *ISIJ Int.* **2018**, *58*, 333–339. [CrossRef]

8. Zhang, S.S.; Li, M.Q.; Liu, Y.G.; Luo, J.; Liu, T.Q. The growth behavior of austenite grain in the heating process of 300M steel. *Mater. Sci. Eng. A-Struct.* **2011**, *528*, 4967–4972. [CrossRef]

9. Gladman, T. *The Physical Metallurgy of Microalloyed Steels*; CRC Press: London, UK, 2002; pp. 236–254. ISBN 978-1902653815.

10. Manohar, P.A.; Ferry, M.; Chandra, T. Five decades of the Zener equation. *ISIJ Int.* **1998**, *38*, 913–924. [CrossRef]

11. Bhadeshia, H.K.D.H.; Honerycombe, R. *Steels Microstructure and Properties*, 2nd ed.; Butterworth-Heinemann: Oxford, UK, 2006; p. 176. ISBN 978-0750680844.

12. Gladman, T.; Pickering, F.B. Grain-coarsening of austenite. *Iron Steel Inst. J.* **1967**, *205*, 653–664.

13. Li, Y.M.; Sun, X.J.; Li, Z.D.; Yong, Q.L.; Wang, X.J.; Zhang, K. Stability of reversed austenite in Mn-Ni steel. *Heat Treat. Met.* **2016**, *41*, 111–116.

14. Himeno, M.; Shibata, K.; Fujita, T. Effects of the Widmanstatten-like Reversed Austenite on Mechanical Properties in Fe-Ni Alloys. *ISIJ Int.* **1979**, *65*, 235–244.

15. Li, J.H.; Wang, H.H.; Luo, Q.; Li, L.; Sun, C.; Misra, R.D.K. Correlation between microstructure and impact toughness of weld heat affected zone in 5 wt.% manganese steel. *J. Iron Steel Res. Int.* **2018**, in press.

16. Law, N.C.; Edmonds, D.V. The formaiton of austenite in a low-alloy steel. *Metall. Mater. Trans. A* **1980**, *11*, 33–46. [CrossRef]

17. Zhang, W.Y. *Welding Metallurgy*; China Machine Press: Beijing, China, 2014; p. 183. ISBN 978-7111044123.

18. Majka, T.F.; Matlock, D.K.; Krauss, G. Development of microstructural banding in low-alloy steel with simulated Mn segregation. *Metall. Mater. Trans. A* **2002**, *33*, 1627–1637. [CrossRef]

19. Senuma, T. Physical Metallurgy of Modern High Strength Steel Sheets. *ISIJ Int.* **2001**, *41*, 520–532. [CrossRef]

Article

Effect of Welding Current on Weld Formation, Microstructure, and Mechanical Properties in Resistance Spot Welding of CR590T/340Y Galvanized Dual Phase Steel

Xinge Zhang [1,2], Fubin Yao [3], Zhenan Ren [2,*] and Haiyan Yu [4]

[1] School of Mechanical and Aerospace Engineering, Jilin University, Changchun 130025, China; zhangxinge@jlu.edu.cn
[2] College of Materials Science and Engineering, Jilin University, Changchun 130025, China
[3] Doosan Infracore China Co., Ltd., Yantai 264006, China; yaofbjlu@126.com
[4] Sanyou Automobile Parts Manufacturing Co., Ltd., Changchun 130022, China; yuhy_sy@163.com
*   Correspondence: renzhenan@yeah.net; Tel.: +86-431-8509-4376

Received: 20 September 2018; Accepted: 13 November 2018; Published: 17 November 2018

Abstract: During resistance spot welding, the welding current is the most important process parameter, which determines the welding heat input and then has a great influence on the welding quality. In present study, the CR590T/340YDP galvanized dual phase steel widely used as automobile material was carried out using resistance spot welding. The effect of welding current on the weld formation, microstructure, and mechanical properties was studied in detail. It was found that the quality of weld appearance decreased with the increase of welding current, and there was a Zn island on the weld surface. The microstructure of the whole resistance spot welded joint was inhomogeneity. The nugget zone consisted of coarse lath martensite and a little of ferrite with the columnar crystal morphology, and the microstructure of weld nugget became coarser when the welding current was higher. There was an optimum welding current value and the tensile strength reached the maximum. This investigation will provide the process guidance for automobile body production.

Keywords: Resistance spot welding; galvanized dual phase steel; microstructure; mechanical properties; weld formation

---

1. Introduction

In the automotive field, reducing the body weight is one of the most fundamental ways to save energy and reduce pollution. Therefore, more and more lightweight and high-strength materials have been developed for body manufacturing [1–3]. Galvanized dual phase steel has the advantages of low yield ratio, good formability, high tensile strength, good matching of strength and plasticity, corrosion resistance, and so on, which has been used in automobile body manufacturing [4–6]. The wide application potential of galvanized dual phase steel in the automotive field mainly depends on the welding methods and its welding quality.

In previous literature, several welding methods, such as laser welding [7–10], gas metal arc welding (GMAW) [11–13], and friction stir welding (FSW) [14–16], have been employed to weld the galvanized dual phase steel, and these welding methods have their own advantages and disadvantages, as well as being applicable to different base metal materials and joint shapes. Resistance spot welding has the advantages of high efficiency and low cost, which is an important welding method for manufacturing the automobile sheet structures [17,18]. Many scholars have studied the microstructure, softening zone characteristics, mechanical behavior, and welding spatter defect of resistance spot welding of dual phase steels [19–23]. The CR590T/340YDP galvanized dual phase steel sheet used as an automobile manufacturing material not only has good corrosion resistance, but also high strength, which can effectively reduce the weight of the car body, and some research on resistance spot welding have been carried out [24–26]. Wang et al. [27] investigated the effect of base material chemical compositions on the properties of the resistance spot welding joint of DP590 steel, and the results indicated the tensile strength and toughness of welded joints were affected by the chemical compositions of the base material, especially the carbon content. Namely, for the same grade DP590 steel, the weld formation, microstructure, and mechanical properties of resistance spot welding will be different if the chemical compositions of the base material are not the same. During the resistance spot welding process, it is well known that the welding heat generation can be expressed as $Q = I^2 R t$ (I is the welding current; R is the resistance; and t is the welding time). In general, the welding time and welding current are 10^{-1} s level and kA level, respectively. Therefore, the welding current is considered as the key factor to determine the welding heat input and influence of the welding quality [28]. Wang et al. [29] established a finite element model for resistance spot welding of DP590 steel, and the nugget formation process was investigated. The simulative result for nugget size was obviously bigger than that of the experiment under a large welding current, although they were well fitted under a suitable welding current. Therefore, the experimental investigation on the effect of the welding current on properties in resistance spot welding of DP590 steel has an important significance.

To explore the weldability, and provide the process with guidance for automobile body production, the CR590/340Y galvanized dual phase steel sheets employed for the automobile front longitudinal beam part were carried out by resistance spot welding, and the influence of the welding current on weld formation, microstructure, microhardness, and tensile strength was studied in detail.

2. Materials and Methods

In this study, the base metal is galvanized dual phase steel (Trademark: CR590T/340Y) sheet, which is always used to manufacture the automobile's front longitudinal beam part. The thickness of the base metal is 2 mm, and its .hemical compositions are listed in Table 1. The XRD pattern result indicates that the base metal consisted of ferrite (86.2 vol.%) and martensite (13.8 vol.%), as shown in Figure 1. The galvanized dual phase steel sheets were conducted with double-side hot galvanizing and the weight of the Zn layer is 80 g/m^3. The yield strength and tensile strength of the base metal are 356 MPa and 605 MPa, respectively.

Table 1. Chemical compositions of base metal (wt.%).

Component	C	Si	Mn	S	P	Al	Cr	Nb	Mo	Fe
wt%	0.087	0.008	1.694	0.007	0.001	0.030	0.160	0.012	0.011	Balance

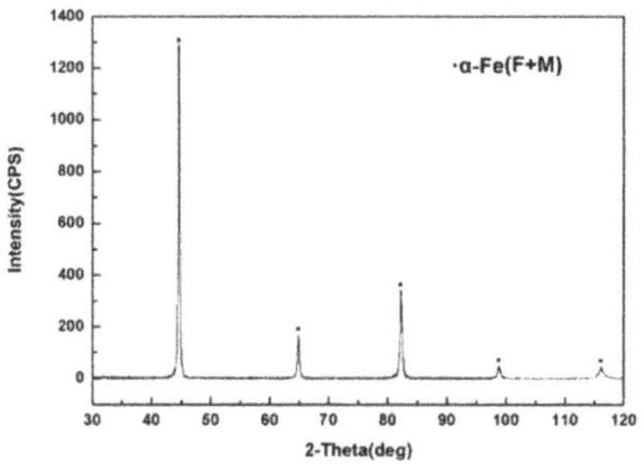

Figure 1. XRD pattern of base metal.

The resistance spot welding robot (Shougang MOTOMAN Robot Co., Ltd., Beijing, China) was employed to weld the overlap joints, and the working face diameter of Cu-Cr-Zr alloy electrodes was 6 mm. The dimensions and configuration of the joint are indicated in Figure 2. The welding time and electrode force were set at 20 cycles and 4.0 kN by preliminary process experiments, and the main experimental parameters are given in Table 2.

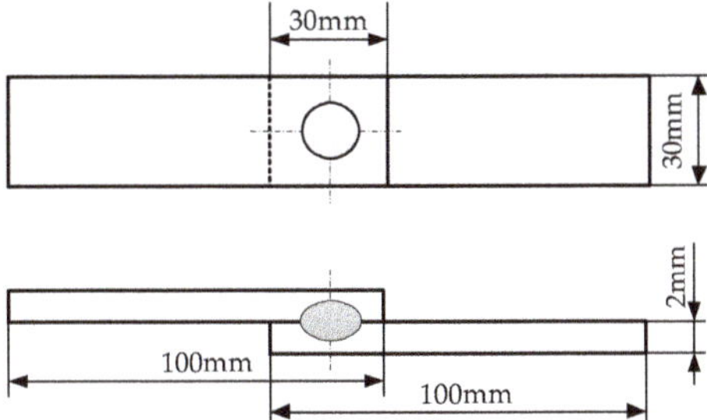

Figure 2. Schematic diagram of dimensions and configuration of resistance spot welding joint.

Table 2. Experimental parameters.

Parameter	Welding Current (kA)	Welding Time (Cycle, 1 Cycle = 0.02 s)	Electrode Force (kN)
Value	8.5~12.0	20	4.0

After welding, the appearance of the resistance spot weld was observed by digital camera (Canon Ixus1000, Cannon, Tokyo, Japan). The resistance spot weld was cut from weld nugget center, and then mounted, polished, and etched for the microstructure observed and analysis. The etching solution was 4 vol.% nitric acid. The microstructure of base metal, heat affected zone, and weld nugget was examined by Evol-18 scanning electron microscopy (SEM) (Carl Zeiss, Jena, Germany) with energy dispersive spectroscope (EDS). The phase structures were identified using the X-ray diffractometer (XRD) (D/MAX-2500PC, Rigaku, Tokyo, Japan) machine with 50kV voltage, 300 mA current, Cu Kα radiation, and 4 °/min scanning rate. The Vickers hardness was measured on the cross section of the welded joint using an MH-3 microhardness test machine (Shanghai Tuming, Shanghai, China), and the test load and load time were 1.961 N and 10 s, respectively. The tensile shear test was performed using CSS-44100 material testing equipment with a 100 kN maximum load (China Mechanical Testing Equipment Co., Ltd., Changchun, China), and the strain speed was 6 mm/min.

3. Results and Discussion

3.1. Weld Formation

3.1.1. Weld Appearance

The resistance spot weld appearance depends on the coupling effect of heat, electric, and load, which will influence the welded joint quality, corrosion resistance, and look. Figure 3 shows the appearances of typical resistance spot welded joints, which were gained with different welding currents. During the resistance spot welding process, while the welding current was 8.5 kA~10.5 kA, there was no welding spatter on the spot weld surface. Until the welding current was 11 kA, the welding spatters were generated, and the welding spatters obviously increased. The welding spatter is caused by the large welding heat input, which results in the faster speed of melting metal than the expansion speed of the plastic ring, and the melted liquid metals fly out of the plastic ring. Because the welding current is the main parameter to determine the heat input ($Q = i^2Rt$, Q is the welding heat input, i is the welding current, and t the is welding time), the welding heat input increases sharply when the welding current is increased, then the high heat input brings about more welding spatters.

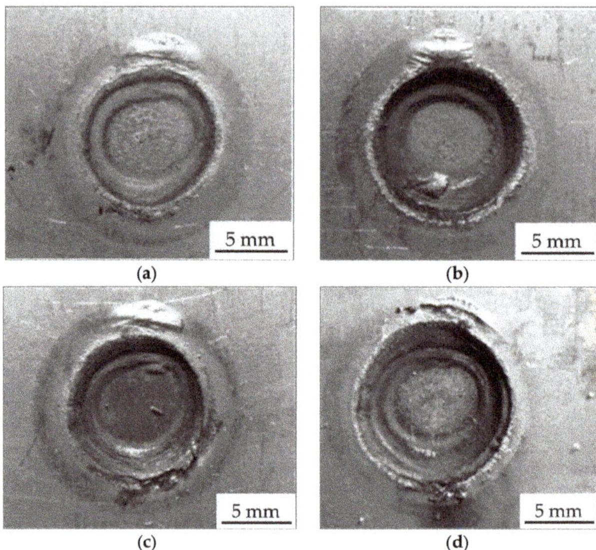

Figure 3. Resistance spot weld appearance gained with different welding currents: (**a**) 8.5 kA; (**b**) 10.0 kA; (**c**) 11.0 kA; (**d**) 12.0 kA.

Figure 4 indicates the three feature zones on the resistance spot weld surface, and it comprises of three circle zones (circle zone I, circle zone II, and circle zone III) from the center to the outside. The circle zone I was the area where the electrodes contacted with the base metal during the resistance spot welding process, and is located in the centre of the spot weld surface. It was produced by various physical factors, such as the heat, electricity, and load, of the electrodes and the binding force of the base metal. Because of the highest temperature in the circle zone I and the low melting point of Zn (692 K), the welding heat caused the Zn layer to melt and be squeezed away by the electrodes, meanwhile the base metal substrate was exposed and oxidized. The circle zone II is located outside the adjacent region of the circle zone I. The profile of the circle zone II depended on the working surface shape of the electrodes and the indentation depth of the spot welded joint. On the circle zone II, the Zn extruded from circle zone I along the edge of the electrodes, and molten Zn on circle zone II was aggregated by the action of gravity and surface tension, then solidified to form the Zn island as shown in Figure 4. The SEM image and element (white spots in Figure 5b–d) map distribution of the Zn island (on the circle zone II in Figure 4 as an example) are shown in Figure 5, and the main components of the Zn island comprised of O (8.57%), Fe (15.44%), and Zn (65.24%). The circle zone III was the heat affected zone of spot welded joints. The base metal in this zone was heated and the Zn layer was oxidized to form ZnO. The Zn island also could generate on the circle zone III.

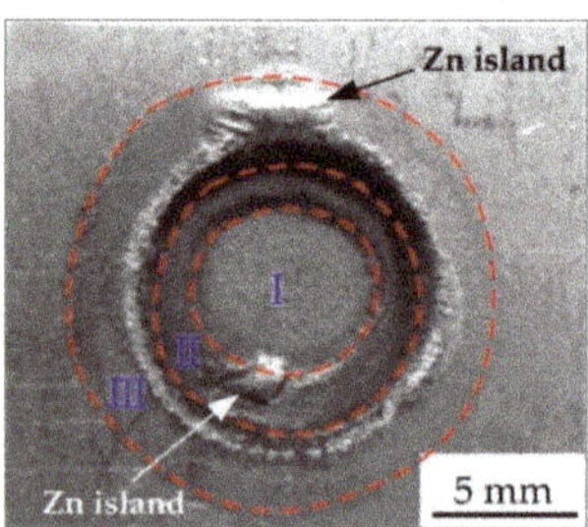

Figure 4. Three feature zones on the resistance spot weld surface.

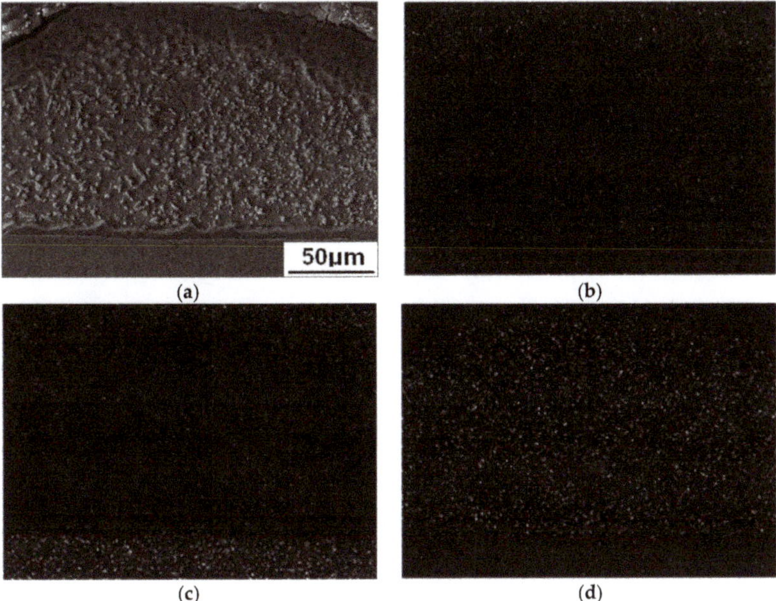

Figure 5. SEM image and element map distribution of Zn island: (**a**) SEM image; (**b**) O Kα 1; (**c**) Fe Kα 1; (**d**) Zn Kα 1.

From Figure 3a, due to the low welding current, the Zn on the surface of circle zone I was partially melted and the color was not much different from that on the base metal; the profile of circle zone II was small with a smooth transition, and the Zn island was generated in the circle zone III. While the welding current was 10.0 kA, the Zn layer on the surface of the circle zone was not seriously damaged, and the inner and outer colors of circle zone II were quite different. The Zn island was formed on the adjacent part of circle zone I, because the Zn layer on the circle zone II was melted and extruded, and then cooled and crystallized on the outside. There was a Zn island on the circle zone III. When the welding current continued to increase to 11.0 kA, the Zn layer on the circle zone I was seriously damaged. There was a Cu-Zn alloy formed by the reaction of the molten Zn layer and copper on the edge of circle zone II, and the electrodes' adhesion occurred. When the welding current reached 12.0 kA, the appearance quality of the spot welded joint decreased obviously and many spatters were generated because of the uneven heat distribution of the electrodes. Figure 6 indicates the relation between the welding current and the diameters of the three feature zones on the weld appearance. When the welding current was 8.5~9.5 kA, the diameters of circle zone I were almost unchanged. With the continued increase of the welding current, the diameters of circle zone I increased to the maximum at first and then decreased gradually. The diameter of the circle zone II changed little with a low welding current, which was the largest with a 10.5 kA current. The diameter of the circle zone III increased with the increase of the current from 8.5 kA to 10.5 kA, but when the welding current was higher than 10.5 kA, the diameter of circle zone III did not change obviously. The results displayed that the welding current had an important influence on the weld appearance, and a low welding current was used on the basis of meeting the strength requirements of the welded joint.

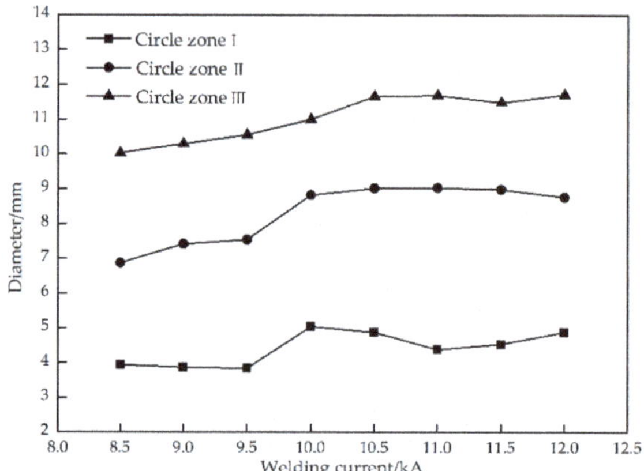

Figure 6. Effect of the welding current on sizes of three feature zones on the weld appearance.

3.1.2. Main Dimensions of Welded Joint Cross-Section

The main dimensions of the resistance spot welded joint cross-section include the indentation (usually expressed by indentation rate, D/δ) and weld nugget width at the overlap surface (W), as shown in Figure 7. The indentation influences the weld appearance smoothness, reduces the welded joint cross-section size, and causes stress concentration, which results in reducing the strength of the welded joint. The tensile strength of the resistance spot welded joint is mainly controlled by the W.

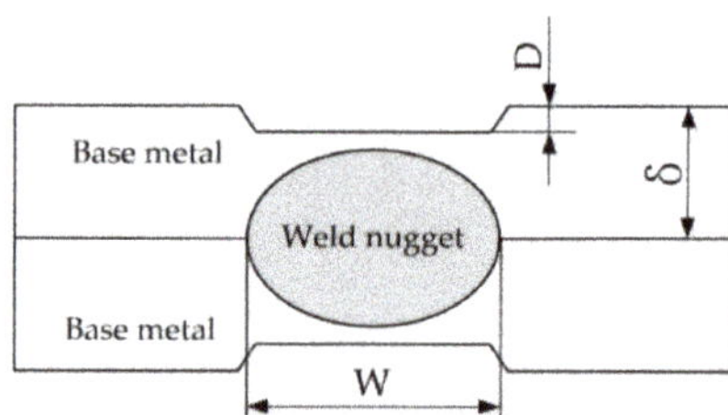

Figure 7. Schematic diagram of the main dimensions of the welded joint cross-section.

The effect of the welding current on the indentation rate is indicated in Figure 8a. The results indicated that the indentation rate was small and increased less with a low welding current because the welding heat input was small, which caused a small amount of base metal melting. When the welding current was between 9.5 kA and 11.0 kA, the welding heat input increased rapidly, so more base metal was melted and the indentation rate increased. If the welding current was greater than 11.0 kA, the indentation rate increased gradually due to welding spatters and other defects, and the indentation was too serious to satisfy the welding quality requirements. Figure 8b displays the relationship between the welding current and the W. The weld nugget width increased rapidly from 7.36 mm to 8.64 mm with an increase of the current from 8.5 kA to 10.0 kA, and the maximum value was 8.75 mm at the 10.5 kA welding current. While the current changed from 10.0 kA to 10.5 kA, the welding heat input reached a quasi-steady state and the change of the weld nugget width was small. When the welding current was greater than 10.5 kA, the current density was higher, and a large number of welding spatters were generated, which reduced the amount of melted base metal in the weld nugget and thus decreased the weld nugget width. If the welding current continued to increase to 12.0 kA, the weld nugget width increased because of the larger welding heat input, but there were many welding spatters, and also some shrinkage and crack defects in the weld nugget.

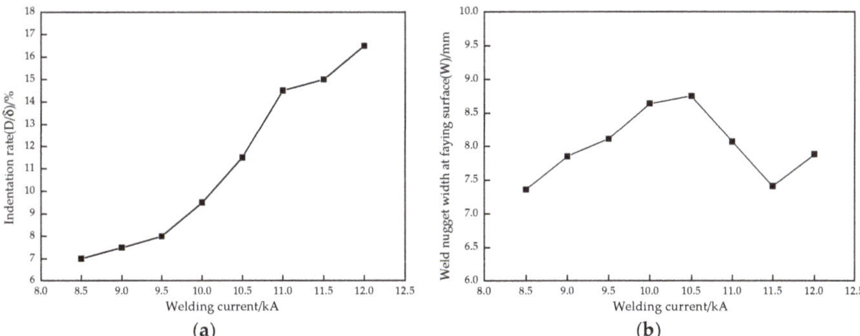

(a)   (b)

Figure 8. Relationships between the welding current and (**a**) indentation rate (D/δ); (**b**) weld nugget width at overlap surface (W).

3.2. Microstructure

Due to the uneven heat input and different cooling conditions, the microstructure of the resistance welded joints was very inhomogeneous. As shown in Figure 9a, the whole resistance spot welded joint comprised of the base metal (a zone), uneven structure zone (b zone), fine grain zone (c zone), superheated zone (d zone), and weld nugget zone (e zone). The microstructure of the base metal consists of ferrite and martensite as shown in Figure 9b. The grain and microstructure in the uneven structure zone were obviously heterogeneous, as shown in Figure 9c. Because the temperature in this zone was between Ac_1 and Ac_3 during the resistance spot welding process, the phase transformation and recrystallization occurred for part of the base metal, and the fine ferrite and martensite formed; meanwhile, the ferrite, which was not austenitized, became the coarse ferrite. Therefore, there were also similarities to the structure of the base metal in this zone. During the welding process, the base metal in the fine grain zone was heated to above Ac_3, and all ferrite and martensite were recrystallized to austenite. The fine and homogeneous ferrite and martensite were obtained after cooling, which were similar to the normalized microstructure of heat treatment, as shown in Figure 9d. The overheated zone consisted of coarse lath martensite and a little of the ferrite shown in Figure 9e. The temperature in this zone was between 1100° and the solidus temperature, and the austenite was overheated. The grain grew up seriously and the chemical compositions in the grain were uniform, hence, the coarse martensite was obtained after rapid cooling. The nugget zone comprised of coarse lath martensite and a little of ferrite with the columnar crystal morphology, as shown in Figure 9f. Because the maximum temperature gradient in the nugget zone was along the axis of the electrodes, the molten liquid metal nucleated and crystallized at the fusion line first, and then formed the columnar austenite along the direction of the higher temperature gradient. Finally, the solid austenite transformed into martensite because of the rapid cooling rate of the welding process and low carbon component in the base metal.

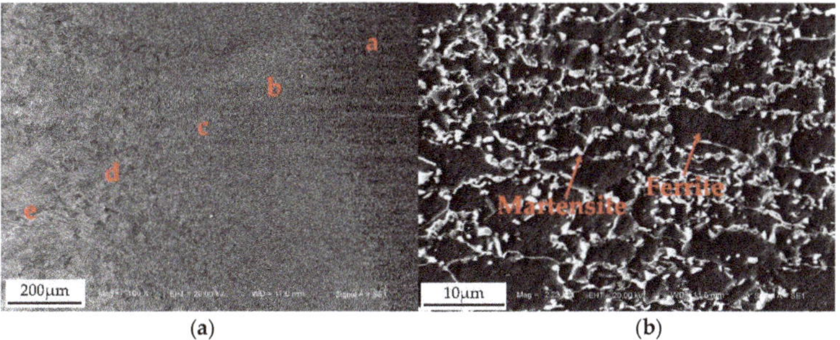

(a)   (b)

Figure 9. *Cont.*

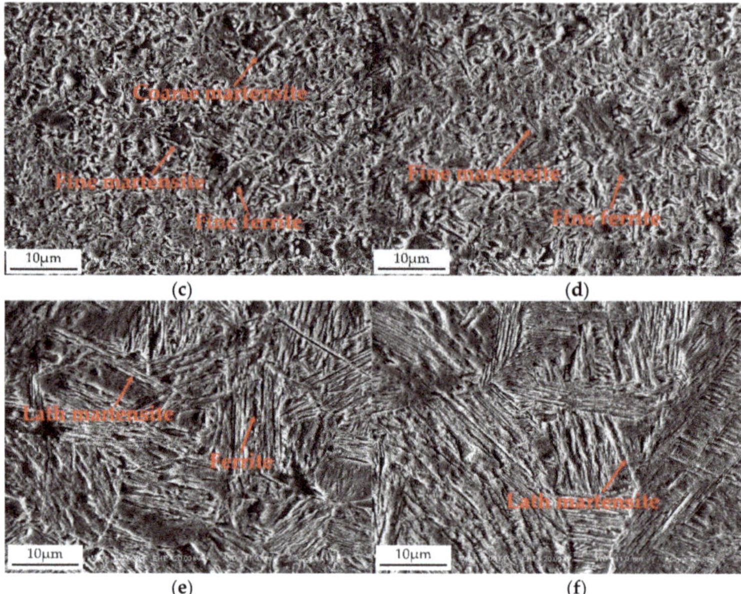

Figure 9. Microstructure of (**a**) different zones of the welded joint; (**b**) base metal; (**c**) uneven structure zone; (**d**) fine grain zone; (**e**) superheated zone; (**f**) weld nugget zone.

The weld nugget is the important zone of resistance spot welded joints, and its microstructure characteristics directly affected the mechanical properties of welded joints. The heat generation and heat transfer in the nugget zone were different under different welding currents, so the microstructure was very different. Figure 10 shows the influence of a typical current on the microstructure of the weld nugget. It can be seen that while the welding current was low, the welding heat input was low, and the weld nugget was mainly lath martensite with a fine and uniform structure. At the same time, the plastic deformation of the weld nugget zone was large, and there was no welding defect in the weld nugget zone. While the welding current was 10.5 kA, the welding heat input increased and the microstructure became coarser. The decrease of the cooling rate resulted in the decrease of martensite and the increase of ferrite. The microstructure of the weld nugget was relatively uniform, and there was no obvious welding defect. If the welding current was the maximum value of 12.0 kA, a large number of spatters caused some heat loss and a rapid cooling rate, so that the weld nugget microstructure consisted of lath martensite. The grains in the weld nugget center grew up and coarsened greatly, as shown in Figure 10c. In short, when the welding current varied from 8.5 kA to 12 kA, the weld nugget columnar structure gradually coarsened, mainly because of the larger heat input and the reduction of the cooling rate.

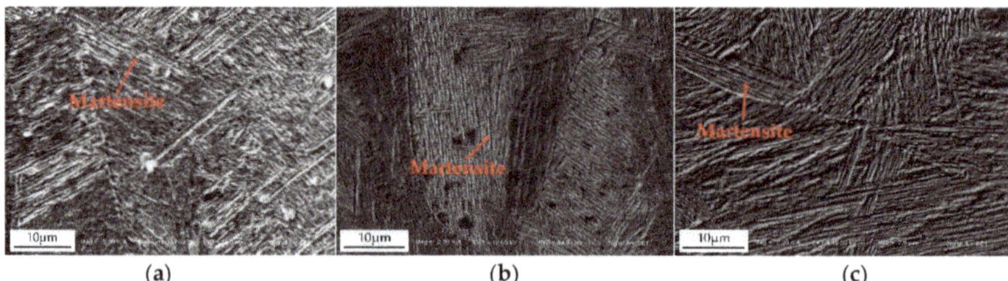

Figure 10. Effect of the typical welding current on the microstructure of the weld nugget: (**a**) 8.5 kA; (**b**) 10.5 kA; (**c**) 12 kA.

3.3. Microhardness Distribution

The microhardness measurement was performed on the cross section of the resistance spot welded joint. The schematic diagram of the measurement location is shown in Figure 11. The distance between two test points in the base metal and weld nugget zone was 0.5 mm. Because the HAZ width was narrow, the two test points distance was 0.25 mm in HAZ.

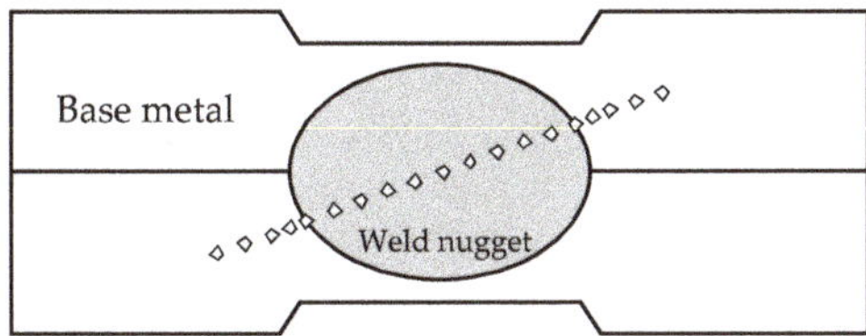

Figure 11. Schematic diagram of the microhardness measurement location of the welded joint.

Figure 12 displays the microhardness distributions of resistance spot welded joints gained with different welding currents. The microhardness of the base metal was 198 $HV_{0.2}$, and the microhardness in HAZ was obviously higher than the base metal. The microhardness of the weld nugget zone was the highest, which was above 350 $HV_{0.2}$. The microhardness at the edge of the weld nugget was slightly higher than that of the weld nugget center. Due to the uneven distribution of the current density at the welding joint, the current density at the edge of the weld nugget was greater than the average current density, which generated a greater resistance heat. Therefore, during the resistance spot welding process, the solidification and crystallization first occurred at the edge of the weld nugget. Meanwhile, the temperature gradient was large and the cooling rate was fast, so the martensite was large and coarse and the microhardness was higher in this zone. With the increase of the welding current, the average microhardness in the weld nugget zone decreased. While the welding current was 8.5 kA, the microhardness of the weld nugget was higher because of the faster cooling rate, and the microstructure was lath martensite, which was uniform and fine, so the hardness was higher. While the welding current was 10.5 kA, the cooling rate should decrease. The columnar crystals of the weld nugget grew up, mainly composed of lath martensite and acicular ferrite, which resulted in the decrease of hardness. When the welding current continued to increase to 12.0 kA, the microhardness of most zones of the weld nugget was higher than that of 10.5 kA due to the large heat input and coarse structure, but there were welding defects in the center of the weld nugget, so that the average value of microhardness decreased.

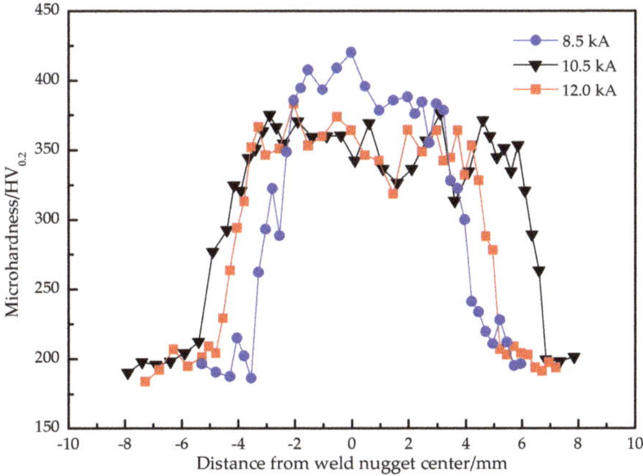

Figure 12. Microhardness distributions of welded joints with different welding currents.

3.4. Tensile Shear Strength

The tensile shear force is often used to characterize the welded joint strength. The larger the tensile shear force is, the better the strength is. Figure 13 indicates the effect of the welding current on tensile shear force. Compared with Figure 8b, it can be found that the variation trend of the W and tensile shear force with the welding current is basically the same; that is, the larger the weld nugget width, the greater the tensile shear force. With the first increase of welding current, the weld nugget width and tensile shear force all obviously increased. While it was 10.5 kA, the weld nugget width and tensile shear force all reached the maximum value of 8.75 mm and 24.20 kN, respectively. Subsequently, the weld nugget width dropped rapidly and an inflection point appeared while the welding current was 11.5 kA, but the tensile force decreased continuously. With the large welding current, the melting amount of the base metal increased gradually, which caused the increase of the weld nugget width. Thus, the bonding strength of the resistance spot welded joint increased and the tensile shear force increased. However, when the welding current was 12.0 kA, the welding heat input was too large, resulting in a large number of spatters, shrinkage cracks, and other defects in the welded joint. Although the weld nugget width increased, the effective bonding width of the welded joint decreased, so the tensile shear force decreased.

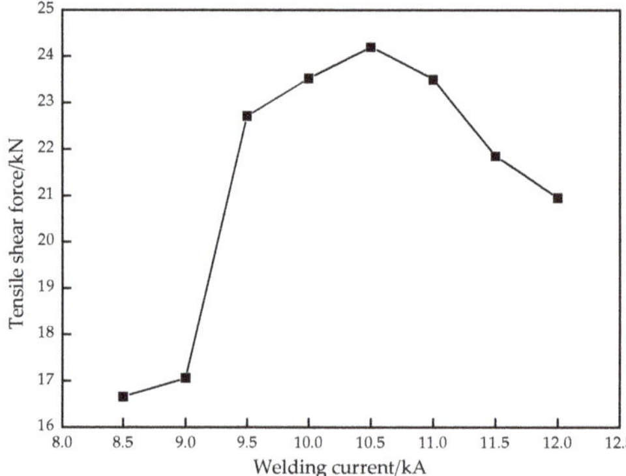

Figure 13. Effect of the welding current on the tensile shear force of welded joint.

During the tensile shear experiment, there were two failure modes: Interface failure and pullout failure, as shown in Figure 14. The SEM images of fracture surfaces were indicated in Figure 15. The welded joints, which were gained with the lower welding current ($\leq$9.5 kA) and higher welding current ($\geq$11 kA), always ruptured along the overlap surface, as shown in Figure 14a. While the welding current was low ($\leq$9.5 kA), the W was narrow, as well as the strength of the base metal was high. Therefore, under the tensile shear force, the crack produced on the weld nugget edge at the overlap surface at first and then extended along the overlap surface until the welded joint failed with the interface failure mode. While the welding current was high ($\geq$11 kA), the bonding strength of welded joints was small due to the welding defects, such as spatter, shrinkage, and cracks, in the weld nugget, which resulted in the smaller tensile shear force for the welded joint, and its tensile shear specimen also ruptured along the overlap surface. As shown in Figure 15a, the river pattern was obvious on the fracture surface, which illustrated the brittle fracture characteristics. While the welding current was 10 kA, the fracture pattern of the welded joint was the pullout failure mode, as shown in Figure 14b. It can be found that the fracture surfaces are mainly dimples and a small amount of cleavage steps from Figure 15b. Under the tensile shear force, the tensile specimen first produced necking in the HAZ of the welded joint. With the increase of tensile shear force, the dimples grew up and converged, and then the tensile specimen broke down in the base metal. In the present study, the weld nugget width and tensile shear force were all maximum with a 10.5 kA welding current.

During the tensile shear test for welded joints produced with a 10.5 kA welding current, because of the large weld nugget and few welding defects, the welded joint was not easy to rupture from the overlap surface. The tensile stress on the edge of the weld nugget increased gradually, owing to the angle between the overlap surface and tensile force. The HAZ was the weakest zone due to the inhomogeneous microstructure, coarse grains, and low plasticity and toughness. With the increase of tensile force, the necking occurred first in the HAZ, and a number of micro-holes began to form in the center of the necking. The micro-holes grew and formed the dimples, which converged to form the crack. Finally, the crack was torn along the HAZ to form the pullout tear failure, as show in Figure 14c. Figure 15c shows the pullout tear fracture surface, which was mainly composed of small and uniform dimples.

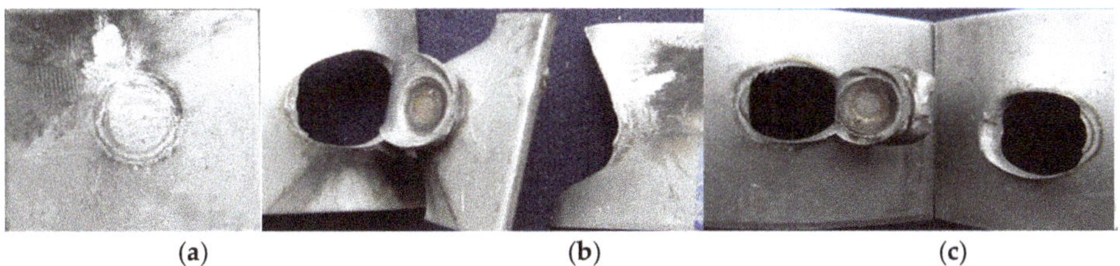

(a)   (b)   (c)

Figure 14. Three failure modes of the resistance spot welded joints: (**a**) Interface failure; (**b**) pullout failure (base metal tear fracture); (**c**) pullout failure (pullout tear fracture).

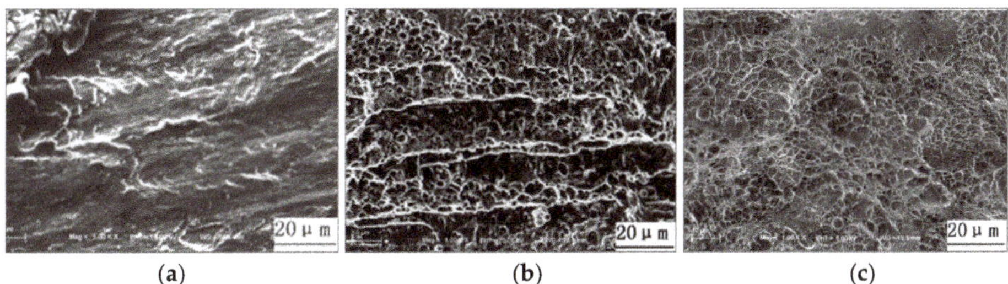

(a)   (b)   (c)

Figure 15. SEM images of the fracture surface: (**a**) Interface failure; (**b**) pullout failure (base metal tear fracture); (**c**) pullout failure (pullout tear fracture).

4. Conclusions

(1)   The Zn island was generated on the resistance spot weld surface because of the melting and aggregating of the Zn layer under the action of heated and squeezed by the electrodes. While the welding current increased from 8.5 kA to 12.0 kA, the indentation rate kept growing to 16.5% due to the increase of the welding heat input. However, the weld nugget width obviously increased at first, which reached the maximum when the welding current was 10.5 kA, and then decreased.

(2)   The whole resistance spot welded joint comprises of the base metal, uneven structure zone, fine grain zone, superheated zone, and weld nugget zone. The nugget zone was mainly comprised coarse lath martensite and little ferrite with columnar crystal morphology due to the high temperature gradient and rapid cooling rate. With the increase of the welding current, the microstructure of the weld nugget became coarser; meanwhile, the martensite decreased and the ferrite increased.

(3)   The microhardness of the weld nugget zone was highest and the base metal was lowest. With the increase of the welding current, the average microhardness in the weld nugget zone decreased. While the welding current increased from 8.5 kA to 10.5 kA, the tensile shear force was obviously raised, owing to the increase of the weld nugget width. The tensile shear force reached the maximum value of 24.20 kN with a 10.5 kA welding current. If the welding current increased

continuously, the tensile shear force decreased because of a large number of spatters and the high indentation rate. The failure modes mainly depended on the weld nugget width at the overlap surface and welding defect. Therefore, while the CR590T/340YDP galvanized dual phase steel sheets with 2 mm thickness were performed using resistance spot welding, the recommended welding current was 10.0~10.5 kA with a 20 cycles welding time and 4.0 kN electrode force.

Author Contributions: Writing, funding acquisition and investigation, X.Z.; conceptualization, methodology and supervision, Z.R.; validation and investigation, F.Y.; formal analysis and data curation, H.Y.

Funding: This research was funded by Jilin Scientific and Technological Development Program (20160520055JH).

Conflicts of Interest: The authors declare no conflict of interest.

References

1. Tucker, R. Trends in automotive lightweighting. *Met. Finish.* **2013**, *111*, 23–25. [CrossRef]
2. Sakundarini, N.; Taha, Z.; Abdul-Rashid, S.H.; Ghazila, R.A.R. Optimal multi-material selection for lightweight design of automotive body assembly incorporating recyclability. *Mater. Des.* **2013**, *50*, 846–857. [CrossRef]
3. Behrens, B.-A.; Klose, C.; Chugreev, A.; Heimes, N.; Thürer, S.E.; Uhe, J. A Numerical Study on Co-Extrusion to Produce Coaxial Aluminum-Steel Compounds with Longitudinal Weld Seams. *Metals* **2018**, *8*, 717. [CrossRef]
4. Xue, X.; Pereira, A.B.; Amorim, J.; Liao, J. Effects of Pulsed Nd:YAG Laser Welding Parameters on Penetration and Microstructure Characterization of a DP1000 Steel Butt Joint. *Metals* **2017**, *7*, 292. [CrossRef]
5. Lee, S.G.; Patel, G.R.; Gokhale, A.M.; Sreeranganathan, A.; Horstemeyer, M.F. Quantitative fractographic analysis of variability in the tensile ductility of high-pressure die-cast ae44 mg-alloy. *Mater. Sci. Eng. A* **2006**, *427*, 255–262. [CrossRef]
6. Pathak, N.; Butcher, C.; Worswick, M.J.; Bellhouse, E.; Gao, J. Damage Evolution in Complex-Phase and Dual-Phase Steels during Edge Stretching. *Materials* **2017**, *10*, 346. [CrossRef] [PubMed]
7. Bandyopadhyay, K.; Panda, S.K.; Saha, P.; Baltazar-Hernandez, V.H.; Zhou, Y.N. Microstructures and failure analyses of dp980 laser welded blanks in formability context. *Mater. Sci. Eng. A* **2016**, *652*, 250–263. [CrossRef]
8. Cui, Q.L.; Parkes, D.; Westerbaan, D.; Nayak, S.S.; Zhou, Y.; Liu, D.; Goodwin, F.; Bhole, S.; Chen, D.L. Effect of coating on fiber laser welded joints of dp980 steels. *Mater. Des.* **2016**, *90*, 516–523. [CrossRef]
9. Jia, Q.; Guo, W.; Li, W.; Zhu, Y.; Peng, P.; Zou, G. Microstructure and tensile behavior of fiber laser-welded blanks of dp600 and dp980 steels. *J. Mater. Process. Technol.* **2016**, *236*, 73–83. [CrossRef]
10. Dong, D.; Liu, Y.; Wang, L.; Yang, Y.; Jiang, D.; Yang, R. Microstructure and deformation behavior of laser welded dissimilar dual phase steel joints. *Sci. Technol. Weld. Joi.* **2015**, *21*, 75–82. [CrossRef]
11. Koganti, R.; Angotti, S.; Joaquin, A.; Jiang, C.; Karas, C. Gas metal arc welding (GMAW) process optimization for uncoated dual phase 600 material combination with aluminized coated and uncoated boron steels for automotive body structural applications. In Proceedings of the ASME 2007 International Mechanical Engineering Congress & Exposition, Seattle, WA, USA, 11–15 November 2007.
12. Yan, B.; Lalam, S.H.; Zhu, H. *Performance Evaluation of GMAW Welds for Four Advanced High Strength Steels*; SAE Technical Paper Series; SAE International: Warrendale, PA, USA, 2005.
13. Koganti, R.; Jiang, C.; Karas, C. Gas metal arc welding (GMAW) process optimization of 1.5 mm uncoated dual-phase 780 (DP780) joint for automotive body structural applications. In Proceedings of the ASME 2009 International Mechanical Engineering Congress & Exposition, Lake Buena Vista, FL, USA, 13–19 November 2009.
14. Khan, M.I.; Kuntz, M.L.; Su, P.; Gerlich, A.; North, T.; Zhou, Y. Resistance and friction stir spot welding of dp600: A comparative study. *Sci. Technol. Weld. Joi.* **2013**, *12*, 175–182. [CrossRef]
15. Santella, M.; Hovanski, Y.; Frederick, A.; Grant, G.; Dahl, M. Friction stir spot welding of dp780 carbon steel. *Sci. Technol. Weld. Joi.* **2010**, *15*, 271–278. [CrossRef]

16. Lee, H.; Kim, C.; Song, J.H. An Evaluation of Global and Local Tensile Properties of Friction-Stir Welded DP980 Dual-Phase Steel Joints Using a Digital Image Correlation Method. *Materials* **2015**, *8*, 8424–8436. [CrossRef] [PubMed]

17. Aslanlar, S.; Ogur, A.; Ozsarac, U.; Ilhan, E. Welding time effect on mechanical properties of automotive sheets in electrical resistance spot welding. *Mater. Des.* **2008**, *29*, 1427–1431. [CrossRef]

18. Zhang, X.; Zhang, J.; Chen, F.; Yang, Z.; He, J. Characteristics of Resistance Spot Welded Ti6Al4V Titanium Alloy Sheets. *Metals* **2017**, *7*, 424. [CrossRef]

19. Ma, C.; Chen, D.L.; Bhole, S.D.; Boudreau, G.; Lee, A.; Biro, E. Microstructure and fracture characteristics of spot-welded dp600 steel. *Mater. Sci. Eng. A* **2008**, *485*, 334–346. [CrossRef]

20. Hernandez, V.H.B.; Nayak, S.S.; Zhou, Y. Tempering of martensite in dual-phase steels and its effects on softening behavior. *Metall. Mater. Trans. A* **2011**, *42*, 3115. [CrossRef]

21. Nayak, S.S.; Hernandez, V.H.B.; Zhou, Y. Effect of chemistry on non-isothermal tempering and softening of dual-phase steels. *Metall. Mater. Trans. A* **2011**, *42*, 3242. [CrossRef]

22. Khan, M.I.; Kuntz, M.L. Microstructure and mechanical properties of resistance spot welded advanced high strength steels. *Mater. Trans.* **2008**, *49*, 1629–1637. [CrossRef]

23. Senkara, J.; Zhang, H.; Hu, S.J. Expulsion prediction in resistance spot welding. *Weld. J.* **2004**, *83*, 123–132.

24. Wang, B.; Hua, L.; Wang, X.; Song, Y.; Liu, Y. Effects of electrode tip morphology on resistance spot welding quality of DP590 dual-phase steel. *Int. J. Adv. Manuf. Technol.* **2016**, *83*, 1917–1926. [CrossRef]

25. Banerjee, P.; Sarkar, R.; Pal, T.K.; Shome, M. Effect of nugget size and notch geometry on the high cycle fatigue performance of resistance spot welded DP590 steel sheets. *J. Mater. Process. Technol.* **2016**, *238*, 226–243. [CrossRef]

26. Zhao, D.W.; Wang, Y.X.; Zhang, L.; Zhang, P. Effects of electrode force on microstructure and mechanical behavior of the resistance spot welded dp600 joint. *Mater. Des.* **2013**, *50*, 72–77. [CrossRef]

27. Wang, M.; Wu, Y.X.; Pan, H.; Lei, M. Effect of base metal chemical composition on properties of resistance spot welding joint of DP590 steel. *Trans. China Weld. Inst.* **2010**, *31*, 33–35.

28. Wan, X.D.; Wang, Y.X.; Zhang, P. Modelling the effect of welding current on resistance spot welding of dp600 steel. *J. Mater. Process. Technol.* **2014**, *214*, 2723–2729. [CrossRef]

29. Wang, M.; Zhang, H.T.; Pan, H.; Lei, M. Numerical simulation of nugget formation in resistance spot welding of dp590 dual-phase steel. *J. Shanghai Jiaotong Univ.* **2009**, *43*, 56–60.

 *materials*

Article

Application of Cold Wire Gas Metal Arc Welding for Narrow Gap Welding (NGW) of High Strength Low Alloy Steel

Rafael A. Ribeiro [1,*], Paulo D. C. Assunção [2], Emanuel B. F. Dos Santos [3], Ademir A. C. Filho [2], Eduardo M. Braga [2] and Adrian P. Gerlich [1]

1   Centre for Advanced Materials Joining (CAMJ), University of Waterloo, 200 University Avenue West, Waterloo, ON N2L 3G5, Canada; agerlich@uwaterloo.ca
2   Metallic Materials Characterization Laboratory (LCAM) Federal University of Pará (UFPA), Rua Augusto Corrêa, 1—Guamá, Belém, PA 66075-110, Brazil; pd2costa@uwaterloo.ca (P.D.C.A.); eng.angelo80@gmail.com (A.A.C.F.); edbraga@ufpa.br (E.M.B.)
3   Liburdi Automation Inc., Liburdi GAPCO, 400 ON-6, Dundas, ON L9H 7K4, Canada; esantos@liburdi.ca
*   Correspondence: rdearauj@uwaterloo.ca; Tel.: +1-226-749-4832

Received: 21 December 2018; Accepted: 18 January 2019; Published: 22 January 2019

Abstract: Narrow gap welding is a prevalent technique used to decrease the volume of molten metal and heat required to fill a joint. Consequently, deleterious effects such as distortion and residual stresses may be reduced. One of the fields where narrow groove welding is most employed is pipeline welding where misalignment, productivity and mechanical properties are critical to a successful final assemblage of pipes. This work reports the feasibility of joining pipe sections with 4 mm-wide narrow gaps machined from API X80 linepipe using cold wire gas metal arc welding. Joints were manufactured using the standard gas metal arc welding and the cold wire gas metal arc welding processes, where high speed imaging, and voltage and current monitoring were used to study the arc dynamic features. Standard metallographic procedures were used to study sidewall penetration, and the evolution of the heat affected zone during welding. It was found that cold wire injection stabilizes the arc wandering, decreasing sidewall penetration while almost doubling deposition. However, this also decreases penetration, and incomplete penetration was found in the cold wire specimens as a drawback. However, adjusting the groove geometry or changing the welding parameters would resolve this penetration issue.

Keywords: GMAW; CW-GMAW; Narrow gap welding; sidewall penetration; high strength steel; X80

1. Introduction

Narrow gap welding (NGW) is a technique used to weld thick joints with the aim of reducing the molten metal deposited volume, ultimately decreasing distortion and residual stresses caused by thermal stresses developed during welding. One of the drawbacks associated with this technique is sidewall penetration caused by the arc wandering during welding. This problem is so critical in heavy duty welding that models to predict it are found in the literature [1].

Various NGW variants were developed to overcome such drawbacks: rotating arc, swing arc and wave-shaped wire systems. Another reason to use these modified systems in narrow gap welding is to improve the wettability of the sidewall by the welding pool and to better distribute the heat across the weld to mitigate distortions. A common setback for these systems is their expense and the use of highly trained personnel in their operation, which considerably increases the costs of manufacturing.

Not only are arc-modified processes employed in NGW, but also increased-deposition processed are successfully used, e.g., tandem gas metal arc welding (T-GMAW) [2] and twin GMAW [3]. However, these processes rely on increased deposition, while they increase the amount of heat transferred to the substrate, which might cause increased distortion or residual stresses. To mitigate distortion and other heat-induced detrimental effects, cold metal transfer (CMT) was developed, in which the wire feed motion is reversed at a time synchronized with the pulse current to optimize the metal transfer. A review of the applications of CMT emphasizing the low dilution and the effect of post weld heat treatments (PWHT) in various materials can be found in [4].

One alternative to these sophisticated welding processes is the cold wire gas metal arc welding (CW-GMAW), which consists of the standard gas metal arc welding (GMAW) with an extra cold wire (non-energized) fed into the arc-welding pool system. This process increases the deposition of the welds while maintaining the nominal heat input for the same welding parameters in standard GMAW. Previous research [5] has shown that the feeding of cold wire causes a slight increase in current without a corresponding increase in penetration. A feature that distinguishes GMAW and CW-GMAW is the reduced dilution caused by cold wire feed rates, which decreases penetration and melting of the base metal. Ultimately, the two processes differ by range of admissible parameters, since the cold wire can stabilize metal transfer [6]. Moreover, this difference in dilution can be linked to the thermal signature, as reported in [7], showing that cold wire welds result in lower heat-induced distortion compared to standard GMAW welds.

Previous research [8] demonstrates the possibility of welding U-grooves using CW-GMAW with pure carbon dioxide as shielding gas, and this method was primarily developed to be applied in shipbuilding. The feasibility of this process to weld a 5 mm wide groove was recently studied on ASTM A131 grade A steel, revealing that, due to the pinning of the arc to the cold wire, the sidewall erosion was considerably reduced in comparison to welds manufactured with standard GMAW [9]. Subsequent research [10] studied the effect of the CW-GMAW on process-induced residual stresses, concluding that welds manufactured by CW-GMAW have lower levels of residual stresses. This decrease in residual stresses might explain the improvement in fatigue life reported in [11].

The present work reports the further development on the work of prior research [9], and reports preliminary results regarding the application of CW-GMAW to weld high strength pipeline steels in narrow gap configuration. The feasibility of NGW employing CW-GMAW was assessed by comparing the severity of sidewall erosion in the welds and the presence of defects in the root pass. The results point to the general feasibility of NGW using CW-GMAW pipeline applications, and illuminate possible future work to avoid certain defects found during welding. The arc attachment to sidewalls can be ascribed to the arc shortest electron path according to Zhang et al. [12].

2. Experimental Set-Up and Materials

NGW were fabricated using both GMAW and the cold wire gas metal arc welding (CW-GMAW) process. Figure 1 shows a schematic of the geometry of the narrow groove used in this work and the detail of cold wire positioning regarding the wire electrode and cold wire feeding angle. The grooves were welded using a Lincoln R500 welding power source linked to a Fanuc ArcMate 120i robotic arm. The size of the joints was 140 mm (length) $\times$ 115 mm (width) $\times$ 15 mm (thickness). Figure 1 shows groove gap and the root face. To determine the reproducibility, three replicates were manufactured for each welding condition. It is important to mention that the welds were performed in constant voltage mode, where no synergic controls were used during welding to adaptively control the arc dynamics.

To manufacture the welds, ER100S-G in the diameter of 1.2 mm was used as electrode, while the cold wire had a diameter of 1.0 mm. API X80 [9] was selected as base metal. The nominal compositions of the electrodes and of the base metal are given in Table 1. Moreover, no weaving or preheating was used during welding.

Cold wire feed rates are expressed here as a fraction of the electrode mass feed rate, due to the fact that two wires of different diameters were used for the electrode and cold wire, respectively. To quantify the cold wire feed rate as function of the electrode wire, the mass percentage as a fraction of the electrode was used. The mass feed rate was calculated from the wire density and cross-sectional area.

To evaluate the welding process, a NGW joint design used in heavy welding applications such as pipeline welding was considered. This configuration was chosen to demonstrate an immediate application of the process, which is critical to welding of thick structures, since it decreases the heat-induced detrimental effects of welding passes.

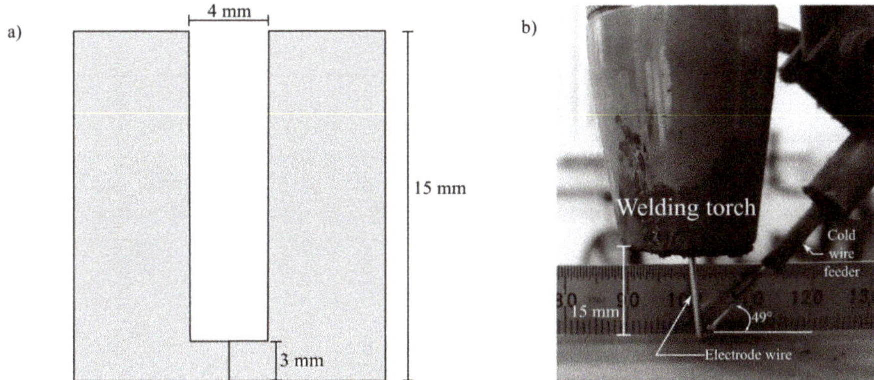

Figure 1. Schematics of the groove geometries used in this work: (**a**) schematic of the narrow groove, showing the cross-section; and (**b**) detail of CW positioning and feeding angle.

Table 1. Nominal chemical composition of the welding wires and the base metal.

	C	Mn	P	S	Ti	Mo	Ni	Cr	Fe
ER100S-G	0.08	1.25–1.80	0.20–0.55	0.01	0.10	0.25–0.55	1.20–2.10	1.25–1.80	Balance
API X80	0.22	1.85	0.025	0.015	0.06	-	-	-	Balance

During welding, the current and voltage signals were acquired at the sampling frequency of 20 kHz for 2 s with synchronized high speed imaging at 5000 fps with shutter speed of 25 ms, an aperture of f/22, and a narrow band pass filter of 900 ± 10 nm. The videography was performed with the camera in parallel to the groove longitudinal line to record the metal transfer inside the groove. The high speed images shown in this work were selected to adequately represent the arc dynamics and metal transfer, when in a stable condition. The welding parameters used are reported in Table 2. For all welds, the shielding gas mixture used was Ar-15%CO_2 at a flow rate of 17 L/min, and the contact tip to work-piece distance (CTWD) was constant and equal to 17 mm. The welding parameters were set to apply the same heat input using both processes. The quantities of average voltage ($U_{average}$), average current ($I_{average}$), and average power ($P_{average}$) were calculated using the average instantaneous algorithm according to Equations (1)–(3):

$$U_{average} = \sum_{1}^{i} U_i \tag{1}$$

$$I_{average} = \sum_{1}^{i} I_i \tag{2}$$

$$P_{average} = \sum_{1}^{i} U_i \times I_i \tag{3}$$

The arc arc stability was also assessed through cyclogrammes which are voltage versus current plots, and are used to study the events occurring in the arc electric. They are useful since they show the amount of short-circuits and account for the general arc stability of the process [13]. For a more thorough discussion on cyclogrammes and their respective zones, the reader should refer to prior work [14].

Table 2. Welding parameters.

Welding Process	WFS (m/min)	Voltage (V)	Travel Speed (m/min)	CWFR (%)	Deposition (kg/h)
GMAW	7.62	25	0.41	-	4.11
CW-GMAW	7.62	25	0.41	80	7.39

Once the experiments were completed, the specimens were subjected to standard metallographic procedures and etched with 5% Nital to reveal the macrostructure. The cross-sections showing the passes sequence were taken from the start, middle, and end of the joints. The cross-sections showing the complete joint were taken from the middle of the joints. The hardness map was performed with a 200 g load and 10 s dwell time. The distance between indentations was 0.3 mm and the distance between lines was 0.3 mm.

3. Results

3.1. Electrical Data

Table 3 presents the actual electrical data for current, voltage and power probed for the welding conditions employed in this work. For GMAW and CW-GMAW, the heat input per pass was similar.

Table 3. Average electrical parameters sampled during welding and nominal resulting heat input as response of the welding power source.

Welding Process	Pass	Average Voltage (V)	Average Current (A)	Average Power (W)	NHI (kJ/mm)
GMAW	root	25.13	299.69	7520.13	1.11
	fill	25.13	270.64	6801.29	1.00
	cap	25.12	253.27	6427.07	0.95
CW-GMAW	root	25.14	292.20	7344.18	1.08
	fill/cap	25.16	277.95	6991.18	1.03

3.1.1. Oscillograms

Figure 2 shows the oscillograms for the standard GMAW condition. For the root pass, one can notice a periodic repetition with frequency of approximately 3 Hz (Figure 3a). By correlating the electrical signals to the high speed images, it was observed that this repetition is a consequence of arc attachment to the groove sidewalls. As a consequence of the arc attachment to the sidewall, the power source promotes an instantaneous increase in the arc current, as shown in detail in Figure 3b. It is possible that this was caused by the auto-regulation system of the power source, which interprets the attachment of the arc to sidewalls as an increase in wire feed speed, consequently increasing the current to keep the melting rate constant while the arc length is reduced due to the arc attachment to the groove walls.

On the other hand, one can notice that this repetitive behavior did not occur in the fill and cap passes (Figure 2c,d), likely due to a reduced degree of arc constriction, and the shortest path to the electron conduction being the bottom of the groove and not the sidewalls. On the other hand, the severity of short-circuits in the cap pass was higher when compared to the filler pass. This suggests that the distance between the droplet (as shown in the high speed frames, Figure 6) and the substrate is lower, favoring short-circuits.

Figure 3 shows the oscillograms for the CW-GMAW specimens. It is noted that the periodic pattern observed in the oscillograms of GMAW for root pass was not observed for the entire sampling period in CW-GMAW of 2000 ms. This suggests the sidewall erosion was mitigated during CW-GMAW, since sidewall penetration causes the periodicity in the observed signal. Moreover, the short-circuit severity in the fill/cap pass is less prominent than that observed in the GMAW specimen.

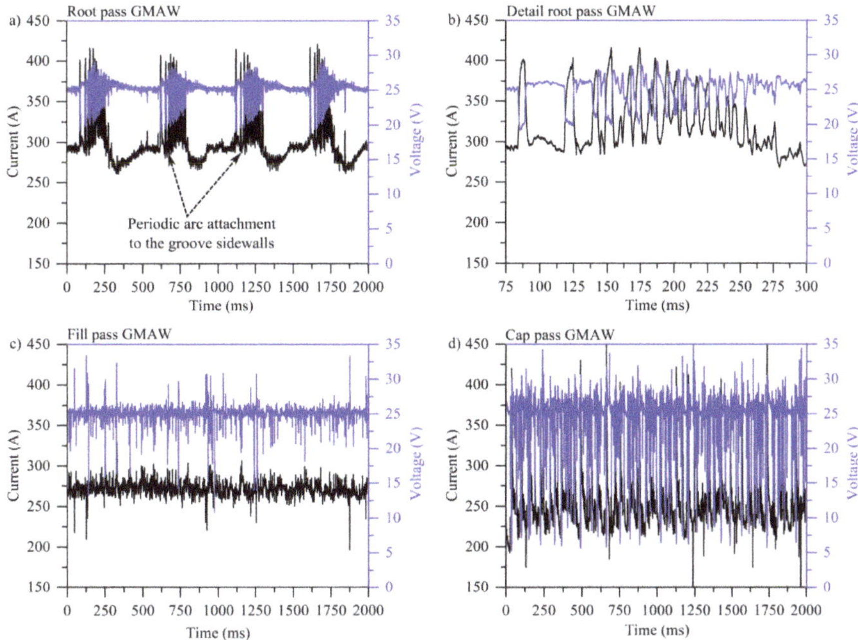

Figure 2. Typical oscillograms for the GMAW condition: (**a**) root pass; (**b**) detail of the root inside the period of repetition; (**c**) fill pass; and (**d**) cap pass.

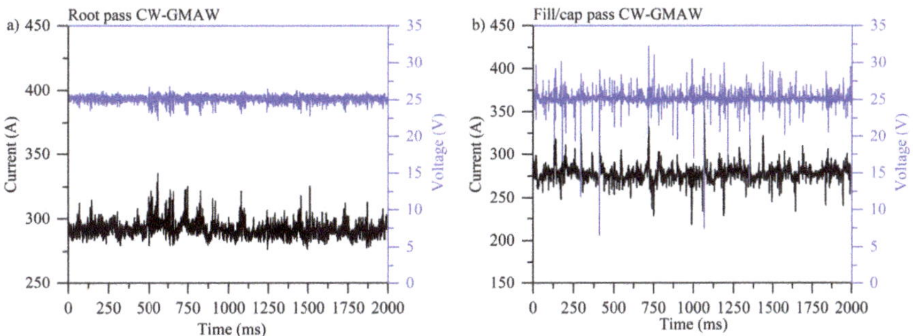

Figure 3. Typical oscillograms for the CW-GMAW condition: (**a**) root pass; and (**b**) fill/cap pass.

3.1.2. Cyclogrammes

Figure 4 shows typical cyclogrammes for GMAW specimens where one can notice that in the root pass (Figure 4a) the number events corresponding to short-circuits (region inside the dashed rectangle) are larger than the arc burning area (darker region, upper left-side of the dashed square). Ultimately, this cyclogramme points out an unstable condition where the arc burning region is smaller compared to the perturbed region.

As reflected by oscillograms, the fill pass was more stable, with some short-circuits and high voltage points, indicating large variations in arc length (see Figure 4b). In the cap pass (Figure 4c), one can notice a larger short-circuit region with slight variation in arc length. The results also suggest that there was a high variation in current, probably due to an increased quantity of metal to melt, interpreted by the power source as an increase in wire feed speed (higher melting speed).

Figure 5 shows the cyclogrammes for the CW-GMAW condition. One observes that, by comparing the cyclogrammes for root pass between standard GMAW and CW-GMAW, the arc burning operation range for CW-GMAW is shorter, indicating that this was more stable than the root in GMAW condition. One observes the complete absence of short-circuits (see Figure 5a). The cause of such stabilization should be attributed to the cold wire feed. Meanwhile, comparing the fill/cap condition (Figure 5b) between CW-GMAW and GMAW, one finds short-circuits events where voltage dropped below 20 V, along with variations in arc length, suggested by high values of voltage for the same range of current.

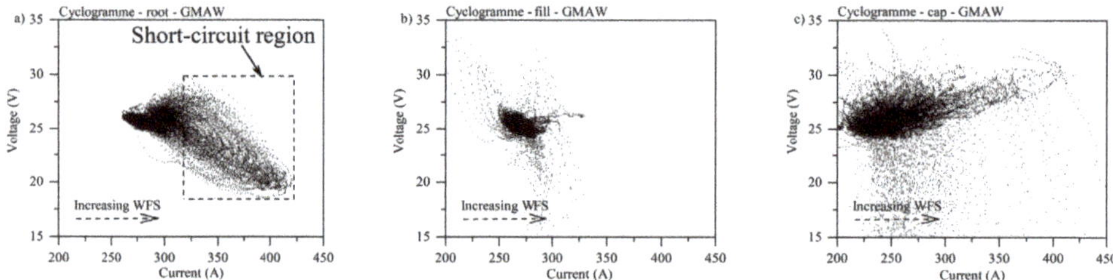

Figure 4. Cyclogrammes for the GMAW specimens: (**a**) root pass; (**b**) fill pass; and (**c**) fill pass.

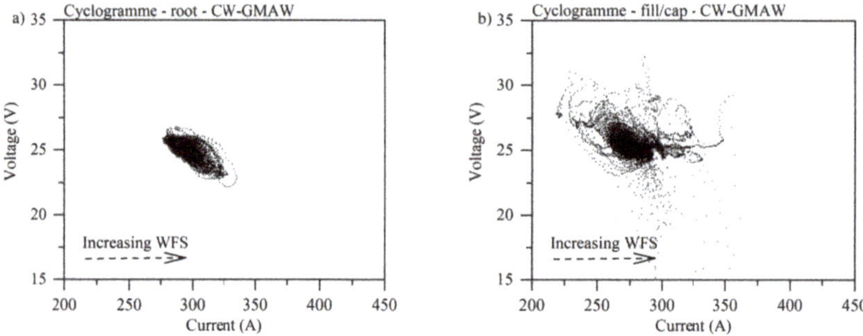

Figure 5. Cyclogrammes for the GMAW specimens: (**a**) root pass; (**b**) fill pass; and (**c**) fill pass.

3.2. High Speed Imaging

Figure 6 shows the high speed frames for the standard GMAW and CW-GMAW inside the grooves. As suggested by the electrical signals, it was possible to verify that the arc for the GMAW often attaches to the sidewalls, thus eroding it. The same author claims that the arc attachment was avoided when welding in constant current mode, since, in these sources, to maintain and almost constant melting speed, the voltage highly varies. Meanwhile, in CW-GMAW, the arc attachment was to the cold wire, which prevents sidewall erosion. In Figure 6a, one can discern that the arc is pinned to the rear (cold wire) and not to the sidewalls.

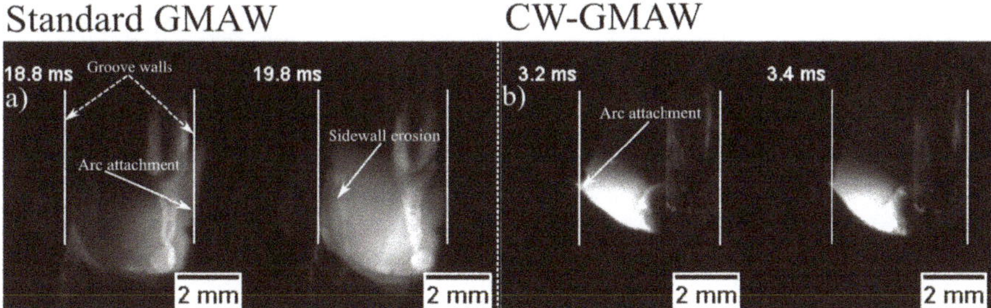

Figure 6. High speed frames, arc dynamic behavior inside the groove: (**a**) standard GMAW; and (**b**) CW-GMAW.

3.3. Sidewall Penetration

The erosion of the sidewall was caused by the arc self regulation dynamics, which tried to establish the shortest path to electron flow, thereby increasing the melting rate. However, as the arc moved through the groove, the attachment point was continually changing, melting multiple points across the groove walls. This multiple erosion points might have detrimental effects on the mechanical properties of the joint.

Figure 7 shows the sidewall penetration caused by the root pass during welding. One observes that the standard GMAW erodes the sidewalls consistently with the oscillograms shown in Figure 2a. There, every time the arc attached to the sidewalls, intense and fast short-circuiting caused the current to increase abruptly (to fuse the extra metal and restore the compatible voltage (arc length, in constant voltage) to the current pre-set before welding). Conversely, as in the CW-GMAW, the arc was attached to the cold wire, thus there is effectively no sidewall penetration, as can be seen in Figure 6a.

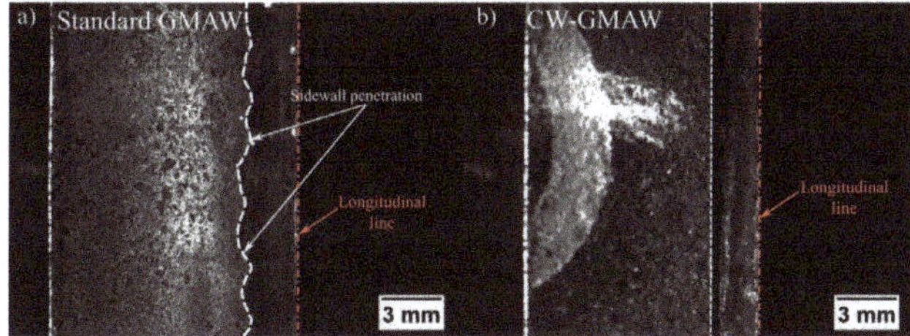

Figure 7. Top view of the sidewall penetration caused by the root pass; the specimens were cut along the longitudinal line (dashed red line): (**a**) standard GMAW; and (**b**) CW-GMAW. Once can note that the specimens were cut along the longitudinal axis of the joint. Note: In (**b**), the mark on its front is a written identification of the sample that could not be removed.

3.4. Evolution of the HAZ in Root Pass

Figure 8 shows the evolution of the heat affected zone (HAZ) in the root pass for the standard GMAW welds in three different locations of the bead: start, middle and end. One can find that the arc wandering resulted in welds that are rather non-symmetrical. This was likely due to the inconsistency in penetration, associated with is incomplete fusion defect. Moreover, the size of the root face also contributed to this issue. In addition, one can note in Figure 8 that, as the weld progressed, the plate became hotter and consequently penetration was slightly increased in the end of the weld.

Figure 9 shows the root pass in three locations across the bead for CW-GMAW; no discontinuities (based on cross-sections inspection) were likely to occur in the root pass, and, generally, the sidewall penetration in the root was much more symmetrical due to the arc pinning to the cold wire. One finds that the increase in deposition decreased the penetration of the weld, causing incomplete penetration.

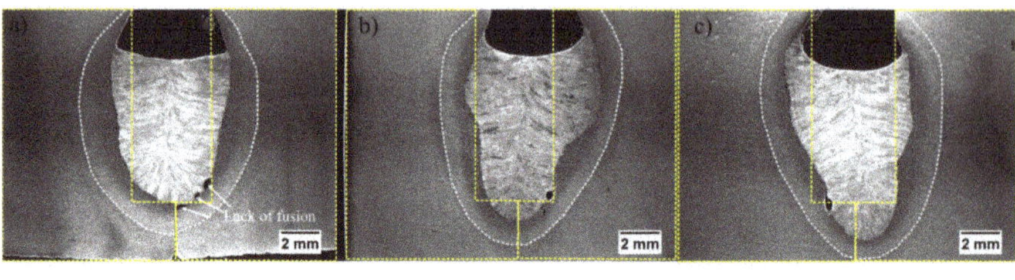

GMAW - Start   GMAW - Middle   GMAW - End

Figure 8. Root pass macrographs for the standard GMAW process: (**a**) start of the joint; (**b**) middle of the joint; and (**c**) end of the joint.

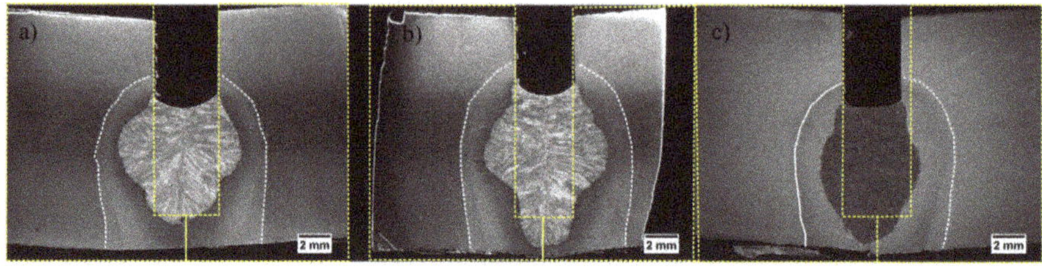

CW-GMAW - Start   CW-GMAW - Middle   CW-GMAW - End

Figure 9. Root pass macrographs for the CW-GMAW process: (**a**) start of the joint; and (**b**) middle of the joint; and (**c**) end of the joint.

3.5. Macrographs

Figure 10 shows cross-sections from the middle of the bead for both the standard GMAW and CW-GMAW. One notices that the middle cross-section of the standard GMAW has an acceptable morphology without discontinuities, while achieving suitable penetration in the root face. Conversely, incomplete penetration persists in CW-GMAW. Due to the cold wire feed rates, more mass was deposited in CW-GMAW, which caused a decrease in penetration. This accounts for the incomplete penetration. Another technicality in CW-GMAW is the presence of inclusions, most likely oxides, due to the increased level of titanium in the weld metal due to the cold wire feeding, as the same wire was used as electrode and cold wire.

The presence of inclusions points to the need for more careful grinding after the root to clean the silicates formed during welding pool solidification. Another feature that differs between the cross-sections is the quantity of passes to fill them out. Standard GMAW required three passes and CW-GMAW only two. Ultimately, the overall heat input in CW-GMAW was lower due this difference in number of passes.

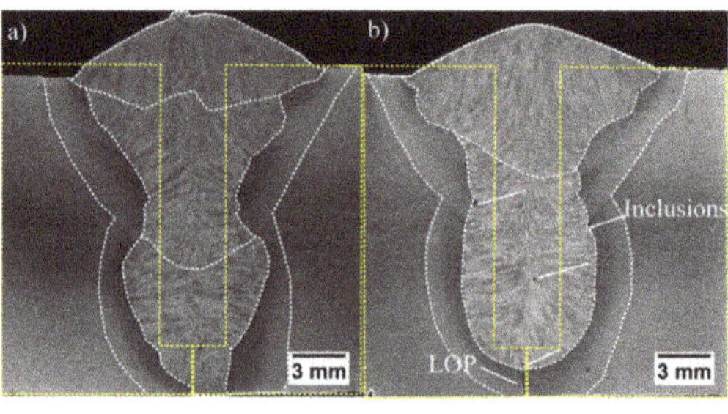

Standard GMAW   CW-GMAW

Figure 10. Middle of the bead cross-section showing the final weld morphology: (**a**) standard GMAW; and (**b**) CW-GMAW. The arrows indicate discontinuities such as inclusions and incomplete fusion.

To mitigate incomplete fusion, two alternatives might be used: increase the current by means of wire feed speed to increase penetration, or decrease the root face height to facilitate higher penetration.

3.6. Micrographs

The HAZ width of the standard GMAW and CW-GMAW specimens were compared. The intercritical heat affected zone (ICHAZ), which is basically formed between 800 and 500 °C upon cooling, is narrower in CW-GMAW welds compared to GMAW.

Weld Interface—Root Pass

Figure 11a shows the HAZ from the weld interface to base metal in the conventional GMAW specimen. Figure 11b shows the HAZ from the interface in the CW-GMAW specimen. In both specimens, the images were taken at the root pass, in a one pass weld (see Figures 8b and 9b, respectively).

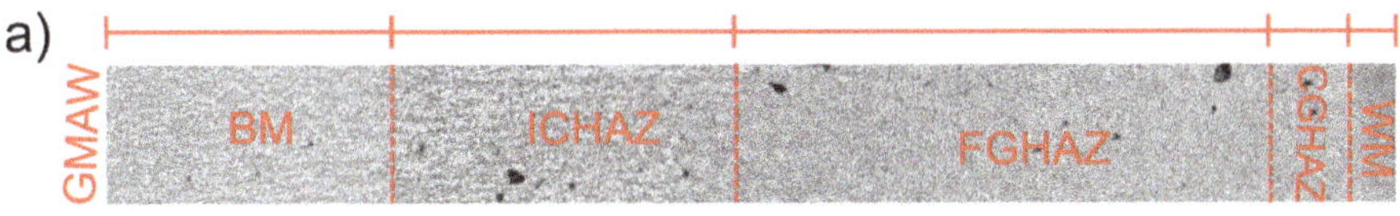

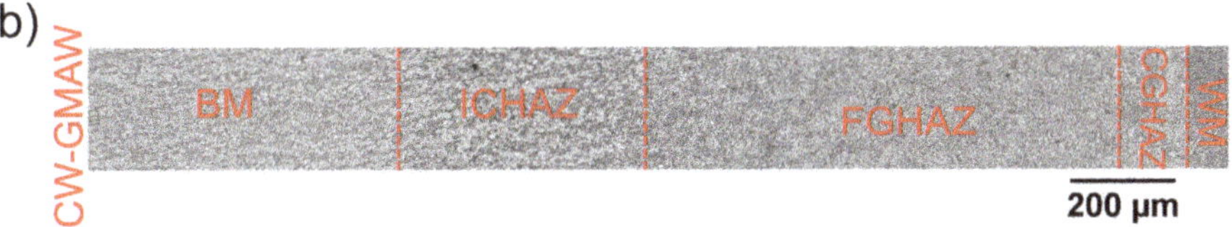

Figure 11. Heat affected zone (HAZ) width in the GMAW specimen showing the Weld metal (WM), Coarse grain heat affected zone (CGHAZ), Fine grain heat affected zone (FGHAZ), Intercritical heat affected zone (ICHAZ), and Base metal (BM): (**a**) standard GMAW; and (**b**) CW-GMAW.

One can see that the ICHAZ is slightly narrower in CW-GMAW than in the standard GMAW. This is consistent with previous research. This seems to indicate that CW-GMAW specimens had a shorter cooling time between 800 and 500 °C compared to the conventional GMAW specimens. This seems to indicate that less heat was actually applied across the weld joint.

3.7. Vickers Hardness

To characterize the strength of the of the joint via cross-sections, maps of the Vickers hardness values are reported here. Figure 12 shows the hardness map of the completed joint for standard GMAW and CW-GMAW specimens.

One can see that the Vickers hardness in the weld for the CW-GMAW is higher, which indicates that the cooling rate of the weld metal in CW-GMAW was faster than the standard GMAW. For some conditions, the melting efficiency of CW-GMAW was higher than in conventional GMAW. The melting efficiency is the amount of heat that actually transfers to the melting pool over the welding total power, and the difference in this value might explain slower cooling rates.

Figure 13 shows the hardness map of the root pass in cross-section extracted from the middle of the joint. Regarding the root pass, one can see that the values are similar in standard GMAW and CW-GMAW. However, comparing Figures 12 and 13, one can see that the bottom of the CW-GMA welds is harder than the bottom of conventional GMAW. Given that the hardness patterns of the roots were similar, one has to consider that the cooling rate near the root was faster in CW-GMAW. However, the exact mechanism behind this fact is still unclear, but may be due to a faster heat transfer due to a larger contact area (the area of contact between the weld metal and the base metal) as well as a higher gradient between the weld and base metals in CW-GMA welds.

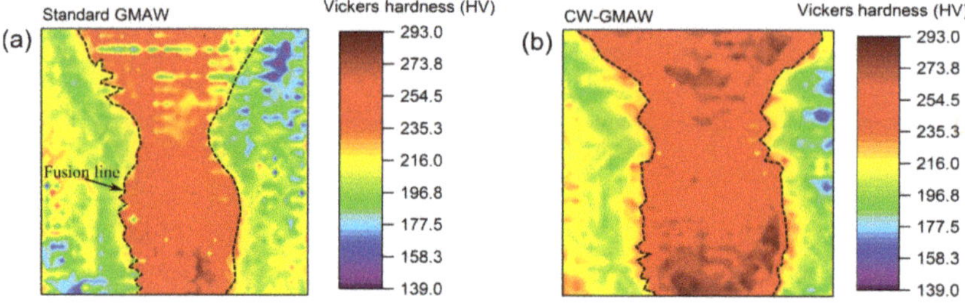

Figure 12. Hardness map of the cross-sections: (**a**) standard GMAW; and (**b**) CW-GMAW, with weld interface marked by dashed lines.

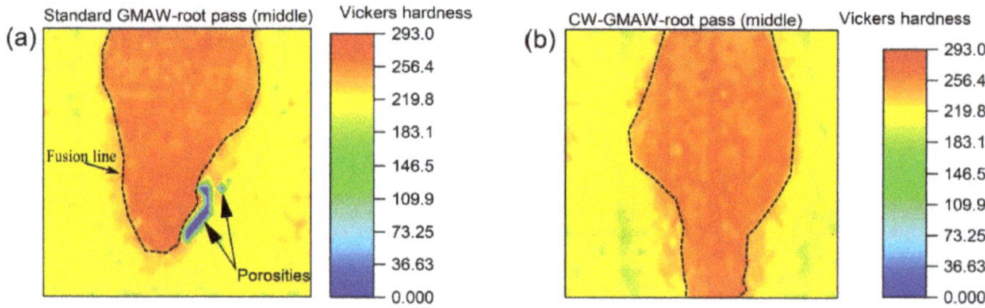

Figure 13. Hardness map of the root pass, middle cross-section: (**a**) standard GMAW; and (**b**) CW-GMAW, with weld interface marked by dashed lines.

4. Discussions

The electrical oscillograms in Figure 2 show that the standard GMAW experiences some sort of oscillating pattern. It was noticed that this pattern corresponds to the instant the arc eroded the sidewalls. Moreover, the author pointed out that the arc attached to sidewall such that the current flows in the shortest path, thus the arc potential in the arc column remains at the minimum possible level.

This phenomenon in constant voltage (CV) leads ultimately to the melting of the contact tip and interruption of the welding processes. However, it was noted that, in constant current (CC), this process did not occur since the reduction in arc length, induced by the arc attachment to the sidewall, caused a small reduction in arc length in comparison the reduction in current, thus the power source reached a new equilibrium point.

On the contrary, in CW-GMA welding, the current path was shortest to the cold wire, which caused the arc to climb to it. This phenomenon prevented the arc attachment to the sidewalls, and consequently prevented sidewall erosion. Moreover, one observes that the arc was much more stable in CW-GMA welding (Figure 3) than in standard GMAW (Figure 2). The cyclogrammes for the the two cases confirmed this assertion. The root pass performed using CW-GMAW (Figure 5a) was more stable than in the conventional GMAW (Figure 4a), which had a tail that points to short-circuits caused by the sidewall erosion (dashed square).

Regarding the cap pass, one can observe that, in conventional GMAW, the arc tended to climb towards the contact, as indicated by the higher values of voltage (increase in arc length) and current in Figure 4c. However, in the CW-GMAW, such phenomenon did not occur as systematically, with only occasional values of current and voltage tending to the high voltage region. High speed images were used to investigate the arc dynamical behavior. The images show that the arc attached to the sidewalls in standard GMAW because of the internal regulation of the source in CV, as discussed above (Figure 6). In CW-GMAW, as expected from the electrical signals, no sidewall erosion was detected, as can be seen in Figures 6 and 7.

Figure 8 shows the macros of the joint manufactured using conventional GMAW. One can observe that some fusion points are lacking. These were caused by the arc climbing causing over-melts in one direction, but leaving a gap and causing the lack of fusion. On the contrary, in CW-GMAW, one can notice that, as the arc is pinned to the cold wire, this causes a more stable melting pattern avoiding incomplete fusion across the joint (Figure 9).

The macros, taken from the middle of joints (Figure 10), show the difference in productivity. The standard GMAW joint was completely filled with three passes while the the CW-GMAW was filled with two passes. However, one observes that, in relation to the cold wire joint, there was a lack of penetration due to the arc pinning to the cold wire, which limits the penetration and dilution.

Figure 11 compares the HAZ for the two processes: standard GMAW and CW-GMAW. The difference in their size can be attributed to differences in thermal signature of CW-GMAW in comparison to GMAW. The difference in ICHAZ might be linked to higher thermal gradient in CW-GMAW than in conventional GMAW. This thermal gradient might result from an improved melting efficiency in CW-GMAW for some conditions.

Figures 12 and 13 show the hardness maps over the complete macro and the root pass for both conventional GMAW and CW-GMAW, respectively. The Vickers hardness values in CW-GMAW (Figure 12) point to shorter cooling time (higher cooling rate), which might be linked to the higher thermal gradient caused by the possibly higher melting efficiency in CW-GMAW. This might explain the higher hardness in CW-GMAW compared to the standard GMAW. Regarding the root pass, the difference in hardness was likely due to the larger gradient formed by the larger joined area in CW-GMAW root. This led to higher cooling rates in the reheated root of the weld metal, with higher hardness.

5. Conclusions

Narrow groove welds in API X80 were fabricated using the standard GMAW and the CW-GMAW to assess the process feasibility using CW-GMAW for a joint with 4 mm gap. Taking into account the results discussed, the following conclusions can be drawn:

1. The welds using GMAW process presented serious sidewall penetration that might compromise their integrity.
2. The welds fabricated using CW-GMAW did not present sidewall erosion during root pass, conversely the increase in deposition compromised the penetration in the root region.
3. The amount of inclusions found in the the cold wire welds might be linked to the grade of the welding wires used.

Author Contributions: R.A.R. conceived, designed, and performed the experiments; P.D.C.A. assisted with measurements and data collection; E.B.F.D.S., A.A.C.F. and E.M.B. contributed to the discussion of the experimental data; R.A.R. wrote the manuscript; and A.P.G. interpreted results and revised the manuscript.

Funding: The authors acknowledge Natural Science and Engineering Research Council of Canada (NSERC) and TransCanada Pipelines Ltd. for the financial support.

Acknowledgments: The authors would like to acknowledge the Centre for Advanced Materials Joining (CAMJ) of the University of Waterloo, for the use of its facilities where all the welds were performed.

References

1. Li, W.; Gao, K.; Wu, J.; Wang, J.; Ji, Y. Groove sidewall penetration modeling for rotating arc narrow gap MAG welding. *Int. J. Adv. Manuf. Technol.* **2015**, *78*, 573–581. [CrossRef]
2. Cai, X.Y.; Lin, S.B.; Fan, C.L.; Yang, C.L.; Zhang, W.; Wang, Y.W. Molten pool behaviour and weld forming mechanism of tandem narrow gap vertical GMAW. *Sci. Technol. Weld. Join.* **2016**, *21*, 124–130. [CrossRef]
3. Michie, K. Twin-wire GMAW: Process characteristics and applications. *Weld. J.* **1999**, *78*, 31–34.
4. Selvi, S.; Vishvaksenan, A.; Rajasekar, E. Cold metal transfer (CMT) technology—An overview. *Def. Technol.* **2018**, *14*, 28–44. [CrossRef]
5. Ribeiro, R.A.; dos Santos, E.B.F.; Assunção, P.D.C.; Maciel, R.R.; Braga, E.M. Predicting Weld Bead Geometry in the Novel CW-GMAW Process. *Weld. J.* **2015**, *94*, 301s–311s.
6. Xiang, T.; Li, H.; Wei, H.L.; Gao, Y. Effects of filling status of cold wire on the welding process stability in twin-arc integrated cold wire hybrid welding. *Int. J. Adv. Manuf. Technol.* **2016**, *83*, 1583–1593. [CrossRef]
7. Cabral, T.S.; Braga, E.M.; Mendonça, E.A.M.; Scott, A. Influence of procedures and transfer modes in MAG welding in the reduction of deformations on marine structure panels. *Weld. Int.* **2015**, *29*, 928–936. [CrossRef]
8. Maciel, R.R.; Filho, L.d.S.B.; Assunção, P.D.C.; Braga, E.d.M. Comparative Study Betweenthe Processes Gmaw and Gmaw-Cw With Variations in the Wire Feed To Fill Chamfer in "U". In Proceedings of the 22nd International Congress of Mechanical Engineering (COBEM), Ribeirão Preto, Venue, 3–7 November 2013; Number 1, pp. 1–9.
9. Assunção, P.D.C.; Ribeiro, R.A.; dos Santos, E.B.F.; Gerlich, A.P.; Braga, E.M. Feasibility of narrow gap welding using the cold-wire gas metal arc welding (CW-GMAW) process. *Weld. World* **2017**, *61*, 659–666. [CrossRef]
10. Costa, E.S.; Assunção, P.D.C.; Dos Santos, E.B.F.; Feio, L.G.; Bittencourt, M.S.Q.; Braga, E.M. Residual stresses in cold-wire gas metal arc welding. *Sci. Technol. Weld. Join.* **2017**, *22*, 706–713. [CrossRef]
11. Marques, L.F.N.; dos Santos, E.B.F.; Gerlich, A.P.; Braga, E.M. Fatigue life assessment of weld joints manufactured by GMAW and CW-GMAW processes. *Sci. Technol. Weld. Join.* **2017**, *22*, 87–96. [CrossRef]
12. Zhang, G.; Shi, Y.; Zhu, M.; Fan, D. Arc characteristics and metal transfer behavior in narrow gap gas metal arc welding process. *J. Mater. Process. Technol.* **2017**, *245*, 15–23. [CrossRef]
13. Suban, M.; Tušek, J. Methods for the determination of arc stability. *J. Mater. Process. Technol.* **2003**, *143–144*, 430–437. [CrossRef]
14. Jorge, V.L.; Gohrs, R.; Scotti, A. Active power measurement in arc welding and its role in heat transfer to the plate. *Weld. World* **2017**, *61*, 847–856. [CrossRef]

Article

Effect of Temperature and Hold Time of Induction Brazing on Microstructure and Shear Strength of Martensitic Stainless Steel Joints

Yunxia Chen [1] and Haichao Cui [2,*]

[1]   School of Materials Science and Engineering, Shanghai Dianji University, Shanghai 201306, China; cyx1978@yeah.net

[2]   Shanghai Key Laboratory of Materials Laser Processing and Modification, Shanghai Jiao Tong University, Shanghai 200240, China

*   Correspondence: haichaocui@sjtu.edu.cn; Tel./Fax: +86-21-34202814

Received: 11 August 2018; Accepted: 30 August 2018; Published: 1 September 2018

Abstract: 1Cr12Mo martensitic stainless steel is widely used for intermediate and low-pressure steam turbine blades in fossil-fuel power plants. A nickel-based filler metal (SFA-5.8 BNi-2) was used to braze 1Cr12Mo in an Ar atmosphere. The influence of brazing temperature and hold time on the joints was studied. Microstructure of the joints brazed, element distribution and shear stress were evaluated at different brazing temperatures, ranging from 1050 °C to 1120 °C, with holding times of 10 s, 30 s, 50 s and 90 s. The results show that brazing joints mainly consist of the matrix of the braze alloy, the precipitation, and the diffusion affected zone. The filler metal elements diffusion is more active with increased brazing temperature and prolonged hold time. The shear strength of the brazed joints is greater than 250 MPa when the brazing temperature is 1080 °C and the hold time is 30 s.

Keywords: induction brazing; elements diffusion; microstructural evolution; shear strength; stainless steel

---

1. Introduction

1Cr12Mo stainless steel is a modified material made by appropriately increasing the content of Mo to hold the temper brittleness on the base of AISI 403. As a martensitic heat-resistant stainless steel with good creep strength and moderate corrosion resistance, 1Cr12Mo stainless steel is widely used for intermediate and low-pressure steam turbine blades in fossil-fuel power plants. Its mechanical properties, fatigue resistance, and corrosion resistance have been researched [1–4]. However, most of these studies are limited to traditional processing technology, such as furnace and vacuum brazing [5]. Compared with vacuum brazing, induction brazing is a faster and more effective technique, which provides a fast and controllable method of heating to help elements dissolution, diffusion, and chemical reaction between the base metal and the filler metal. The heating rate of induction brazing can reach 100 °C/s, which is important to avoid liquation of the braze alloy with different solidus and liquidus temperatures [6].

For excellent performance of the brazed joint, nickel-based filler metal is often used in high-temperature alloy induction brazing. B, Si, and other elements are added to the filler metal to lower its melting point temperature and improve its liquid flow rate. However, B and Si in the filler metal can react with some metallic elements and form high hardness brittle intermetallic compounds, usually located in the diffusion affected zone of the welded joints. These intermetallic compounds have adverse effects on joint performance. The brazing temperature and hold time at high temperature have a critical influence on the diffusion of B and Si. Proper brazing temperature and hold time are helpful to the diffusion of B and Si, and the diffusion between the filler metal and the base metal [7]. Compared with BMn50NiCuCrCo and BNi82CrSiBFe filler metals, SFA-5.8 BNi-2 filler metal is the best for stainless steel brazing because of its distinct weldability [8]. In this study, induction brazing of 1Cr12Mo using a nickel-based brazing alloy, BNi-2, was investigated. Both the microstructural evolution and shear strength of the brazed joint are evaluated.

2. Experimental Procedure

1Cr12Mo stainless steel was used as the base metal, with the chemical composition in weight percent of Cr (11.50~13.00), Ni (0.30~0.60), Mo (0.30~0.60), Mn (0.30~0.50), C (0.10~0.15), Si (0.05), P (0.035), S (0.030), and Fe (balance) according to the national (Chinese) standard GB8732. A nickel-based filler metal, BNi-2, containing in weight percent Cr (6.0~8.0), B (2.75~3.5), Si (4.0~5.0), Fe (2.5~3.5), Ni (balance) was chosen. The filler metal was in powder form with the granule size about 400 mesh. The base metal was processed to shear specimens described in the national (Chinese) standard GB11363-89, then cleaned using an ultrasonic bath and acetone solvent, and dried with hot air. Before brazing, the shear specimens were assembled as shown in Figure 1. The gap between brazed materials was 2 μm and sufficient filler metal was put on the packing place. To prevent the powder from running away, some alcohol was used during brazing. Figure 2 shows a schematic illustration of the induction brazing. A HX-GP-25 type high-frequency inductor was used as the heating equipment, and the heating current was 600 A. A high speed infrared temperature measuring instrument (Kleiber KMGA740, Kleiber, Allgäucity, German) was used to measure and record the brazing temperature. Due to the low content of Cr elements in BNi-2, to prevent oxidation the induction brazing was performed in an Ar atmosphere, and the Ar gas flow rate was 25 L/min. The brazing temperatures were 1050, 1080, 1120 and 1150 °C. The holding times were 10 s, 30 s, 50 s, and 90 s respectively.

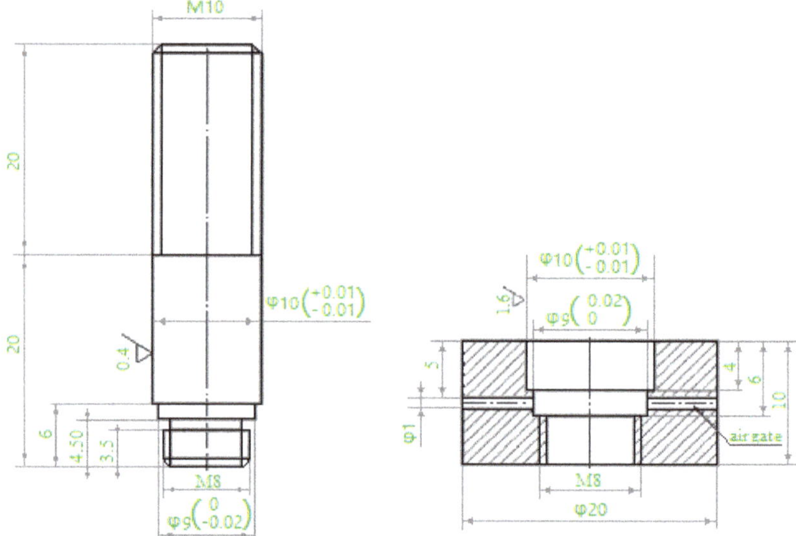

Figure 1. The brazed specimen for shear strength test (units: mm).

The brazed samples were cut using a metallographic sample cutting machine, then executed in accordance with the standard metallographic procedure. The cross section of the brazed joint was examined using the JSM-7600 UHR thermal field emission scanning electron microscope (JEOL, Tokyo, Japan) with an operating voltage of 15 kV. To evaluate the bonding strength of the base metal and the filler metal, the shear test was conducted. The shear test piece was drawn by a universal testing machine (Zwick, Ulm-Einsingen, German) with a constant speed of 1 mm/min at room temperature.

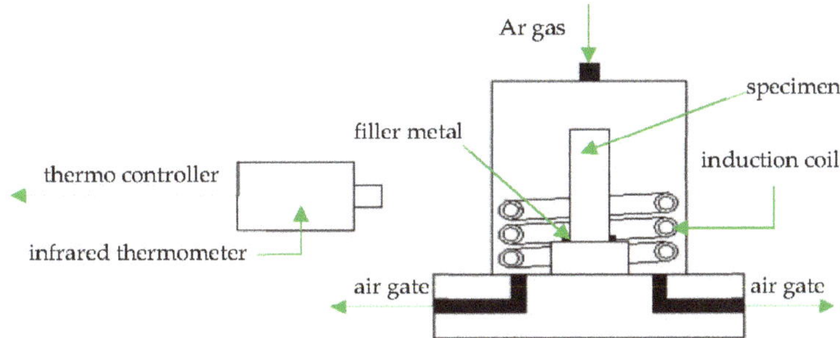

Figure 2. Schematic illustration of the induction brazing.

3. Results and Discussion

Figure 3 shows the SEM backscattered image of BNi-2 brazed at 1120 °C for 10 s, 30 s, 50 s, and 90 s respectively. The distribution of the elements in the joint can be observed in the backscattered image. The specimen areas containing high-atomic number elements appear light, while the areas containing low-atomic number elements appear dark. Based on this information, it is clear that the elements distribution of the joint is not uniform. Due to the rapid heating and cooling rate in the induction brazing, there is not enough time for the elements to distribute to equilibrium. As a result, different phases generated in the joint.

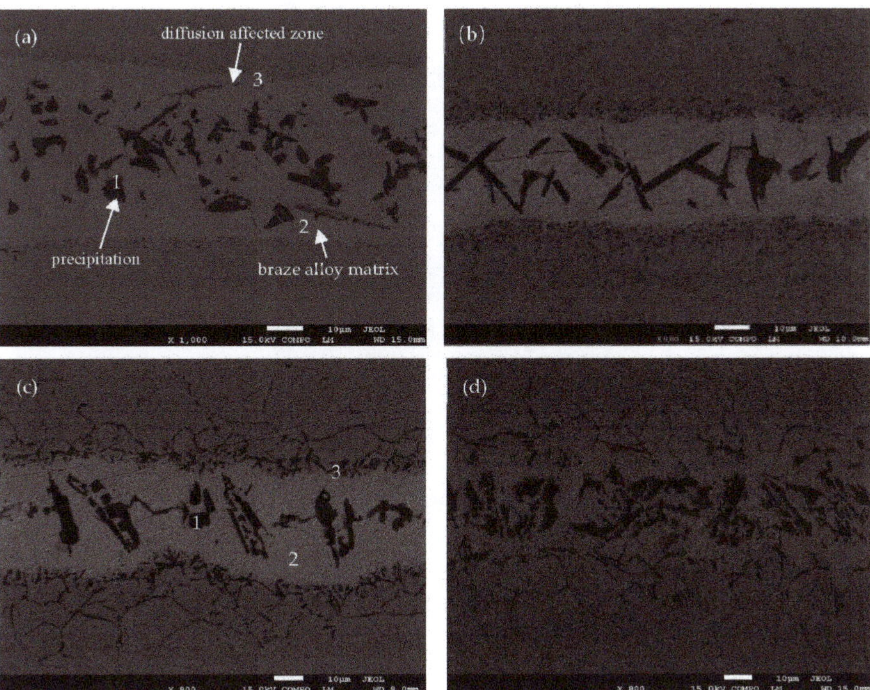

Figure 3. The SEM backscattered image of BNi-2 brazed at 1120 °C for (**a**) 10 s, (**b**) 30 s, (**c**) 50 s, (**d**) 90 s.

As shown in Figure 3, the bond region consists of three parts: the matrix of the braze alloy, the precipitation, and the diffusion affected zone. It has been reported that the matrix of the braze alloy is the isothermally solidified zone (ISZ) formed by isothermal solidification during the holding time [9,10]. The microstructure of the ISZ is γ solid solution (which solutes the rich Ni) and free γ′ precipitates. The precipitation in the middle of the joint is the athermally solidified zone which formed at the end of the solidification and is controlled by added elements to depress the melting point [11]. The diffusion affected zone consists of CrB, due to B diffusion and strong metal compounds for Cr and B.

Figure 3 also proves that the diffusion affected zone is more active as holding time increases. When the holding time is only 10 s, as shown in Figure 3a, there is almost no reaction layer between the base metal and the filler metal. When the holding time is prolonged to 30 s, little reaction layer can be observed (see Figure 3b). When the holding times are 50 s and 90 s, a net structure (see Figure 3c,d) formed in the diffusion affected zone, which has been reported as enhancing the joining strength of the base metal and the filler metal [12]. At the same time, the area of athermally solidified zone decreased.

Figure 4 shows the elements distribution of the joint using line-scan analysis of the brazing temperature at 1120 °C for 10 s. From the base metal to the filler metal, the content of Fe and Cr decrease while the content of Ni and Si increase. The reason is that there is an interdiffusion between the base metal and the filler metal. When the scanning line reaches the precipitation, the content of Fe and Ni decrease sharply, while the content of Cr increases to maximum. Based on the principles of SEM backscattered images, it can be determined that there is B element in the precipitation. Therefore, the precipitation is identified as CrB.

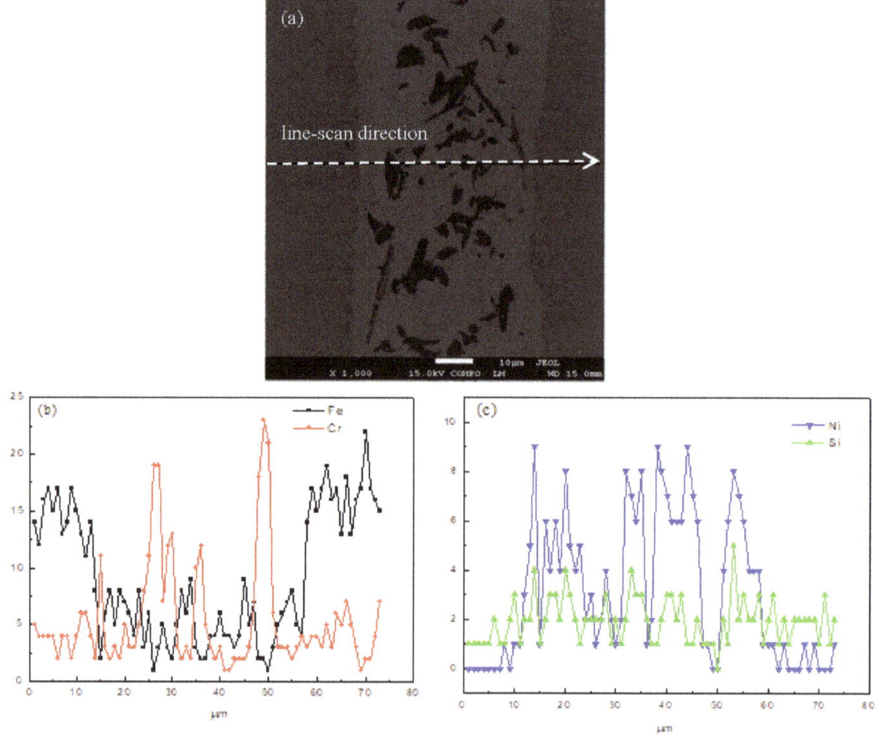

Figure 4. Line-scan analysis of the joint brazed at 1120 °C for 10 s, (**a**) line-scan direction, (**b**) content of Fe and Cr, (**c**) content of Ni and Si.

Figure 3a,c show the chemical analysis of different phases in the joint brazed at 1120 °C for 10 s and 50 s respectively. There are three different phases observed in the joint. Their corresponding chemical compositions are shown in Tables 1 and 2. It suggests that the high brazing temperature will result in some Fe atoms melting from the base metal to the joint, so there are a few Fe atoms detected in the joint. Meanwhile, the different chemical compositions of the phases in the two joints result from the different holding times during the brazing procedure. When the brazing temperature is at 1120 °C for 10 s, the precipitation in the joint (point 1 in Figure 3a) is CrB without other elements. This result is consistent with the line-scan result mentioned above. The probable reason is that the precipitation was formed in the althermal solidification at the end of the isothermal solidification. According to the phase diagram, the solubility of B in Ni decreased at the end of the isothermal solidification, then the B element was left in the liquid and resulted in the formation of CrB. With the prolonging of the holding time, more Fe atoms dissolved into the joint. The precipitation at point 1 in Figure 3c, consists of B, Cr, Fe and Ni (as shown in Table 2), which is different from the precipitation at point 1 in Figure 3a. The differences in the matrix of the joints are the content of Fe, Si and B. When the holding time is prolonged, more Fe melts into the joint, and more Si and B atoms diffuse. Due to the atom size of B being smaller than that of Si, the diffusion rate of B is faster than that of Si, and there is no B element detected at point 2 in Figure 3c. Point 3 in Figure 3a is the diffusion affected zone between the base metal and the filler metal. The composition reveals the diffusion of Ni and Si atoms. Point 3 in Figure 3c is a precipitation that primarily comprises B, Cr and Fe. As reported, B diffuses into the base metal to form the intermetallics along the grain boundaries of the base metal [13].

Table 1. The chemical compositions labeled 1,2,3 in Figure 3a.

Element	1		2		3	
	Wt %	At %	Wt %	At %	Wt %	At %
Ni	0.33	0.11	56.38	49.09	11.08	10.45
Cr	58.07	22.54	2.84	2.79	9.63	10.25
Fe	-	-	34.33	31.42	78.57	77.88
Si	0.29	0.21	4.75	8.65	0.72	1.42
B	41.31	77.13	1.7	8.05	-	-

Table 2. The chemical compositions labeled 1,2,3 in Figure 3c.

Element	1		2		3	
	Wt %	At %	Wt %	At %	Wt %	At %
Ni	3.46	1.05	48.09	45.43	1.74	0.47
Cr	24.32	8.31	4.08	4.36	5.17	1.57
Fe	21.13	6.72	45.06	44.75	31.95	9.01
Si	-	-	2.77	5.46	0.16	0.09
B	51.1	83.93	-	-	60.98	88.86

Figure 5 shows the SEM backscattered images of the joints brazed at 1050, 1080 and 1120 °C for 30 s, respectively. The diffusion affected zone (see Figure 5a) is not obvious, but is quite clear in Figure 5b,c. The shape of the boride phase also varies with the brazing temperature. The higher the brazing temperature, the easier the boride phases achieve phase equilibrium.

Table 3 shows the shear test results of brazed joints for varying brazing parameters. Most shear stress values of brazed joints are above 250 MPa, except for test samples 1 and 3; the former was brazed at a low temperature (1050 °C) and the latter was brazed for a short time (10 s). Figure 6a,b show the variation of shear stress for different brazing parameters. In Figure 6a, when the brazing temperature is 1120 °C, the shear stress of the brazing joint increases with the prolonging of the holding time. However, the increase of shear stress is not obvious when the holding time exceeds 50 s. The shear stress of the brazing joint increases with the increasing brazing temperature when the hold time is 30 s, as shown in Figure 6b. The microstructure of the joint is an indicator for its mechanical properties. The different holding time and brazing temperature that resulted in varying boride phases in the joint, which have a slight effect on the shear stress in the joint. However, the shear stress is heavily dependent on the shape of the diffusion affected zone and whether the chemical composition content in the matrix of the braze joint can easily achieve phase equilibrium [14].

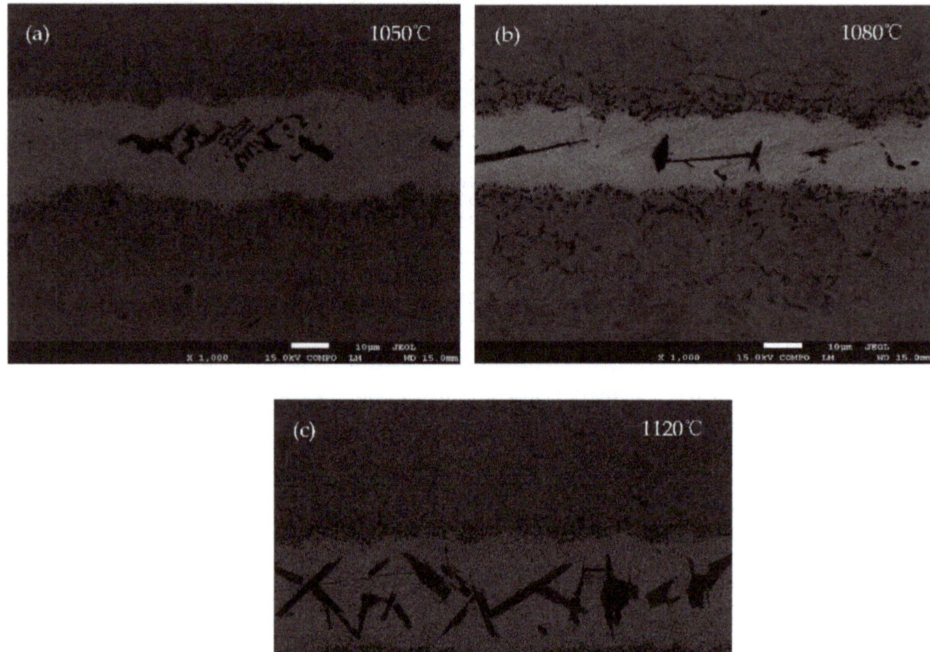

Figure 5. The SEM backscattered images of BNi-2 brazed at (**a**) 1050 °C, (**b**) 1080 °C and (**c**) 1120 °C for 30 s.

Table 3. Shear test results of the brazed joints for different process parameters.

No.	Temperature (°C)	Time (s)	Shear Stress (MPa)
1	1050	30	221.2
2	1080	30	252.1
3	1120	10	235.2
4	1120	30	269.5
5	1120	50	277.8
6	1120	90	285.6
7	1150	30	276.5

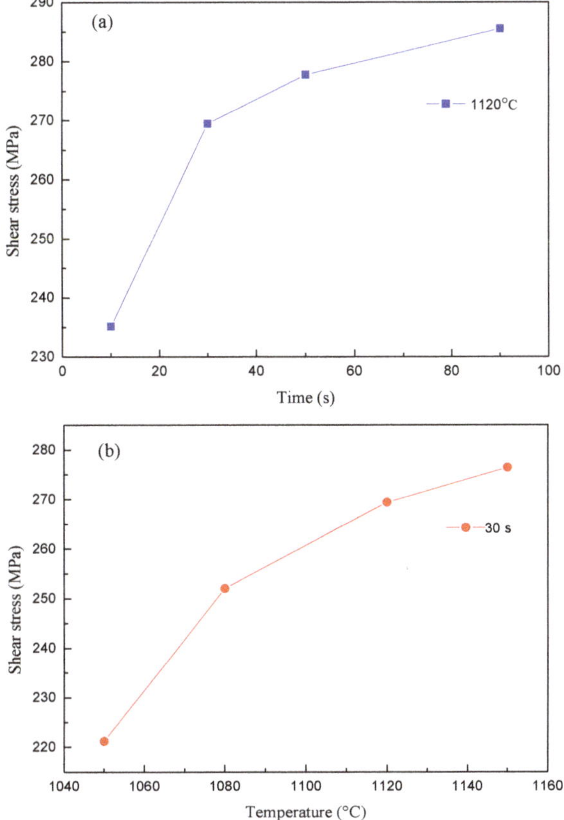

Figure 6. The shear test results of the brazed joints, (**a**) at 1120 °C, (**b**) for 30 s.

4. Conclusions

The induction brazing of 1Cr12Mo martensitic stainless steel with nickel-based filler metal (SFA-5.8 BNi-2) in an Ar atmosphere was performed in this paper. The effect of temperature and hold time of induction brazing on microstructure and shear strength has been discussed. The research results show that the brazed joint consists of three parts, the matrix of the braze alloy, the precipitation and the diffusion affected zone. With increases of temperature and holding time, the strength of the brazed joint was increased. The shear strength of the joints brazed is 285.6 MPa when the Ar gas flow rate is 25 L/min, the heating current is 600 A, the brazing temperature is 1120 °C, and the holding time is 90 s.

Author Contributions: Y.C. conceived and designed the experiments, analyzed the data and wrote the paper; H.C. performed the experiments and contributed analysis tools.

Funding: This research was funded by National Natural Science Foundation of China grant number 51809161, Shanghai Municipal Natural Science Foundation grant number 18ZR1416000.

Acknowledgments: The authors gratefully acknowledge the financial support of the National Natural Science Foundation of China (Grant Nos. U1660101).

Conflicts of Interest: The authors declare no conflict of interest.

References

1. Lemus-Ruíz, J.; Verduzco, J.A.; González-Sánchez, J.; López, V.H. Characterization, shear strength and corrosion resistance of self joining AISI 304 using a Ni Fe-Cr-Si metallic glass foil. *J. Mater. Process Tech.* **2015**, *223*, 16–21. [CrossRef]
2. Wu, N.; Li, Y.; Wang, J. Microstructure of Ni-NiCr laminated composite and Cr18-Ni8 steel joint by vacuum brazing. *Vacuum* **2012**, *86*, 2059–2063. [CrossRef]
3. Jiang, W.; Gong, J.M.; Tu, S.T. Effect of holding time on vacuum brazing for a stainless steel plate-fin structure. *Mater. Des.* **2010**, *31*, 2157–2162. [CrossRef]
4. Jiang, W.; Gong, J.; Tu, S.T. A new cooling method for vacuum brazing of a stainless steel plate-fin structure. *Mater. Des.* **2010**, *31*, 648–653. [CrossRef]
5. Ma, Q.; Li, Y.; Wu, N.; Wang, J. Microstructure of Vacuum-Brazed Joints of Super-Ni/NiCr Laminated Composite Using Nickel-Based Amorphous Filler Metal. *J. Mater. Eng. Perform.* **2013**, *22*, 1660–1665. [CrossRef]
6. Chen, J.; Fu, Y.; Li, Q.; Gao, J.; He, Q. Investigation on induction brazing of revolving heat pipe grinding wheel. *Mater. Des.* **2017**, *116*, 21–30. [CrossRef]
7. Zhou, Y.; Xia, C.; Yang, J.; Xu, X.; Zou, J. Microstructure and properties of W-Cu/1Cr18Ni9 steel brazed joint with different Ni-based filler metals. *Sci. Eng. Compos. Mater.* **2018**, *25*, 463–472. [CrossRef]
8. Chen, Y.; Cui, H.; Lu, B.; Lu, F. The Microstructural Evolution of Vacuum Brazed 1Cr18Ni9Ti Using Various Filler Metals. *Materials* **2017**, *10*, 385. [CrossRef] [PubMed]
9. Ruiz-Vargas, J.; Siredey-Schwaller, N.; Gey, N.; Bocher, P.; Hazotte, A. Microstructure development during isothermal brazing of Ni/BNi-2 couples. *J. Mater. Process Tech.* **2013**, *213*, 20–29. [CrossRef]
10. Ruiz-Vargas, J.; Siredey-Schwaller, N.; Bocher, P.; Hazotte, A. First melting stages during isothermal brazing, of Ni/BNi-2 couples. *J. Mater. Process Tech.* **2013**, *213*, 2074–2080. [CrossRef]
11. Chakraborty, G.; Chaurasia, P.K.; Murugesan, S.; Albert, S.K.; Murugan, S. Effect of brazing temperature on the microstructure of martensitic-austenitic steel joints. *Mater. Sci. Tech.-Lond.* **2017**, *33*, 1372–1378. [CrossRef]
12. Yuan, X.; Kang, C.Y.; Kim, M.B. Microstructure and XRD analysis of brazing joint for duplex stainless steel using a Ni-Si-B filler metal. *Mater. Charact.* **2009**, *60*, 923–931. [CrossRef]
13. Kang, S.W.; Chen, Y.T.; Liu, H.P. Brazing diffusion bonding of micro-nickel cylinders and SUS-316 stainless steel. *J. Mater. Process Tech.* **2005**, *168*, 286–290. [CrossRef]
14. Jiang, W.; Gong, J.; Tu, S.T. Effect of brazing temperature on tensile strength and microstructure for a stainless steel plate-fin structure. *Mater. Des.* **2011**, *32*, 736–742. [CrossRef]

Article

A Primary Study of Variable Polarity Plasma Arc Welding Using a Pulsed Plasma Gas

Zhenyang Lu [1], Wang Zhang [1], Fan Jiang [1,2,*], Shujun Chen [1] and Zhaoyang Yan [1,3]

[1] Engineering Research Center of Advanced Manufacturing Technology for Automotive Components, Ministry of Education, Beijing University of Technology, Beijing 100124, China; lzy@bjut.edu.cn (Z.L.); wang_zhang0731@163.com (W.Z.); sjchen@bjut.edu.cn (S.C.); zhygyan@126.com (Z.Y.)

[2] Beijing Engineering Researching Center of laser Technology, Beijing University of Technology, Beijing 100124, China

[3] Chemical and Materials Engineering, University of Alberta, Edmonton, AB T6G 2R3, Canada

* Correspondence: jiangfan@bjut.edu.cn; Tel.: +86-137-2008-7645

Received: 10 April 2019; Accepted: 20 May 2019; Published: 22 May 2019

Abstract: A process variant of variable polarity plasma arc welding (VPPAW), that is, the pulsed plasma gas VPPAW process, was developed. The pulsed plasma gas was transmitted into the variable polarity plasma arc through a high-frequency solenoid valve to modify the output of the plasma arc. The collection of arc electrical characteristics, arc shapes, and weld formation from VPPAW, double-pulsed VPPAW (DP-VPPAW), and pulsed plasma gas VPPAW (PPG-VPPAW) was carried out to examine if the pulsed plasma gas was able to play a positive role in improving the stability and quality of the VPPAW process. The arc voltage shows that the pulsed plasma gas had a greater influence on the electrode positive polarity voltage. The lower the plasma gas frequency was, the lower the arc voltage fluctuation frequency was and the greater the arc voltage fluctuation amplitude was. From the arc image, it could be observed that the arc core length had a short decrease during the general rising trend after plasma gas was turned on. The arc core width only had a slight change due to the restriction of the torch orifice. Compared with pulsed current wave, the pulsed plasma gas could better enhance the fluidity of the molten pool to reduce porosity during aluminum keyhole welding.

Keywords: variable polarity plasma arc welding (VPPAW); weld formation; pulsed plasma gas; arc voltage

1. Introduction

Variable polarity plasma arc welding (VPPAW) has been widely used in aeronautics, astronautics, and the automobile industry to produce high-quality and high-precision weld joints of aluminum alloy [1,2]. The constrained process in the plasma arc torch leads to high energy density and arc stiffness of the plasma arc, but makes the output of the arc coupled. This characteristic results in a smaller weld lobe curve than for other processes, and leads the keyhole molten pool easily affected by environmental changes in complex welding environments [3]. All of these cause a poor dynamic stability of the keyhole and weld defects, which restrict the application of VPPAW.

Increasing arc energy density is a common method to improve welding quality. The laser hybrid plasma arc welding process [4,5] can improve the energy density and stability of the plasma arc by benefiting from interactions with laser beams. The gas-focusing plasma arc welding process [6], with arc column secondly constricted by focusing gas, can improve the arc restraint degree and stability significantly. The increase in arc stability effectively reduces the disturbance from welding arc to molten pool, but it is difficult to eliminate the other influence on molten pool stability.

Increasing the robustness of the keyhole molten pool is more helpful for the interference resisting. The controlled pulsed plasma arc welding process [7,8], which makes the holes in periodic opening and closing state by adjusting the current output, could effectively improve the robustness of the molten pool. The double-pulsed VPPAW (DP-VPPAW) [9,10] adds additional high-frequency pulsed current into electrode negative (EN) and electrode positive (EP), which makes the molten pool oscillate periodically to improve stability of the welding process and quality. However, the additional current parameters lead the thermal-force synchronous change and cause difficulties of process control.

The vibration-assisted plasma arc welding process [11,12], which uses mechanical vibration to drive molten pool vibration, is an efficient method to decrease the attendant heat fluctuation of the plasma arc, compared with the current wave. These methods produced satisfactory process effectiveness, but the mechanical coupling of vibration system and welding torch reduced the stability and precision of the plasma welding torch structure.

The previous studies [13,14] have shown that the plasma gas flow rate significantly affects the pressure output, but for the heat output, it is insignificant. Based on this, a novel welding process named pulsed plasma gas VPPAW (PPG-VPPAW) was proposed in this study. A specially designed plasma arc torch was used to control plasma gas flow, and the pulsed plasma gas VPPAW system was developed. This paper focuses on the arc behavior and welding process with PPG-VPPAW. The reason for the periodic variation of arc voltage and arc profile during PPG-VPPAW is discussed based on the arc electric signals and arc image acquired from experimental results. Furthermore, the weld-forming experiments were carried out to explore the reason for the improving the fluidity of the molten pool and reducing porosity during aluminum keyhole PPG-VPPAW.

2. Experimental Procedure

2.1. Experimental System

Figure 1 shows the schematic diagram of the PPG-VPPAW system, which included three parts: the pulsed plasma gas control unit, the VPPAW system, and the data acquisition system. The plasma gas control unit consisted of a S7-200 series PLC and a 35A series high-frequency solenoid valve (rated voltage of DC24V, rated power of 5.4 W, conduction response time of 6 ms, outages response time of 2 ms, and the maximum atmospheric flux of 16.2 L/min for argon). The solenoid valve was installed in the plasma gas tube of the plasma arc torch and as close to the torch nozzle as possible to generate the pulsed plasma gas flow. The VPPAW system consisted of a welding power source, a modified plasma arc torch, and the needed accessories. The data acquisition system mainly consisted of a voltage sensor, a current sensor, a high-speed camera, a data acquisition card, and an industrial computer. During the welding process, the welding current and voltage were collected by the current and voltage sensors. The plasma arc was imaged by the high-speed camera (IDT Y4 series), which was set to focus on the fixed region around the nozzle of the unmovable plasma arc torch. The experimental data was displayed and recorded by the data acquisition card in the industrial control computer.

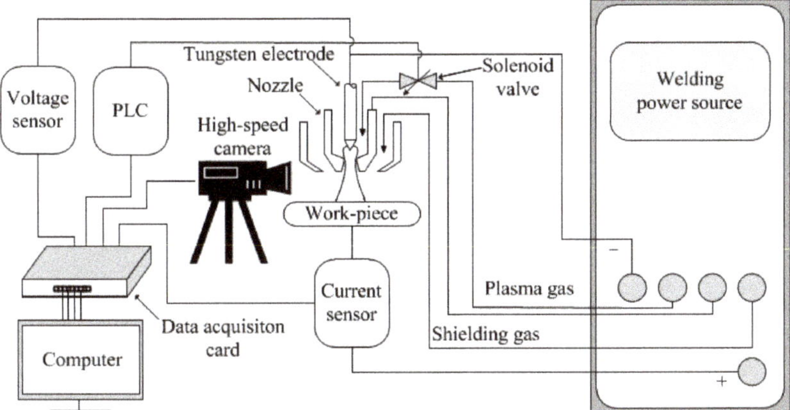

Figure 1. Schematic diagram of the PPG-VPPAW system.

2.2. Experimental Design

The principle of the PPG-VPPAW process is shown in Figure 2. When the solenoid valve was closed, the plasma gas flow was blocked by the solenoid valve and gathered at the entrance of the solenoid valve. When the solenoid valve was opened, the blocked plasma gas flow was released and input into the plasma arc at a velocity greater than the set value at that moment. Then, the plasma gas flow returned to the set value and waited the next closure of the solenoid valve.

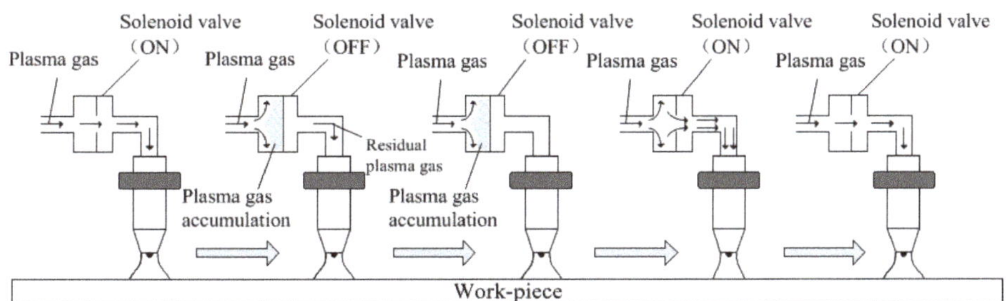

Figure 2. The principle of PPG-VPPAW process.

In order to study the effect of pulsed plasma gas acting on the plasma arc, the arc electric signals and arc image were acquired from PPG-VPPAW and compared with those in VPPAW and DP-VPPAW. The current and plasma gas flow waveforms of the VPPAW, DP-VPPAW, and PPG-VPPAW processes are shown in Figure 3. The VPPAW process has the variable polarity square current wave with unequal straight and reverse polarity time intervals, and the plasma gas flow rate remains constant, as shown in Figure 3a. For the DP-VPPAW process, the current wave has been decreased periodically, the total current wave presents two periodically varying pulses (i.e., a high-frequency variable polarity pulse and a low-frequency pulse), and the plasma gas flow rate is the same as for the VPPAW process, as shown in Figure 3b. In the PPG-VPPAW process, the current waveform is the same as for the VPPAW process while the plasma gas flow rate is input intermittently in the form of a pulse, as shown in Figure 3c. Based on a large number of trials, the process parameters in this study were selected in Table 1 to make the comparison clear. In Table 1, I_{EN} was the electrode negative current; I_{EP} was the electrode positive current; I_B was the basic current in DP-VPPAW; I_P was the pick current in DP-VPPAW. The arc image capture rate was 3000 fps and the signal sampling rate was 10,000 Hz. The plasma gas and the shielding gas both were pure argon.

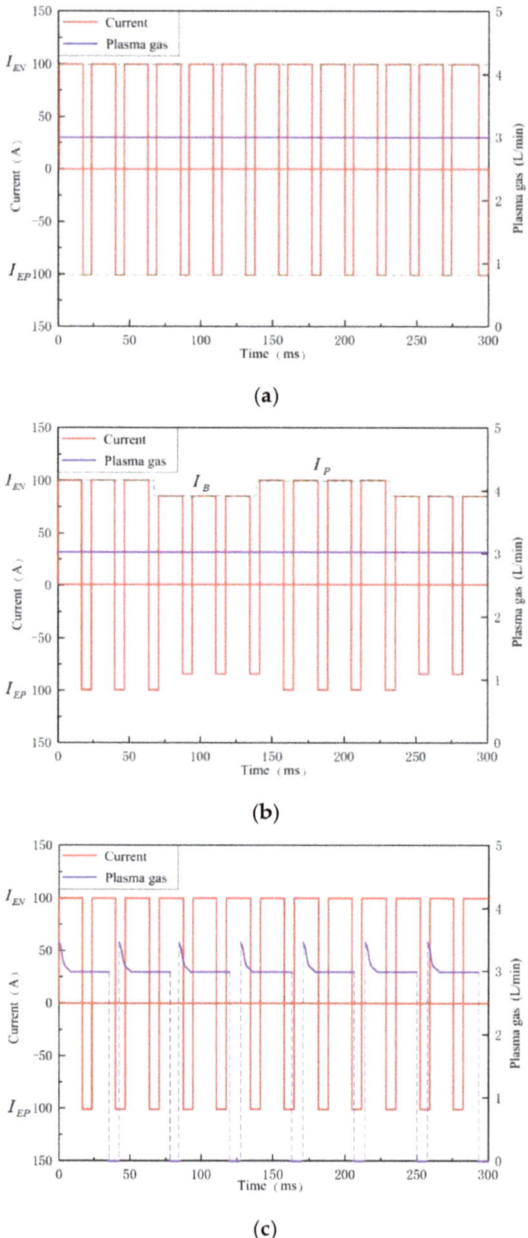

Figure 3. Schematic diagram of current–plasma gas flow waveform. (**a**) VPPAW; (**b**) DP-VPPAW; (**c**) PPG-VPPAW.

Table 1. Parameters for arc image and electrical signal comparison.

Number	Arc Length (mm)	Tungsten Setback (mm)	Shielding Gas Flow Rate (L/min)	Plasma Gas Flow Rate (L/min)	$I_{EN}{:}I_{EP}$ (A)
1-1	6	4	10	3.0	100:100
1-2	6	4	10	3.0	100:100
1-3	6	4	10	3.0	100:100
1-4	6	4	10	3.0	100:100

Number	$I_B{:}I_P$ (A)	Electric Pulse Frequency (Hz)	Electric Pulse Duty Cycle	Solenoid Valve Frequency (Hz)	Solenoid Valve Duty Cycle
1-1	-	-	-	-	-
1-2	80%	3	5:3	-	-
1-3	-	-	-	4	4:1
1-4	-	-	-	20	4:1

In order to study the effect of pulsed plasma gas on the molten pool in the keyhole welding process, the weld-forming experiments were carried out. The weld bead geometries and porosity distribution from the VPPAW, DP-VPPAW, and PPG-VPPAW processes were investigated. In order to study the influence of the pulsed plasma gas on the molten pool, the filling wire was not applied for avoiding the effect of filling material on molten pool behavior and simplifying the experimental model. In these experiments, 5 mm thick 5A06 aluminum alloy was selected as the work piece. The plasma gas and the shielding gas both were pure argon. Based on a large number of trials, the parameters in this part were selected to obtain a good weld bead geometry, as shown in Table 2.

Table 2. Parameters for weld-forming experiments.

Number	Arc Length (mm)	Tungsten Setback (mm)	Shielding Gas Flow Rate (L/min)	Plasma Gas Flow Rate (L/min)	$I_{EN}{:}I_{EP}$ (A)
2-1	5	4	10	3.0	130:150
2-2	5	4	10	3.0	130:150
2-3	5	4	10	3.0	130:150
2-4	5	4	10	3.0	130:150

Number	$I_B{:}I_P$ (A)	Welding Speed (mm/s)	Electric Pulse Frequency (Hz)	Electric Pulse Duty Cycle	Solenoid Valve Frequency (Hz)	Solenoid Valve Duty Cycle
2-1	-	1.83	-	-	-	-
2-2	80%	1.83	3	5:3	-	-
2-3	-	1.83	-	-	20	4:1
2-4	-	1.83	-	-	40	4:1

3. Results and Discussion

3.1. Variation in Arc Electrical Signal

The welding electrical signals of VPPAW (Experiment 1-1) and DP-VPPAW (Experiment 1-2) are shown in Figure 4a,b, respectively. It can be observed that the current waves well fit the preset parameters. The root mean square (RMS) and absolute mean (AM) values of current are shown in Table 3. In comparison with traditional VPPAW, the current of PPG-VPPAW was almost unchanged. Figure 4c,d show the welding electrical signals of PPG-VPPAW under different plasma gas flow pulse frequencies with the same plasma flow rate; the plasma gas flow pulse frequencies were 4 Hz (Experiment 1-3) and 20 Hz (Experiment 1-4), respectively. Compared with Figure 4a,c, it can be observed that the current wave had a negligible effect on the pulsed plasma gas, while the arc voltage fluctuated periodically with the pulsed plasma gas frequency. The electrode negative period voltage (U_{EN}) in the VPPAW process and solenoid valve on state of the PPG-VPPAW process was 24.97 V on average, and the electrode positive period voltage (U_{EP}) was 35.49 V on average. When the solenoid valve was off state and the gas flow pulse frequency was 4 Hz, the U_{EN} and U_{EP} decreased by 5.8 V and 9.8 V, respectively. When the gas flow pulse frequency was 20 Hz, the U_{EN} and U_{EP} decreases were 5.1 V and 8.9 V, respectively. The above results showed that the lower the frequency of plasma gas, the more distinct the influence on the arc voltage waveform. At a certain plasma gas pulse frequency, the pulsed plasma gas has a greater influence on the electrode positive polarity voltage.

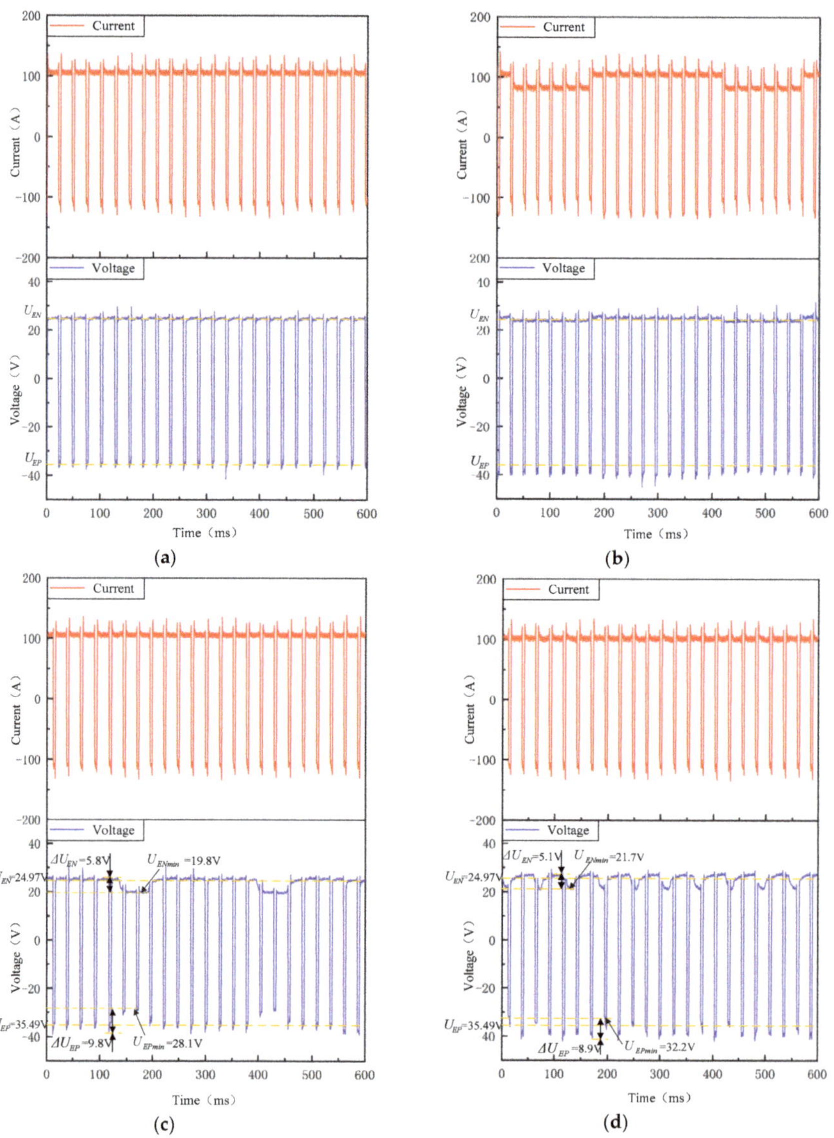

Figure 4. Welding current–voltage synchronizing wave form. (**a**) VPPAW; (**b**) DP-VPPAW; (**c**) PPG-VPPAW (4 Hz); (**d**) PPG-VPPAW (20 Hz).

Table 3. The root mean square (RMS) and absolute mean (AM) values of current.

Welding Methods	RMS (A)	AM (A)
VPPAW	100.95	100.66
DP-VPPAW	93.56	92.39
PPG-VPPAW(4 Hz)	100.98	100.66
PPG-VPPAW(20 Hz)	100.84	100.08

When the plasma gas injects into the arc column, it requires more energy to keep enough ionized particles that let the current through. When the plasma gas is shut off, the required energy which was used to ionize the gas decreases, and the arc voltage also decreases. This decrease relates to the shut-off time of plasma gas and has a maximum value. With the frequency of plasma gas increased, the shut-off time decreases with lack of time to let the arc voltage decrease to the maximum value. In the EP phase of the PPG-VPPAW process, due to the cathodic cleaning phenomenon [15], the size of the arc profile is larger than that in the EN phase, which lets it be more affected by the plasma gas shut-off.

Figure 5 displays more details about the arc voltage of PPG-VPPAW with the plasma gas flow pulse frequencies of 4 Hz and 20 Hz. The pulse signals of "1" and "0" indicate that the solenoid valve is in the on state and off state, respectively. It can clearly be seen that with the plasma gas flow pulse frequency of 4 Hz, the arc voltage is dropped immediately to the minimum voltage when the plasma gas is shut off, which costs 14 ms, and the voltage decrease rate is 0.44 V/ms. When the plasma gas is turned on, it takes 28 ms to return the average voltage, and the voltage increase rate is 0.20 V/ms accordingly. Once the plasma gas flow pulse frequency increases to 20 Hz, the voltage decrease rate is nearly same as that in 4 Hz (0.46 V/ms), but it is hard to find a stable minimum voltage due to the lack of shut-off time of the plasma gas. When the plasma gas is turned on in 20 Hz situation, the voltage recovery time is 22 ms, and the voltage increase rate is 0.23 V/ms accordingly. The above results show that the plasma arc needs more time to recover the effect of plasma gas closure, and the plasma gas flow pulse frequency has less effect on the voltage decrease rate.

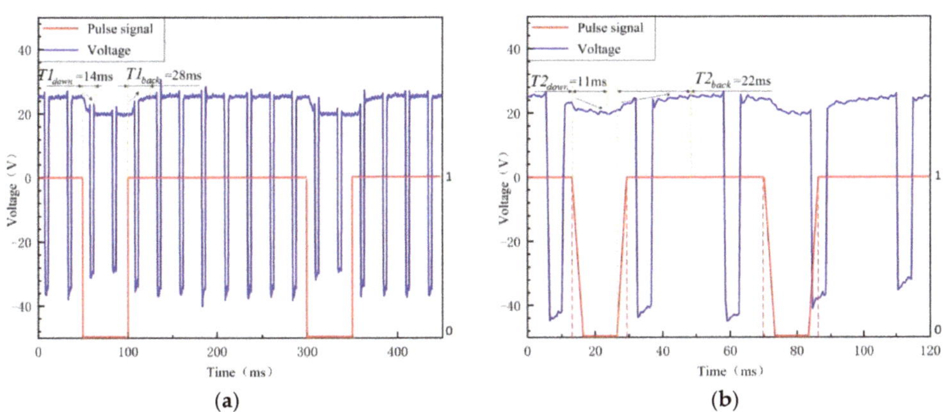

Figure 5. The arc voltage of PPG-VPPAW. (**a**) PPG-VPPAW (4 Hz); (**b**) PPG-VPPAW (20 Hz).

When the plasma gas is shut off, there is no plasma gas injecting immediately and the arc zone could be considered as a relative closed environment and establish balance easily. Compared with it, when the plasma gas is turned on, the plasma gas continuously injects into the arc column and makes it harder to establish balance, so the arc voltage needs more time to stabilize after the plasma gas is turned on. Furthermore, the higher the plasma gas frequency is, the lower the plasma gas accumulation rate is, so the arc voltage needs less time to stabilize after the plasma gas is turned on. The voltage decrease rate is correlated with arc characteristic, but there is no significant correlation with the frequency of plasma gas. However, when the frequency of plasma gas is above a certain value, the plasma gas is turned on again when the arc voltage is not yet decreased to the stable voltage without plasma gas, so the recovery speed of the voltage becomes faster.

3.2. Variation in Arc Profile

In order to better observe and analyze the variation in arc profile collected by the high-speed video camera, it is necessary to divide the arc region for regionalization and measurement. The arc image processing procedure is shown in Figure 6. The original arc image is shown in Figure 6a. Figure 6b is the colored gray-scale image that is transformed from the original arc. According to the intensity of the arc light, the arc can be divided into 256 scale levels. Yang et al. illustrated that the gray scale was larger with higher intensity in the arc gray-scale image, with 90% of the arc intensity as the arc core region [16,17]. The gray-scale image was colored by RGB for the convenience of arc core region [18,19], the red and green region of processed arc image were the core and edge region, respectively. As a result, the arc profile would be measured and analyzed using the arc core region, as presented in Figure 6b. The arc profile, which was defined by arc core length L and arc core width under the nozzle D, was accurately measured by computerized measurement technique.

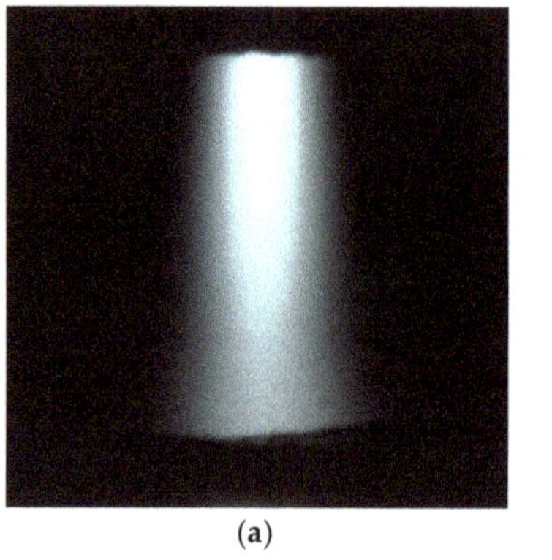
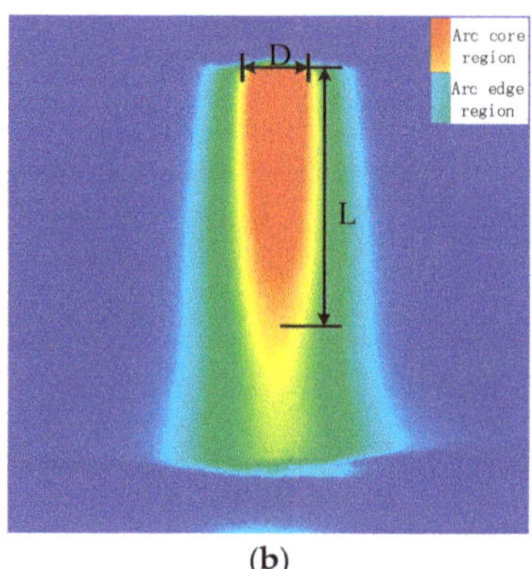

(a) (b)

Figure 6. Edge extraction and regionalization of welding arc. (**a**) Original arc image; (**b**) processed arc image.

Figure 7a displays the variation of arc profile in a complete plasma gas cycle of PPG-VPPAW with the 4 Hz plasma gas pulse frequency in EN phase. Figure 8a shows the corresponding L and D, which were measured using the method mentioned above. From 0 to 30 ms, the arc images were relatively stable, the L and D were about 3.76 mm and 1.05 mm, respectively. When the plasma gas was shut off, from 30 ms to 60 ms, the arc constricted rapidly with the L and D decreased to 0 mm. From 60 ms to 80 ms, the core region of the arc almost disappeared because there was no plasma gas supply. When the plasma gas was turned on, from 80 ms to 100 ms, the core region of the arc significantly increased with L and D increased to 3.18 mm and 1.27 mm, respectively. Then, the L had a decrease to 2.07 mm at 110 ms and increased again to 6 mm in 140 ms. After that, the L gradually decreased to 4.09 mm from 140 ms to 170 ms and then maintained in a stable state. In this period, the D only had a slight change due to the restriction of the torch orifice. The above results show that the arc core length had a short decrease during the general rising trend after plasma gas was turned on. These two peaks of arc core length were due to the initial overshoot phenomenon. Once the plasma gas is turned on, the overshoot velocity is higher than the setting value, which causes the subsequent plasma gas to not keep up, leading the L to decrease.

Figure 7b shows the variation of arc profile in a complete plasma gas cycle of PPG-VPPAW with the 20 Hz plasma gas pulse frequency in the EN phase. The corresponding L and D are shown in

Figure 8b. It can be seen that the change of PPG-VPPAW arc profile with 4 Hz and 20 Hz plasma gas pulse frequencies have the same tendency. As mentioned before, due to the increase of plasma gas pulse frequency, there is not enough shut-off time of plasma gas to affect the arc profile fully, the L cannot decrease to 0 mm from 8 ms to 18 ms, and the first peak value is also less than that in 4 Hz. In the whole cycle with 20 Hz plasma gas pulse frequency, the core region of the arc always exists, and the D keeps balanced at 1.02 ± 0.17 mm.

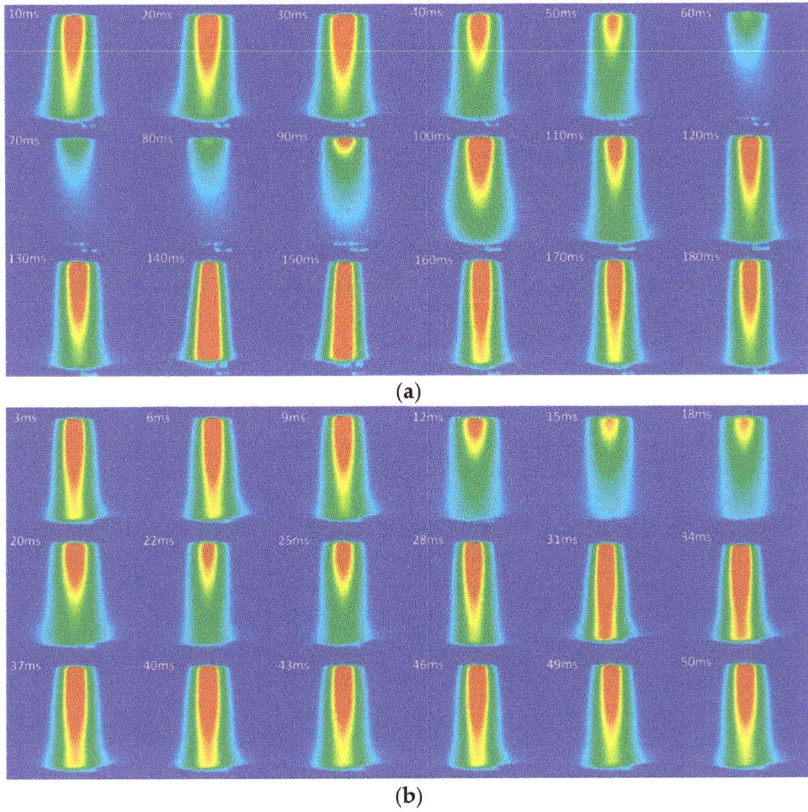

Figure 7. Comparison of arc images at negative polarity stage. (**a**) PPG-VPPAW (4 Hz); (**b**) PPG-VPPAW (20 Hz). (The color bar of this figure is same as Figure 6).

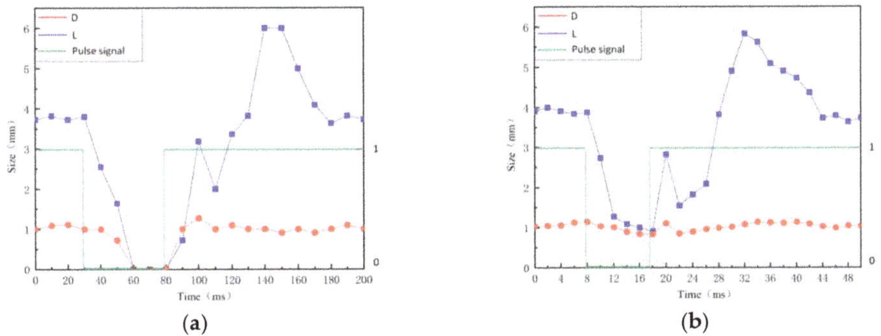

Figure 8. Arc core region size of PPG-VPPAW. (**a**) PPG-VPPAW (4 Hz); (**b**) PPG-VPPAW (20 Hz).

Figure 9 displays the arc profile of EP phase in different VPPAW processes and stages. Figure 9a shows the arc profile in the VPPAW process. Figure 9b,c show shutting-off and turning-on stages, respectively, with 4Hz PPG-VPPAW. Figure 9d,e show shutting-off and turning-on stages, respectively, with 20 Hz PPG-VPPAW. Compared with these arc profiles, it is easy to find that only in the shut-off stage with 4 Hz PPG-VPPAW process, the arc core length L significantly decreased. This phenomenon cannot be found when the plasma gas pulse frequency increased to 20 Hz. That means the influence of plasma gas shutting off in the EP phase of the VPPAW process is much smaller than in the EN phase, especially in a high plasma gas pulse frequency. This is because the arc core region is directly related to the temperature field, which has such a great inertia that the temperature change could not quickly respond to the disturbance. When the plasma gas frequency increases, the effect of pulsed plasma gas on the EP phase is harder to observe.

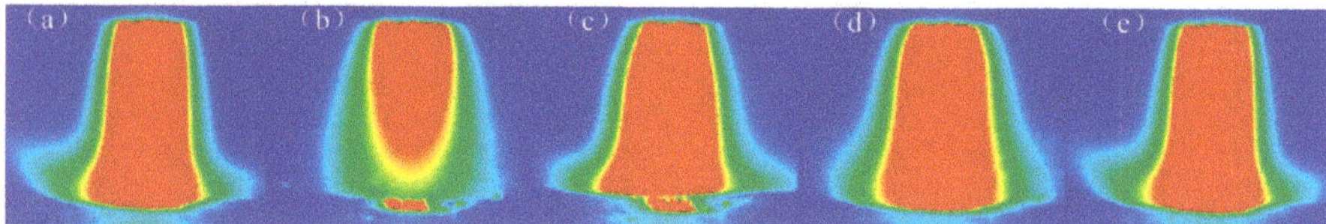

Figure 9. Comparison of arc images at positive polarity stage. (The color bar of this figure is same as Figure 6).

3.3. Variation in Welding Formation

Figure 10 shows the front and back of the PPG-VPPA (20 Hz) weld joint. It can be clearly seen that the keyhole is completely penetrated at the end of the weld joint. The weld joint formed as shown in Figure 10 is due to the lack of metal filling and the molten pool flow toward the front of the weld. Figure 11 shows the appearance of the weld surface with different welding methods. A smooth weld joint profile of VPPAW is shown in Figure 11a. As shown in Figure 11b–d, both DP-VPPAW and PPG-VPPAW form a fish-scale pattern on the surface of the weld joint due to the stirring action on the welding pool caused by the periodical oscillation of the arc pressure. The ripples formed on the weld joint surface show that the fluidity of the welding pool has been enhanced, and benefits the welding quality. Compared with Figure 11b,c, it is clearly demonstrated that the plasma gas frequency has a stronger effect acting on the welding formation than current frequency under the premise of acceptable welding formation. Compared with Figure 11c,d, the denseness of ripple profile on the weld joint surface is directly proportional to the frequency of the plasma gas.

Figure 10. The front and back of the PPG-VPPA (20 Hz) weld joint.

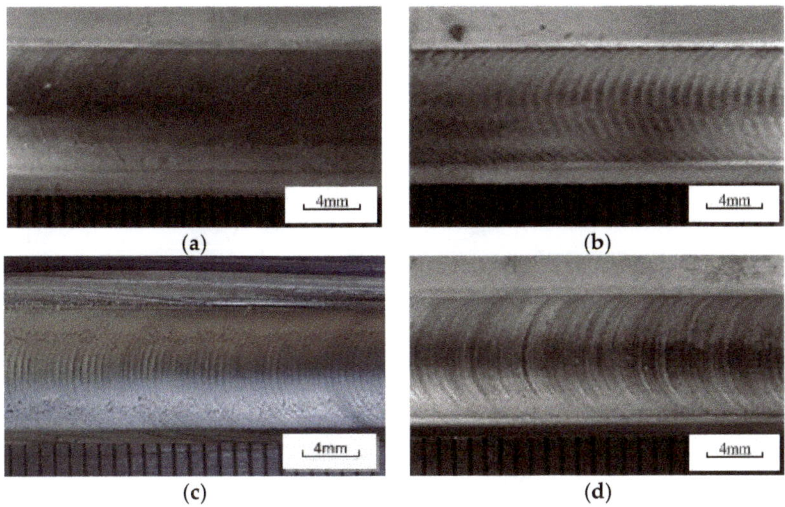

Figure 11. Morphology of weld surface. (**a**) VPPAW; (**b**) DP-VPPAW; (**c**) PPG-VPPAW (20 Hz); (**d**) PPG-VPPAW (40 Hz).

The preparation method of the sample is shown in Figure 12. Three samples were selected in every weld joint. We observed that the characteristics of the three samples of the same weld joint were basically the same. Therefore, a sample of each weld joint was randomly selected for further analysis. The cross-sections of the weld joint with different welding methods are exhibited in Figure 13a. The average weld reinforcement and width are measured as shown in Table 4. The VPPAW process and PPG-VPPAW process have nearly the same current wave, so the weld width from these two processes are similar. For the DP-VPPAW process, due to the decrease of current in the low frequency, the weld width also decreases. The distribution of porosity in the weld fusion line region with different welding methods is shown in Figure 13b, where WZ is weld zone, FZ is fusion zone, and HAZ is heat-affected zone. It shows clearly that the porosity has appeared both in fusion zone and weld zone from the VPPAW process, and the porosity could be observed in the fusion zone from the DP-VPPAW process. In contrast, there is no observable porosity in the cross-section from the PPG-VPPAW process. The results of arc electrical characteristics and arc profile show that PPG-VPPA periodically fluctuates due to the periodic variation of plasma gas flow rate. The molten pool oscillates periodically under the oscillating arc. Therefore, the pulsed plasma gas has a stronger effect than the pulsed current wave which enhances the fluidity of the molten pool and the spillover probability of the porosity from the molten pool increases [20,21].

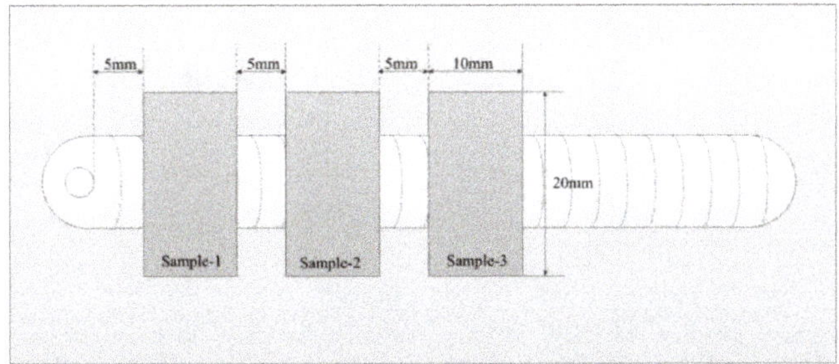

Figure 12. The preparation method of the sample.

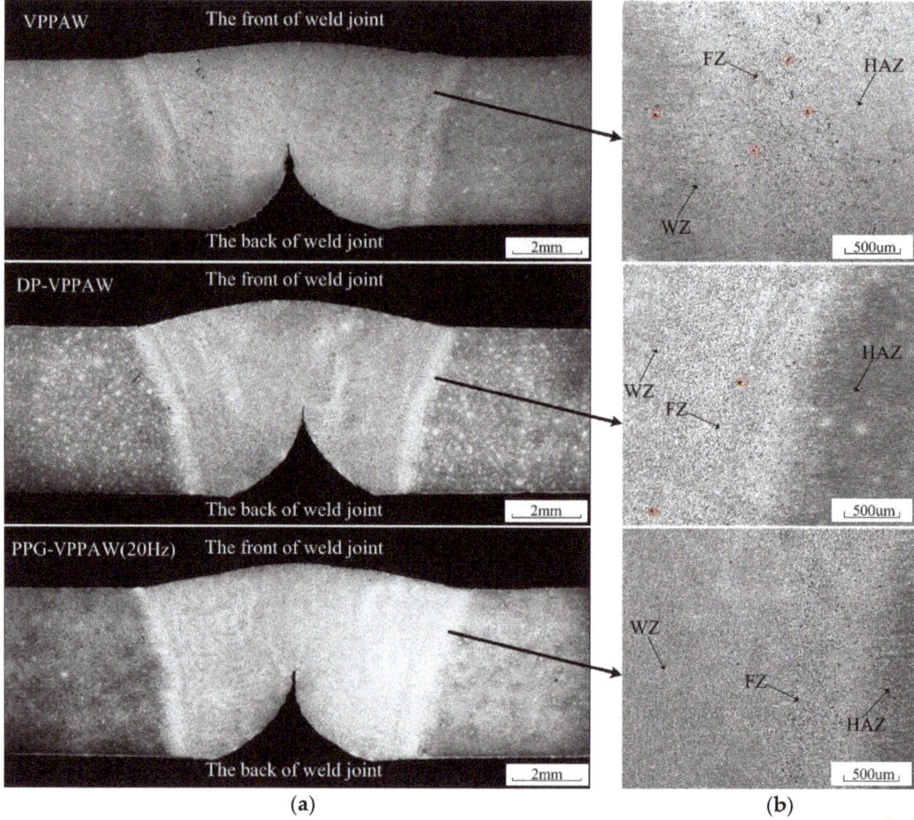

Figure 13. Macrographs of cross-section and the distribution of pores in weld fusion line region. (**a**) Macrographs of cross-section; (**b**) the distribution of pores in weld fusion line region.

Table 4. The weld reinforcement and width.

Welding Methods	Weld Width (mm)	Weld Reinforcement (mm)
VPPAW	10.18	0.64
DP-VPPAW	9.64	0.82
PPG-VPPAW	10.01	0.73

4. Conclusions

This study proposed a novel plasma arc welding process, named pulsed plasma gas variable polarity plasma arc welding process, and investigated the effect of pulsed plasma gas on the arc electrical signal, arc profile, and welding formation. The results can be summarized as follows:

(1) The shut-off time is the key factor affecting arc behavior under the pulsed plasma gas, and this effect is stronger in the EP phase of the variable polarity plasma arc process.
(2) In the EN phase, the length of arc core region is also affected by the shut-off time of plasma gas. Due to the overshoot, two peak values of arc core length appear in the return of plasma gas. The pulsed plasma gas has less effect on arc profile in the EP phase.
(3) Compared with pulsed current wave, the pulsed plasma gas could better enhance the fluidity of the molten pool to reduce porosity during aluminum keyhole welding.

Author Contributions: Z.L., S.C., and F.J. contributed to the project design; W.Z. and Z.Y. performed all experiments and data processing; F.J. wrote this manuscript; all authors participated in the discussion on the results and guided the writing of the article.

Funding: This work is Supported by National Natural Science Foundation of China (No.51875004), Beijing Natural Science Foundation (No.3172004), the China Scholarship Council (CSC) for one-year study in University of Alberta for Fan Jiang, International Research Cooperation Seed Fund of Beijing University of Technology (No.2018B16), the open project of Beijing Engineering Researching Center of Laser Technology (BG0046-2018-09) and start-up funding from Beijing University of Technology.

Conflicts of Interest: The authors declare no conflict of interest.

References

1. Chen, S.J.; Zhang, R.Y.; Jiang, F.; Dong, S.W. Experimental study on electrical property of arc column in plasma arc welding. *J. Manuf. Process.* **2018**, *31*, 823–832. [CrossRef]
2. Jiang, F. Arc Behavior and Weld Pool Stability in Keyhole Plasma arc Welding. Ph.D. Thesis, Beijing University of Technology, Beijing, China, 2014.
3. Cong, B.Q.; Su, Y.; Qi, B.J. Development of pulsed arc welding technology for aluminum Alloy. *Adv. Aviat. Weld. Technol.* **2016**, *11*, 41–46.
4. Mahrle, A.; Schnick, M.; Rose, S. Process characteristics of fibre-laser-assisted plasma arc welding. *J. Phys. D-Appl. Phys.* **2011**, *44*, 345502. [CrossRef]
5. Mahrle, A.; Rose, S.; Schnick, M. Stabilisation of plasma welding arcs by low power laser beams. *Sci. Technol. Weld. Join.* **2013**, *18*, 323–328. [CrossRef]
6. Vredeveldt, H.L. Increased Power Density Plasma arc Welding: The Effect of an Added Radial Gas Flow around the arc Root. Master's Thesis, Delft University of Technology, Delft, The Netherlands, 2014.
7. Wu, C.S.; Jia, C.B.; Chen, M.A. A Control system for keyhole plasma arc welding of stainless steel plates with medium thickness. *Weld. J.* **2010**, *89*, 225–231.
8. Jia, C.B.; Wu, C.S.; Zhang, Y.M. Sensing controlled pulse key-holing condition in plasma arc welding. *Trans. Nonferrous Met. Soc. China* **2009**, *19*, 341–346. [CrossRef]
9. Wang, Y.; Qi, B.; Cong, B. Arc Characteristics in Double-Pulsed VP-GTAW for Aluminum Alloy. *J. Mater. Process. Technol.* **2017**, *249*, 89–95. [CrossRef]
10. Yang, M.; Li, L.; Qi, B. Arc force and shapes with high-frequency pulsed-arc welding. *Sci. Technol. Weld. J.* **2017**, *22*, 580–586. [CrossRef]
11. Chen, S.J.; Yang, X.; Yu, Y. Study on Low Frequency Vibration Pulse Plasma arc Welding of Ti-Alloy. *Weld. J.* **2011**, *10*, 1–6.
12. Wu, C.S.; Zhao, C.Y.; Zhang, C. Ultrasonic Vibration-Assisted Keyholing Plasma arc Welding. *Weld. J.* **2017**, *96*, 279–286.
13. Chen, S.J.; Jiang, F.; Zhang, J.L. Measurement and Analysis of Plasma arc Components. *J. Manuf. Sci. Eng.* **2015**, *137*, 011006. [CrossRef]
14. Chen, S.J.; Zhang, R.Y.; Jiang, F. Measurement and Application of arc Separability in Plasma arc. *Weld. J.* **2016**, *95*, 219–228.
15. Sarrafi, R.; Kovacevic, R. Cathodic cleaning of oxides from aluminum surface by variable-polarity arc. *Weld. J.* **2010**, *89*, 1–10.
16. Yang, M.Y.; Zheng, H.; Li, L. Arc shapes characteristics with ultra-high-frequency pulsed arc welding. *Appl. Sci.* **2017**, *7*, 45. [CrossRef]
17. Yang, M.Y.; Zheng, H.; Qi, B.J.; Yang, Z. Effect of arc behavior on Ti-6Al-4V welds during high frequency pulsed arc welding. *J. Mater. Process. Technol.* **2017**, *243*, 9–15. [CrossRef]
18. Gu, W.P.; Xiong, Z.Y.; Wan, W. Autonomous seam acquisition and tracking system for multi-pass welding based on vision sensor. *Int. J. Adv. Manuf. Technol.* **2013**, *69*, 451–460. [CrossRef]
19. Nele, L.; Sarno, E.; Keshari, A. An image acquisition system for real-time seam tracking. *Int. J. Adv. Manuf. Technol.* **2013**, *69*, 2099–2110. [CrossRef]

20. Cong, B.; Ouyang, R.; Qi, B. Influence of Cold Metal Transfer Process and Its Heat Input on Weld Bead Geometry and Porosity of Aluminum-Copper Alloy Welds. *Rare Metal Mater. Eng.* **2016**, *45*, 606–611. [CrossRef]

21. Liu, A.; Tang, X.; Lu, F. Study on welding process and prosperities of AA5754 Al-alloy welded by double pulsed gas metal arc welding. *Mater. Des.* **2013**, *50*, 149–155. [CrossRef]

Article

Calculation of the Intermetallic Layer Thickness in Cold Metal Transfer Welding of Aluminum to Steel

Zahra Silvayeh [1,*], **Bruno Götzinger** [2], **Werner Karner** [2], **Matthias Hartmann** [3] and **Christof Sommitsch** [1]

1   Institute of Materials Science, Joining and Forming (IMAT), Graz University of Technology (TU Graz), Kopernikusgasse 24/I, 8010 Graz, Austria; christof.sommitsch@tugraz.at
2   Magna Steyr Fahrzeugtechnik AG & Co KG, Liebenauer Hauptstraße 317, 8041 Graz, Austria; bruno.goetzinger@magna.com (B.G.); werner.karner2@magna.com (W.K.)
3   Austrian Institute of Technology (AIT), Light Metals Technologies Ranshofen GmbH (LKR), P.O. Box 26, 5282 Ranshofen, Austria; matthias.hartmann@ait.ac.at
*   Correspondence: zahra.silvayeh@tugraz.at; Tel.: +43-316-873-1658; Fax: +43-316-873-7187

Received: 2 December 2018; Accepted: 17 December 2018; Published: 22 December 2018

Abstract: The intermetallic layer, which forms at the bonding interface in dissimilar welding of aluminum alloys to steel, is the most important characteristic feature influencing the mechanical properties of the joint. In this work, horizontal butt-welding of thin sheets of aluminum alloy EN AW-6014 T4 and galvanized mild steel DC04 was investigated. In order to predict the thickness of the intermetallic layer based on the main welding process parameters, a numerical model was created using the software package Visual-Environment. This model was validated with cold metal transfer (CMT) welding experiments. Based on the calculated temperature field inside the joint, the evolution of the intermetallic layer was numerically estimated using the software Matlab. The results of these calculations were confirmed by metallographic investigations using an optical microscope, which revealed spatial thickness variations of the intermetallic layer along the bonding interface.

Keywords: aluminum-steel blanks; intermetallic layer; cold metal transfer; welding simulation; dissimilar welding; multimaterial car body

---

1. Introduction

Dissimilar joining of aluminum alloy sheets to steel sheets is an indispensable key process for producing multimaterial car bodies, offering both high crash safety and low vehicle weight. Fusion welding processes in particular have marked advantages regarding the efficient joining of hybrid parts of complex shapes. However, thermal joining of aluminum- (Al) to iron- (Fe) based materials is known to be associated with the formation of intermetallic (IM) Al_xFe_y phases at the bonding interface [1–3]. The formation of these phases is mandatory for bonding of the dissimilar materials, however, excessive formation results in brittleness and therefore in poor mechanical properties of the joints. Thus, controlling the thermodynamically unavoidable interfacial reaction between iron and aluminum is a critical issue regarding the performance of dissimilar joints.

Laboratory experiments have identified that in most cases two main IM phases form at the interface between solid iron or steel and liquid aluminum or its alloys: Al_5Fe_2 as the major η-phase [4–8], together with Al_3Fe (also referred as $Al_{13}Fe_4$) as the minor θ-phase [9–34]. Some researchers have found additional Al_xFe_y phases, e.g., $AlFe_3$ or $AlFe$ [7,14–16]. Dybkov [12] reported 'paralinear' growth of the IM phases, meaning that with increasing time the thickness of the Al_5Fe_2 phase tends to grow towards a certain limit, while the thickness of the Al_3Fe phase grows almost linearly after a non-linear initial period. Bouayad et al. [17] also observed that growth of the Al_5Fe_2 phase follows a parabolic relationship, but the growth of the Al_3Fe phase follows a linear relationship.

However, according to Bouché et al. [18], both the Al_5Fe_2 phase as well as the Al_3Fe phase exhibit parabolic growth after an initial non-parabolic transient period. The Al_5Fe_2 crystals are assumed to have a much higher growth rate than the Al_3Fe crystals, since the tongue-like Al_5Fe_2 sub-layer is generally observed to be markedly thicker than the serrated Al_3Fe sub-layer.

At least one, or even both, of these two phases was also found to form in dissimilar cold metal transfer (CMT) welding/brazing of aluminum alloys to steel [35–48]. In comparison to conventional gas metal arc (GMA) welding processes, the CMT process is operated with significantly reduced heat input [47,48], which restricts the growth of the Al_xFe_y phases and therefore limits the thickness of the IM layer [42,43]. The process temperature is high enough to melt the aluminum base material and the aluminum-based filler, but the steel base material remains solid. Thus, dissimilar joining is achieved by a combination of aluminum welding and steel brazing. The single-sided CMT process in particular offers high potential regarding flexible and efficient butt-welding of aluminum alloy sheets to zinc-coated steel sheets, which is of utmost interest in the automotive industry. However, note that the steel sheet can be used in the as-cut condition, i.e., the cutting edge of the steel sheet is uncoated [49].

The growth of the thickness of the IM layer, x_{IM} (m), can be expressed as a function of time, t (s). Diffusion-controlled layer growth, which is assumed as dominant in low temperature and solid state welding processes (e.g., CMT), is commonly expressed using a power-law function:

$$x_{IM} = (kt)^n \tag{1}$$

For parabolic growth, $n = 0.5$. The temperature-dependent growth rate coefficient k (m^2/s) is expected to follow an Arrhenius relationship, where k_0 (m^2/s) is the growth constant, Q (J/mol) is the activation energy, T (K) is the absolute temperature, and $R = 8.314$ J/molK is the gas constant:

$$k = k_0 \exp\left(-\frac{Q}{RT}\right) \tag{2}$$

Table 1 contains values of Q and k_0 as reported in the literature for calculation of the time-dependent parabolic growth of the major η-phase or of the IM layer, respectively. Both of these constants are usually determined by fitting experimental data captured at different temperatures. Obviously, considerable variations exist between the reported values, which can be attributed to differences in the materials investigated, the experimental conditions, and in formulating the growth equation. Note that most of the experiments have been conducted at laboratory conditions within comparatively narrow temperature ranges. Therefore, the influence of transient or non-uniform temperature fields—as occur, for instance, in most industrial welding processes—on the formation of the IM layer is not considered. Furthermore, if iron or aluminum of technical pureness are used for experiments, the growth constants do not consider the influence of alloying elements, which are known to influence the growth of the IM layer and which are normally present in industrial processes. In particular, increasing the silicon content of aluminum alloys retards IM layer growth [4,10,27–33,50,51], but increasing the zinc content promotes IM layer growth [4,34,51]. Increasing the carbon content of steels also retards the growth [52,53]. Note that the constants given in References [54–59] were determined in experiments where both iron and aluminum were solid (s).

During recent years, different methods have been applied by researchers to model the IM layer, since this layer represents a critical feature influencing the mechanical properties of aluminum-steel joints. Rong et al. [60] conducted thermophysical simulations to clarify reaction mechanisms and growth kinetics at the interface between solid steel and liquid aluminum, and to predict the average thickness of the IM layer. Das et al. [61] proposed a combined theoretical–experimental method, including a numerical model and a set of measured results, to estimate the thickness of the IM layer as a function of key process parameters in a lap joint configuration. Zhang et al. [62] used the Monte Carlo

(MC) method to model the growth of IM compounds, and validated their results with bead-on-plate welding of aluminum alloy onto galvanized mild steel.

Table 1. Activation energies and growth constants, as reported in the literature.

Ref.	Researcher	T (°C)	Material Combination	Q (kJ/mol)	k_0 (m²/s)
[5]	Heumann and Dittrich	700–960	pure Fe (s) pure Al (l)	55	-
[8]	Denner and Jones	673–826	0.05 wt % C steel (s) pure Al (l)	170	-
			0.17 wt % C steel (s) pure Al (l)	195	-
[11]	Eggeler, Auer and Kaesche	670–800	low C steel (s) pure Al (l)	134 * 155 **	-
			low C steel (s) Fe-saturated Al (l)	87 * 104 **	-
[17]	Bouayad, Gerometta, Belkebir and Ambari	700–900	pure Fe (s) pure Al (l)	73 * 74 **	-
[30]	Springer, Kostka, Payton, Raabe, Kaysser-Pyzalla and Eggeler	600–675	0.08 wt % C steel (s) pure Al (s,l)	190	-
			0.08 wt % C steel (s) Al + 5 wt % Si (s,l)	17	-
[34]	Springer, Szczepaniak and Raabe, includes data from [4,30,54,57]	400–750	0.08 wt % C steel (s) pure Al (s,l)	190	-
			0.08 wt % C steel (s) Al + 2.5 wt % Zn (s,l)	165	-
[32]	Lemmens, Springer, De Graeve, De Strycker, Raabe and Verbeken	670–725	0.01 wt % C steel (s) pure Al (l)	224	-
			0.01 wt % C steel (s) Al + 1 wt % Si (l)	142	-
			0.01 wt % C steel (s) Al + 3 wt % Si (l)	149	-
			0.01 wt % C steel (s) Al + 10 wt % Si (l)	-72	-
[23]	Tanaka and Kajihara	780–820	pure Fe (s) pure Al (l)	248	1.26×10^2
[29]	Yin, Zhao, Liu, Han and Li	700–800	pure Fe (s) pure Al (l)	207	1.10
			pure Fe (s) Al + 1 wt % Si (l)	169	3.68×10^{-3}
			pure Fe (s) Al + 2 wt % Si (l)	167	1.46×10^{-3}
			pure Fe (s) Al + 3 wt % Si (l)	172	1.38×10^{-3}
[54]	Shibata, Morozumi and Koda	605–655	pure Fe (s) pure Al (s)	226	-
[55]	Jindal, Srivastava, Das and Ghosh	500–600	IF steel (s) pure Al (s)	85	3.82×10^{-8}
[56] [57]	Kajihara Naoi and Kajihara	550–640	pure Fe (s) pure Al (s)	281	1.32×10^2
[58]	Zhe, Dezellus, Gardiola, Braccini and Viala	535	0.03 wt % C steel (s) Al + 7 wt % Si (s)	153	4.37×10^{-4}
[59]	Xu, Robson, Wang and Prangnell	400–480	0.08 wt % C steel (s) Al + 0.8 wt % Si (s)	116	-
		480–570		248	-
		400–570		160	-

* maximum thickness of the IM layer, ** mean thickness of the IM layer, (s) solid, (l) liquid.

This work presents a numerical method, which allows fast estimation and three-dimensional (3D) visualization of the IM layer, because the highly irregular microscopic interface between the IM layer and the weld seam is approximated as a smooth surface. 3D visualization of the layer enables the identification of critical weld seam regions where comparatively thin (possibility of insufficient bonding) or thick (brittleness of the joint) IM layers occur. The presented method, which includes (i) calculation of the temperature at the bonding interface between the steel sheet and the aluminum weld by means of finite element (FE) simulation, (ii) validation of the obtained numerical results with temperature curves measured in CMT welding, (iii) prediction of the thickness of the IM layer based on the calculated temperature field, and finally (iv) validation of the predicted thickness with micrographs of weld cross-sections, has already been presented by the authors of this article [63]. This method was also successfully applied by Borrisutthekul et al. [64] for estimating the effect of different heat flow control measures on the thickness of the IM layer in laser welding, and by Mezrag et al. [65] for the indirect determination of the process efficiency in CMT welding.

2. Experimental Methods

In this study, sheets of aluminum alloy EN AW-6014 T4 were joined with sheets of galvanized mild steel DC04 by means of the single-sided cold metal transfer [47,48] process. The metal sheets were clamped gap-free in horizontal butt-joint position, as schematically illustrated in Figure 1. The dimensions were 250 mm × 150 mm × 0.80 mm for the steel sheet, and 250 mm × 150 mm × 1.15 mm for the aluminum alloy sheet.

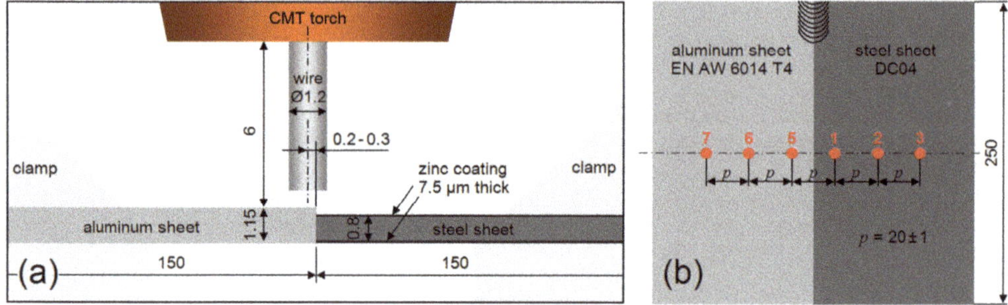

Figure 1. Schematic illustrations of the welding configuration (dimensions in mm): (**a**) front view of the weld butt [46], (**b**) top view showing the positions and the numbering of six thermocouples.

The welding equipment included a Fronius CMT Advanced 4000 power source and a Fronius CMT Braze + torch (Pettenbach, Austria), mounted to the arm of a KUKA KR 30-2 robot (Augsburg, Germany). A filler wire of non-commercial aluminum alloy Al-0.3Mg-0.5Sc-0.4Zr was used. Both feeding of the filler wire and supply of the shielding gas were achieved through the welding torch. The main process parameters applied in the welding experiments are summarized in Table 2, and the nominal compositions (wt %) of the materials are listed in Table 3.

Table 2. Parameters of the welding process.

Parameter	Symbol	Value
Welding current (mean value)	I	71 A
Welding voltage (mean value)	U	8.1 V
Welding speed	v	0.4 m/min
Feeding rate of the filler wire	w	3.9 m/min
Diameter of the filler wire	d_w	1.2 mm
Distance between the torch and the workpiece	d	6 mm
Angle between the torch and the workpiece	δ	90°
Flow rate of the argon shielding gas	$\dot{V}_{Ar}$	12 l/min

Table 3. Nominal compositions (wt %) of the materials used in the experiments [46].

Material	Al	Fe	Mg	Mn	Si	Cu	Zn	Ti	Cr	V	C	P	S	Zr	Sc
Mild steel sheet DC04 (1.0338)	-	bal.	-	max. 0.4	-	-	-	-	-	-	max. 0.08	max. 0.03	max. 0.03	-	-
Aluminum alloy sheet EN AW-6014	bal.	max. 0.35	0.4–0.8	0.05–0.2	0.3–0.6	max. 0.25	max. 0.1	max. 0.1	max. 0.2	0.05–0.2	-	-	-	-	-
Aluminum filler wire Al-0.3Mg-0.5Sc-0.4Zr	bal.	0.02–0.04	0.2–0.3	0.03–0.05	0.03–0.05	-	-	-	-	-	-	-	-	0.3–0.5	0.4–0.6

3. Numerical Methods

The numerical model for calculating the temperature field was created using the ESI Welding Simulation Suite (R 13.5, ESI Group, Paris, France). This software package is based on the Visual-Environment platform, which includes the pre-processing module Visual-Mesh for generating the three-dimensional geometry and the mesh of the model, the module Visual-Weld for defining the welding process, and the post-processing module Visual-Viewer for visualizing the results. The numerical calculation was performed with the finite element (FE) software SYSWELD (R 2017.5, ESI Group, Paris, France).

3.1. Geometry and Mesh

As shown in Figure 2, the model included three parts: the aluminum sheet, the steel sheet, and the weld seam, with dimensions according to the experimental welding configuration. The complete model consisted of approximately 504,000 nodes and 425,000 elements. The interface between the weld seam and the steel sheet was meshed with comparatively fine elements, since the temperature at this interface was of primary interest for calculating the thickness of the IM layer. Figure 2 also shows that the mesh was coarser with increasing distance to the weld seam, in order to limit the required calculation time. A cross-section of the meshed model is shown in Figure 3.

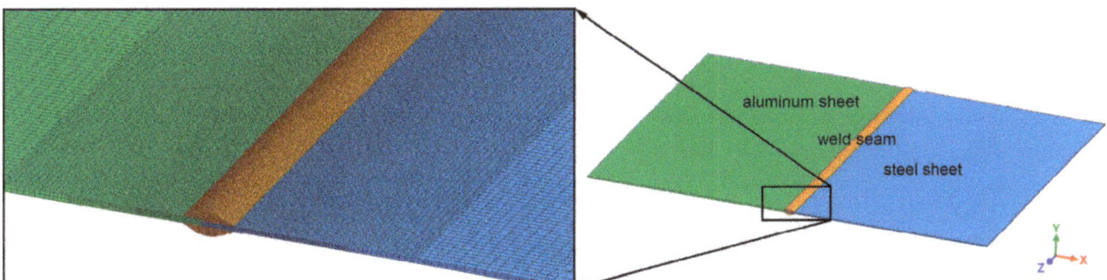

Figure 2. Meshed three-dimensional finite element model.

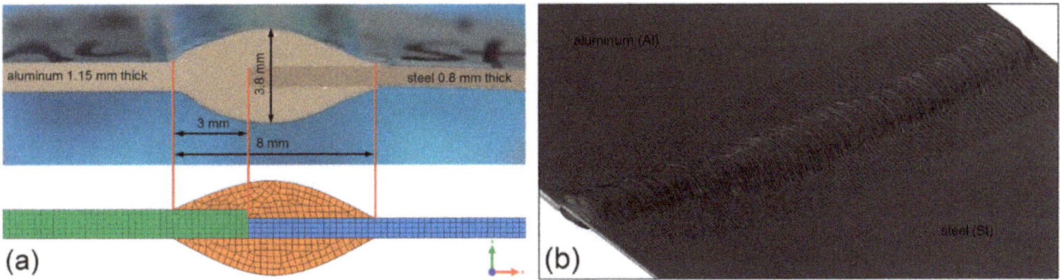

Figure 3. Cross-section of the meshed model based on (**a**) the micrograph of sample 63 and on (**b**) the joint surface reconstruction of sample 62.

According to the results of the welding experiments, the maximum thickness of the weld seam was about 3.8 mm, and the width between the base corners of the weld seam was about 8 mm.

These dimensions were estimated from micrographs of the weld cross-section, as exemplarily shown in Figure 3a. The micrographs were captured using a Zeiss Axio Observer.Z1m (Jena, Germany) optical microscope equipped with a Zeiss Axio-Cam MRc5 camera. The dimensions of the weld seam were validated with a three-dimensional reconstruction of the joint surface as shown in Figure 3b, which was captured using the optical 3D scanning system GOM ATOS III Triple Scan (Braunschweig, Germany). Neither small variations of the actual weld seam geometry nor thermal distortion of the sheets were considered in the model, since their effects on the growth and on the thickness of the IM layer are negligible.

3.2. Material Properties

As shown in Figure 4, temperature-dependent thermal conductivity, specific heat capacity, and density were considered for both sheets and for the weld seam. These thermo-physical properties were obtained by adapting predefined material data of the available Visual-Weld database (R 13.5, ESI Group, Paris, France).

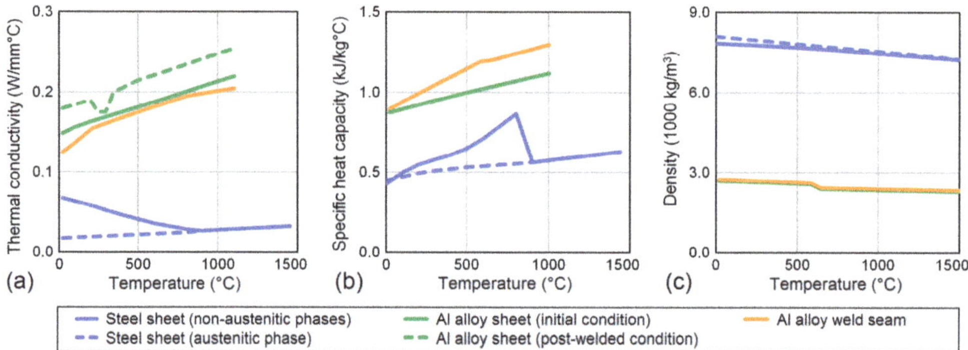

Figure 4. Thermo-physical material properties of the steel sheet, the aluminum alloy sheet and the weld seam: (**a**) thermal conductivity, (**b**) specific heat capacity, and (**c**) density.

3.3. Process Definition and Boundary Conditions

The main process parameters applied in the welding experiments are summarized in Table 2. Based on the mean welding voltage, U (V), the mean welding current, I (A), and the welding speed, v (m/min), the nominal energy input, E (J/mm), was calculated:

$$E = 0.06\frac{UI}{v} \tag{3}$$

The efficiency of the welding process, η, is in the range of about 0.8–0.9 for energy-reduced GMA welding processes with controlled dip transfer [65–67]. However, η tends to increase with decreasing arc power [67]. Since the modelled welding process was operated with comparatively low energy input or arc power, η was approximated as unity in the present model. The heat density distribution, q (W/mm³), describes the time-dependent movement of the heat source along the pre-defined welding trajectory. It was calculated according to the double-ellipsoidal heat source model of Goldak et al. [68,69]:

$$q(x,y,z,t_w) = \frac{6\sqrt{3}}{\pi\sqrt{\pi}}\frac{f\eta Ev}{abc}\exp\left(-\frac{3x^2}{a^2}\right)\exp\left(-\frac{3y^2}{b^2}\right)\exp\left(-\frac{3(z-vt_w)^2}{c^2}\right) \tag{4}$$

In Equation (4), x, y and z (mm) are the coordinates of the fixed coordinate system of the model, and t_w (s) is the welding time. The linear welding trajectory is oriented in z-direction. The heat source moves with constant welding speed, v (mm/s), along the welding trajectory. The heat source center is located at the initial coordinates $x_0 = y_0 = z_0 = 0$ when welding starts at $t_{w0} = 0$. In order to calculate q_f and q_r, the ellipsoidal heat density distributions at the front and rear sections of the weld pool, the factor f is replaced by f_f or by f_r, and the length c is replaced by c_f or by c_r, respectively. The heat fractions f_f and f_r are deposited at the front and rear sections of the weld pool, with $f_f + f_r = 2$. The lengths of the front and rear sections are c_f and c_r. Accordingly, the total length of the weld pool is $c_f + c_r$, with $c_f{:}c_r = 1{:}2$. The total width of the weld pool is $2a$ and the penetration depth is b.

The radiative heat transfer coefficient at the surfaces of the weld and of the sheets, h_r (W/m²K), was calculated based on the Stefan-Boltzmann constant, σ, the thermal emission coefficient, ε, the predefined ambient temperature, T_∞ (K), and the local surface temperature, T (K):

$$h_r = \sigma \, \varepsilon \, (T + T_\infty)\left(T^2 + T_\infty^2\right) \tag{5}$$

In order to quantify the total thermal losses the total heat transfer coefficient, h (W/m²K), was then calculated by adding both the radiative heat transfer coefficient, h_r, and the convective or conductive heat transfer coefficients, h_c:

$$h = h_r + h_c \tag{6}$$

The basic input parameters used for the simulation of the welding process are summarized in Table 4. They basically correspond to the conditions of the welding experiments conducted for validating the results of the simulations. However, in order to model the heat losses it was necessary to make some feasible assumptions. Since the temperature dependence of ε was actually unknown, $\varepsilon = \varepsilon(T)$ was approximated as unity. This is particularly suitable for elevated temperatures, because ε of metals is known to increase markedly with rising temperature, and at elevated temperatures radiative thermal losses become also dominant. The convective and the conductive heat transfer coefficients, h_c (W/m²K), were assumed to be constant over the entire temperature range. Since the metal sheets were clamped with massive metal bars on both sides of the weld butt, the conductive heat transfer between the sheets and the clamps was certainly higher than the convective heat transfer between the weld seam region and the ambient air.

Table 4. Input parameters used for the simulation of the welding process.

Parameter	Symbol	Value
Nominal energy input	E	86.3 J/mm
Efficiency of the welding process	η	1
Width of the weld pool	$2a$	2.0 mm
Penetration depth of the weld pool	b	1.5 mm
Length of the weld pool	$c_f + c_r$	3.0 mm
Duration of the welding process	Δt_w	37.5 s
Duration of the post-welding cooling period	Δt_c	22.5 s
Time increment at time step i	Δt_i	0.25 s
Ambient temperature (= initial sheet temperature)	T_∞	20 °C
Stefan-Boltzmann constant	σ	5.67×10^{-8} W/m²K⁴
Thermal emission coefficient	ε	1
Convective heat transfer coefficient at the weld seam	h_c	10 W/m²K
Conductive heat transfer coefficient at the metal sheets	h_c	200 W/m²K

The dimensions of the weld pool, a, b, c_f and c_r, were calibrated based on micrographs of the joint, as exemplarily shown in Figure 10. Note that in the present work, only the narrow region with the locally molten aluminum sheet was considered for calibration, because the steel sheet remains solid in dissimilar CMT welding/brazing of aluminum alloys to steel. This reasonable simplification allowed the utilization of Goldak's symmetric heat source model according to Equation (4), even though the aluminum-steel joint had an asymmetric shape.

3.4. Calculation of the IM Layer Thickness

The thickness of the IM layer was calculated using the software Matlab (R 2013, MathWorks, Natick, MA, USA). During the non-isothermal welding process, the temperature inside the weld seam and inside the heat affected zone varied strongly with time, $T = T(t)$. Accordingly, the growth rate coefficient varied with time too, $k = k(t)$. Thus, the simple power law function of Equation (1), which is valid for isothermal conditions, cannot describe the growth of the IM layer over the entire welding cycle. However, since assuming T and therefore k as constant is feasible within small time increments, dt, the corresponding growth increment, dx_{IM}, can be calculated based on the first derivation of Equation (1):

$$dx_{IM} = n\, k^n t^{n-1} dt \tag{7}$$

For discrete steps, and if parabolic growth ($n = 0.5$) of the IM layer is assumed, Equation (7) can be transformed into Equation (8). For each of these steps, $i = 1, 2, 3...m$, one can calculate the growth increment, $\Delta x_{IM,i}$, based on the actual time of growth, t_i, the time increment, Δt_i, and the growth rate coefficient, k_i, as follows:

$$\Delta x_{IM,i} = \sqrt{\frac{k_i}{4t_i}}\, \Delta t_i \tag{8}$$

Figure 5 illustrates schematically the relationships between the thickness of the IM layer, x_{IM}, and the time of growth, t, as well as between the growth increment, $\Delta x_{IM,i}$, and the time increment, Δt_i, as expressed by Equation (8).

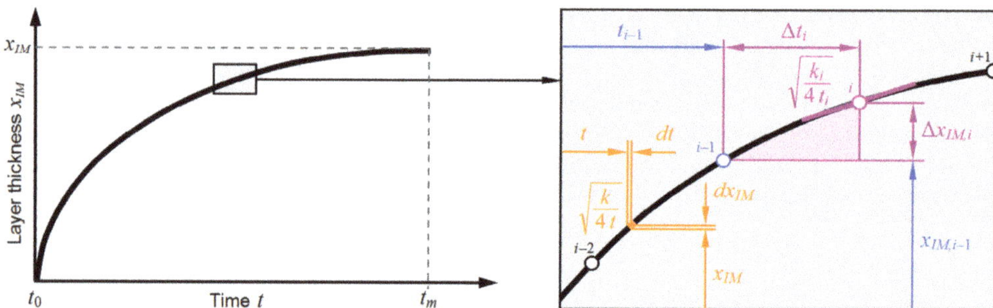

Figure 5. Schematic illustration of the relationship between IM layer thickness and growth time.

Using Equation (9), k_i was calculated with the constants $Q = 190$ kJ/mol and $k_0 = 1.5$ m^2/s, which lay within the range of the values given in Table 1 for the combination of liquid technically pure (i.e., particularly silicon-free) aluminum and solid low-carbon steel:

$$k_i = k_0 \exp\left(-\frac{Q}{RT_i}\right) \tag{9}$$

Keep in mind that k_0 is the general growth constant, which is not related to $t_0 = 0$. By contrast, the growth rate coefficient k_i is certainly related to the time t_i. The temperature T_i was obtained from the finite element simulation, utilizing the constant time increment Δt_i. Note that $t_0 \neq t_{w0}$, because $t_{w0} = 0$ when the welding process starts, but $t_0 = 0$ when the local temperature exceeds the limit of $T_0 = 400\,^\circ$C for the growth of the IM layer. At temperatures below this limit diffusion is quite slow and therefore reactions between aluminum and steel are more or less negligible [5]. According to Equation (10) the total thickness of the IM layer, x_{IM}, can be finally calculated by adding all growth increments, $\Delta x_{IM,i}$:

$$x_{IM} = \sum_{i=1}^{m} \Delta x_{IM,i} \tag{10}$$

4. Results and Discussion

4.1. Temperature Field

The numerical simulation provided the time-dependent temperature field inside the metal sheets and inside the weld seam during the welding process, as well as during the post-welding cooling period. In particular, the evolution of the temperature field next to the weld butt at the top surfaces of the weld seam and of the two metal sheets is illustrated in Figure 6. Though the cooling conditions were equal for both sheets, the non-symmetric temperature field reveals that the majority of the welding heat flowed into the aluminum sheet. This is confirmed by the temperature curves illustrated in Figure 7, which show that the peak temperatures captured at equal distances from the weld butt were almost twice as high in the aluminum sheet as in the steel sheet. Obviously, comparing the temperature curves at particular positions reveals good qualitative and quantitative agreements between the welding experiments (Figure 7a,b) and the welding simulation, (Figure 7c).

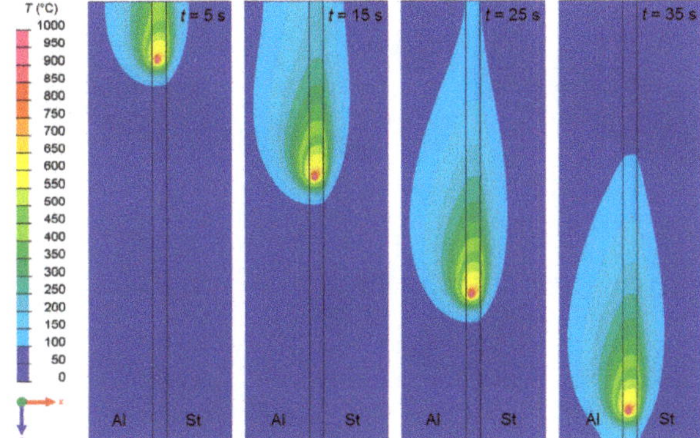

Figure 6. Temperature field at the top surfaces of the aluminum sheet (Al), the steel sheet (St) and the weld seam at 5 s, 15 s, 25 s, and 35 s after starting the welding process.

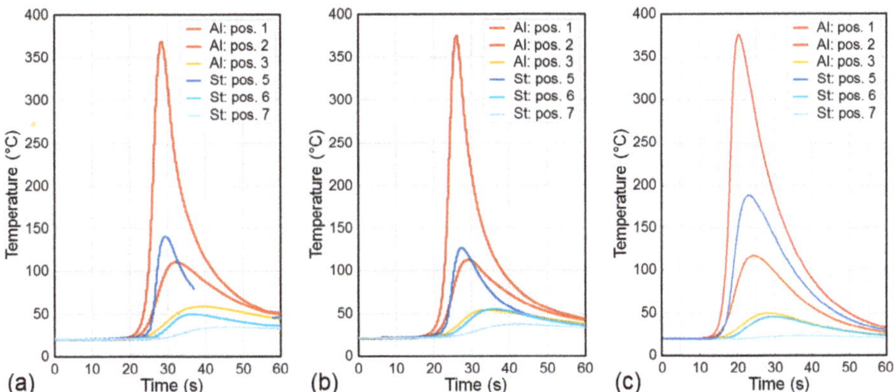

Figure 7. Comparison between temperatures measured using thermocouples during CMT welding of (**a**) sample 62 and (**b**) sample 63, and (**c**) temperatures calculated in the finite element simulation.

The experimentally validated simulation model was used to calculate the temperature not only at the surface, but also at the interface between the weld seam and the steel sheet, which was not accessible for direct temperature measurements. Figure 10a shows that the weld seam contacted the steel sheet at three sections: (i) at the top section, (ii) at the side section, and (iii) at the bottom section.

The idealized dimensions of both the top and the bottom sections were 250 mm × 5 mm (i.e., length of the weld seam × mean width of the weld seam contacting the steel sheet), but the dimensions of the side section were 250 mm × 0.8 mm (i.e., length of the weld seam × thickness of the steel sheet). Detailed information regarding the time-dependent temperature field occurring at the three sections of the bonding interface was required to calculate the time-dependent thickness of the IM layer formed during the welding process.

4.2. Thickness of the Intermetallic Layer

Figure 8 shows the calculated temperature, T, (left column) and the corresponding thickness of the IM layer, x_{IM}, (right column) at the top section of the bonding interface between the steel sheet and the weld seam at the welding times t_w = 10 s, 20 s, 30 s, and 40 s.

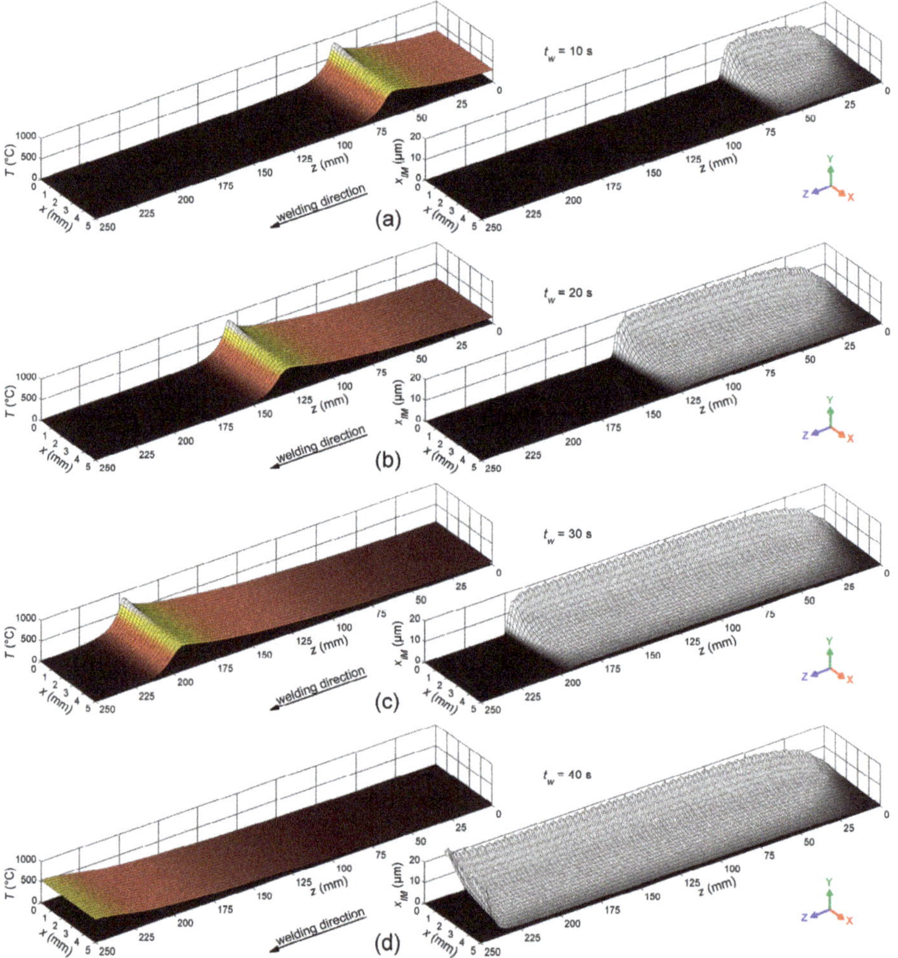

Figure 8. Temperature and thickness of the IM layer, calculated at the top section of the bonding interface between the steel sheet and the weld seam at (**a**) 10 s, (**b**) 20 s, (**c**) 30 s, and (**d**) 40 s after starting the welding process.

As illustrated in Figure 8, the temperature wave related to the movement of the welding torch propagated in welding direction. The peak of this wave was located close to the actual position of the torch. The growth of the IM layer was initiated when the temperature at the wave front exceeded 400 °C, and it proceeded as long as the temperature stayed above this limit. The thickness of the IM layer in welding direction was constant from $z \approx 50$ mm to $z \approx 230$ mm, but it varied distinctly perpendicular to the welding direction. Obviously, the thickness did not decrease when the weld cooled down. However, note that the thickness of the IM layer was overestimated at the end of the weld line, since the thermal weld penetration observed in the experiment (i.e., excessive melting of the sheets due to overheating at the end of the weld line) was not considered in the numerical model. Figure 9 illustrates the time-dependent evolution of both temperature and IM layer thickness at ten different positions (z-coordinates) along the welding direction, and at three different distances (x-coordinates) from the weld butt at the top section of the bonding interface.

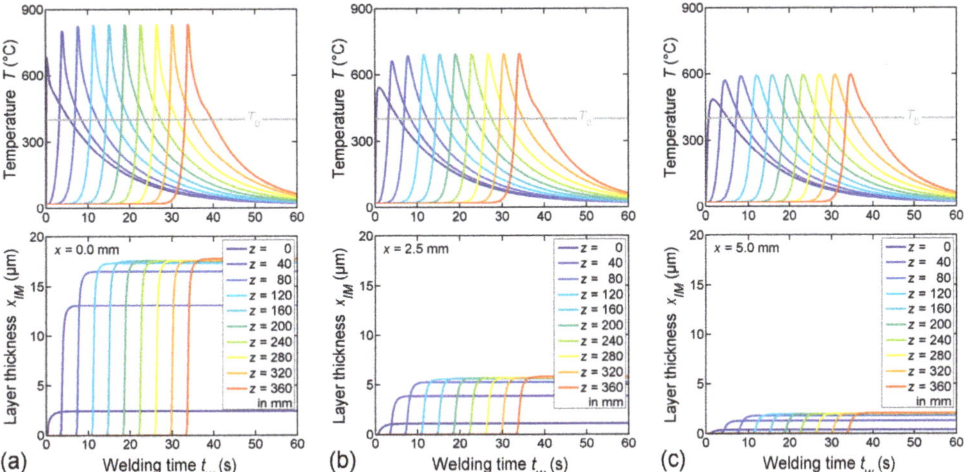

Figure 9. Time-dependent evolution of temperature and IM layer thickness at the top section of the bonding interface (**a**) directly at the weld butt, (**b**) at the center of the interface, and (**c**) at the base corner of the weld seam.

The grey horizontal line shown in the diagrams of the upper row marks the temperature T_0 at which the growth of the IM layer was assumed to start in the present numerical model. The diagrams show (a) peak temperatures of approximately 800 °C directly at the weld butt, and (b) peak temperatures of approximately 600 °C at the base corner of the weld seam. Nevertheless, at all positions the temperature decreased about 90% within 60 s. However, the peak temperature strongly influenced the thickness of the IM layer, as shown in the diagrams of the lower row. The calculated thickness was about 17 µm directly at the weld butt, but it was less than 2 µm at the base corner of the weld seam.

4.3. Experimental Validation

In order to validate the results of the calculations, a sample was extracted from the center of the welded blanks at $z \approx 125$ mm. This sample was embedded, ground, polished, and prepared for metallographic investigations by means of an optical microscope. The merged micrograph of the entire weld cross-section (both unetched and etched using Barker's reagent) is shown in Figure 10. As marked in Figure 10a, three sections of the bonding interface between the weld seam and the steel sheet can be distinguished: (i) the top section, (ii) the side section, and (iii) the bottom section. The surrounding micrographs illustrate the varying thickness of the IM layer (dark grey seam between the light grey aluminum weld and the medium grey steel sheet) at different positions on the three sections of the bonding interface.

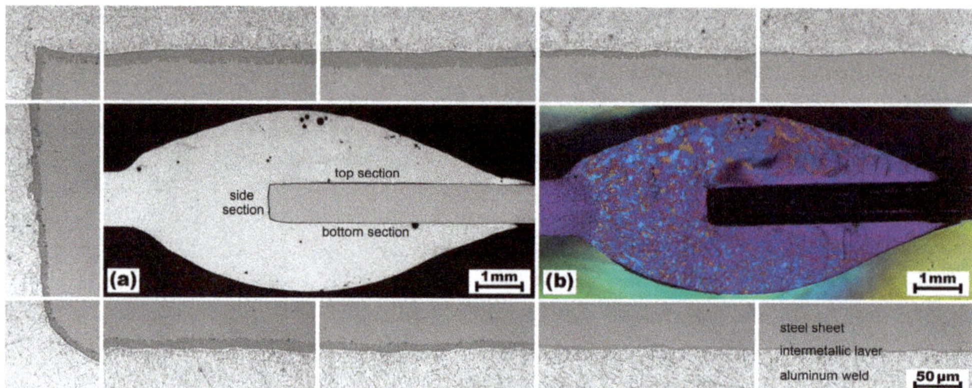

Figure 10. Weld seam cross-section of sample 63, (**a**) unetched and (**b**) etched with Barker's reagent. The surrounding micrographs illustrate the thickness variations of the IM layer.

According to the micrographs shown in Figure 10, the mean thickness of the intermetallic layer was 10.9 µm (SD = 4.2 µm) at the top section, 4.8 µm (SD = 2.1 µm) at the side section, and 7.8 µm (SD = 2.2 µm) at the bottom section. The maximum thickness of approximately 20 µm was found close to the weld butt, at the position where the peak temperatures occurred and where the majority of the filler was deposited. From this point, the thickness of the IM layer decreased bidirectionally: just slightly toward the cutting edge of the steel sheet, but markedly toward the base corner of the weld seam. This was also predicted by the numerical model, as visualized in Figures 8 and 9. However, lack of fusion—which was not included in the present numerical model—occurred at the base corners of the weld seam. The etched micrograph shown in Figure 10b allowed identification of the width of the fusion zone (i.e., the width of the weld pool) between the aluminum sheet and the steel sheet. This width was about 2 mm.

5. Conclusions

This work presents a numerical method for predicting the thickness of the intermetallic (IM) layer formed at the bonding interface in dissimilar welding of aluminum alloy EN AW-6014 T4 sheets to galvanized steel DC04 sheets. The method allows fast estimation of the layer thickness and visualization of the layer growth, because the highly irregular microscopic interface between the IM layer and the weld seam is approximated as a smooth surface. Based on the numerical results, which were also validated with single-sided cold metal transfer (CMT) welding experiments, the following conclusions can be drawn:

(1) Due to the steep temperature gradient perpendicular to the welding direction, the thickness of the IM layer varies distinctly. A thickness of about 17 µm was calculated directly at the weld butt, where the peak temperature was about 800 °C, but a thickness of less than 2 µm was calculated at the base corner of the weld seam, where the peak temperature was about 600 °C. These results corresponded well with experimental microstructure investigations.

(2) Excessive formation of the layer was observed at the end zone of the weld seam, where a thermal hot spot occurred. Hence, this region of the weld seam was identified to be critical regarding brittleness of the joint. In comparison, suppressed formation of the layer was observed at the start zone of the weld seam, where the molten filler alloy is deposited onto the initially cold metal sheets. Hence, this weld seam region was identified to be critical regarding insufficient bonding.

(3)   In order to obtain a more constant layer thickness, increasing the heat input at the start position of the welding trajectory and decreasing the heat input at the end position of the welding trajectory is advisable. For this reason, automated in-line control of main process parameters, in particular of welding current and welding voltage, is required.

(4)   The presented numerical method is based on the temperature field occurring directly at the aluminum-steel interface where the IM layer forms. Thus, the exact experimental or numerical determination of this temperature field is crucial for the subsequent calculation of the layer thickness.

(5)   This method represents an engineering approach which is of practical interest for designing dissimilar aluminum-steel weldments, because it enables estimation and/or validation of the basic growth parameters (i.e., Q and k_0) required for calculating the thickness of the IM layer even for weldments with complex shapes.

Author Contributions: Z.S. did the administration and conceptualization, developed the experimental and numerical methodologies, performed the literature review and the numerical simulations, evaluated the experimental data, validated the experimental and the numerical results, as well as prepared and edited the manuscript. W.K. was responsible for performing the welding experiments, M.H. for producing the filler alloy. B.G. and C.S. reviewed the manuscript and supervised the work.

Funding: The authors would like to thank the Austrian Research Promotion Agency (FFG) for the financial support of this study. Open access funding is provided by the Graz University of Technology.

Acknowledgments: Thanks go to Helmut Treven for assisting the welding experiments as well as to Lukas Potgorschek for doing the metallographic preparations and for operating the optical geometry scanning system.

References

1. Achar, D.R.G.; Ruge, J.; Sundaresan, S. Verbinden von Aluminium mit Stahl, besonders durch Schweißen (I), (II), (III). *Aluminium* **1980**, *56*, 147–293.

2. Atabaki, M.M.; Nikodinovski, M.; Chenier, P.; Ma, J.; Harooni, M.; Kovacevic, R. Welding of aluminum alloys to steels: An overview. *J. Manuf. Sci. Prod.* **2014**, *14*, 59–78. [CrossRef]

3. Jank, N.; Staufer, H.; Bruckner, J. Schweißverbindungen von Stahl mit Aluminium-eine Perspektive für die Zukunft. *BHM* **2008**, *153*, 189–192. [CrossRef]

4. Gebhardt, E.; Obrowski, W. Reaktion von festem Eisen mit Schmelzen aus Aluminium und Aluminium-legierungen. *Z. Metallk.* **1953**, *44*, 154–160.

5. Heumann, T.; Dittrich, S. Über die Kinetik der Reaktion von festem und flüssigem Aluminium mit Eisen. *Z. Metallk.* **1959**, *50*, 617–625.

6. Yeremenko, V.N.; Natanzon, Y.V.; Dybkov, V.I. The effect of dissolution on the growth of the Fe_2Al_5 inter-layer in the solid iron-liquid aluminium system. *J. Mater. Sci.* **1981**, *16*, 1748–1756. [CrossRef]

7. Kobayashi, S.; Yakou, T. Control of intermetallic compound layers at interface between steel and aluminum by diffusion-treatment. *Mater. Sci. Eng. A* **2002**, *338*, 44–53. [CrossRef]

8. Denner, S.G.; Jones, R.D. Kinetic interactions between aluminum$_{liquid}$ and iron/steel$_{solid}$ for conditions applicable to hot-dip aluminizing. *Met. Technol.* **1977**, *4*, 167–174. [CrossRef]

9. Eggeler, G.; Vogel, H.; Friedrich, J.; Kaesche, H. Target preparation for the transmission electron micro-scopic identification of the Al_3Fe (θ-Phase) in hot-dip aluminised low alloyed steel. *Prakt. Metallogr.* **1985**, *22*, 163–170.

10. Eggeler, G.; Auer, W.; Kaesche, H. On the influence of silicon on the growth of the alloy layer during hot dip aluminizing. *J. Mater. Sci.* **1986**, *21*, 3348–3350. [CrossRef]

11. Eggeler, G.; Auer, W.; Kaesche, H. Reactions between low alloyed steel and initially pure as well as iron-saturated aluminium melts between 670 and 800 °C. *Z. Metallk.* **1986**, *77*, 239–244.

12. Dybkov, V.I. Interaction of 18Cr-10Ni stainless steel with liquid aluminium. *J. Mater. Sci.* **1990**, *25*, 3615–3633. [CrossRef]

13. Bahadur, A.; Mohanty, O.N. Structural studies of hot dip aluminized coatings on mild steel. *Mater. Trans. JIM* **1991**, *32*, 1053–1061. [CrossRef]

14. Bahadur, A.; Mohanty, O.N. Aluminium diffusion coatings on medium carbon steel. *Mater. Trans. JIM* **1995**, *36*, 1170–1175. [CrossRef]

15. Shyam, A.; Suwas, S.; Bhargava, S. Microstructural features of iron aluminides formed by the reaction between solid iron and liquid aluminium. *Prakt. Metallogr.* **1997**, *34*, 264–277.

16. Shin, D.; Lee, J.-Y.; Heo, H.; Kang, C.-Y. Formation procedure of reaction phases in Al hot dipping process of steel. *Metals* **2018**, *8*, 820. [CrossRef]

17. Bouayad, A.; Gerometta, C.; Belkebir, A.; Ambari, A. Kinetic interactions between solid iron and molten aluminium. *Mater. Sci. Eng. A* **2003**, *363*, 53–61. [CrossRef]

18. Bouché, K.; Barbier, F.; Coulet, A. Intermetallic compound layer growth between solid iron and molten aluminium. *Mater. Sci. Eng. A* **1998**, *249*, 167–175. [CrossRef]

19. Shahverdi, H.R.; Ghomashchi, M.R.; Shabestari, S.; Hejazi, J. Microstructural analysis of interfacial reaction between molten aluminium and solid iron. *J. Mater. Process. Technol.* **2002**, *124*, 345–352. [CrossRef]

20. Cheng, W.-J.; Wang, C.-J. Growth of intermetallic layer in the aluminide mild steel during hot-dipping. *Surf. Coat. Technol.* **2009**, *204*, 824–828. [CrossRef]

21. Cheng, W.-J.; Wang, C.-J. Study of microstructure and phase evolution of hot-dipped aluminide mild steel during high-temperature diffusion using electron backscatter diffraction. *Appl. Surf. Sci.* **2011**, *257*, 4663–4668. [CrossRef]

22. Tanaka, Y.; Kajihara, M. Morphology of compounds formed by isothermal reactive diffusion between solid Fe and liquid Al. *Mater. Trans.* **2009**, *50*, 2212–2220. [CrossRef]

23. Tanaka, Y.; Kajihara, M. Kinetics of isothermal reactive diffusion between solid Fe and liquid Al. *J. Mater. Sci.* **2010**, *45*, 5676–5684. [CrossRef]

24. Pasche, G.; Scheel, M.; Schäublin, R.; Hébert, C.; Rappaz, M.; Hessler-Wyser, A. Time-resolved X-Ray microtomography observation of intermetallic formation between solid Fe and liquid Al. *Metallogr. Mater. Trans. A* **2013**, *44*, 4119–4123. [CrossRef]

25. Van Alboom, A.; Lemmens, B.; Breitbach, B.; De Grave, E.; Cottenier, S.; Verbeken, K. Multi-method identification and characterization of the intermetallic surface layers of hot-dip Al-coated steel: $FeAl_3$ or Fe_4Al_{13} and Fe_2Al_5 or Fe_2Al_{5+x}. *Surf. Coat. Technol.* **2017**, *324*, 419–428. [CrossRef]

26. Takata, N.; Nishimoto, M.; Kobayashi, S.; Takeyama, M. Crystallography of Fe_2Al_5 phase at the interface between solid Fe and liquid Al. *Intermetallics* **2015**, *67*, 1–11. [CrossRef]

27. Takata, N.; Nishimoto, M.; Kobayashi, S.; Takeyama, M. Morphology and formation of Fe-Al intermetallic layers on iron hot-dipped in Al-Mg-Si alloy melt. *Intermetallics* **2014**, *54*, 136–142. [CrossRef]

28. Takata, N.; Nishimoto, M.; Kobayashi, S.; Takeyama, M. Growth of Fe_2Al_5 phase on pure iron hot-dipped in Al-Mg-Si alloy melt with Fe in solution. *ISIJ Int.* **2015**, *55*, 1454–1459. [CrossRef]

29. Yin, F.; Zhao, M.; Liu, Y.; Han, W.; Li, Z. Effect of Si on growth kinetics of intermetallic compounds during reaction between solid iron and molten aluminum. *Trans. Nonferrous Met. Soc. China* **2013**, *23*, 556–561. [CrossRef]

30. Springer, H.; Kostka, A.; Payton, E.J.; Raabe, D.; Kaysser-Pyzalla, A.; Eggeler, G. On the formation and growth of intermetallic phases during interdiffusion between low-carbon steel and aluminum alloys. *Acta Mater.* **2011**, *59*, 1586–1600. [CrossRef]

31. Lemmens, B.; Springer, H.; Duarte, M.J.; De Graeve, I.; De Strycker, J.; Raabe, D.; Verbeken, K. Atom probe tomography of intermetallic phases and interfaces formed in dissimilar joining between Al alloys and steel. *Mater. Charact.* **2016**, *120*, 268–272. [CrossRef]

32. Lemmens, B.; Springer, H.; De Graeve, I.; De Strycker, J.; Raabe, D.; Verbeken, K. Effect of silicon on the microstructure and growth kinetics of intermetallic phases formed during hot-dip aluminizing of ferritic steel. *Surf. Coat. Technol.* **2017**, *319*, 104–109. [CrossRef]

33. Dangi, B.; Brown, T.W.; Kulkarni, K.N. Effect of silicon, manganese and nickel present in iron on the inter-metallic growth at iron-aluminum alloy interface. *J. Alloys Compd.* **2018**, *769*, 777–787. [CrossRef]

34. Springer, H.; Szczepaniak, A.; Raabe, D. On the role of zinc on the formation and growth of intermetallic phases during interdiffusion between steel and aluminium alloys. *Acta Mater.* **2015**, *96*, 203–211. [CrossRef]

35. Agudo, L.; Eyidi, D.; Schmaranzer, C.H.; Arenholz, E.; Jank, N.; Bruckner, J.; Pyzalla, A.R. Intermetallic Fe_xAl_y-phases in a steel/Al-alloy fusion weld. *J. Mater. Sci.* **2007**, *42*, 4205–4214. [CrossRef]

36. Agudo, L.; Weber, S.; Pinto, H.; Arenholz, E.; Wagner, J.; Hackl, H.; Bruckner, J.; Pyzalla, A. Study of microstructure and residual stresses in dissimilar Al/Steels welds produced by cold metal transfer. *Mater. Sci. Forum* **2008**, *571*, 347–353. [CrossRef]

37. Agudo, L.; Jank, N.; Wagner, J.; Weber, S.; Schmaranzer, C.; Arenholz, E.; Bruckner, J.; Hackl, H.; Pyzalla, A. Investigation of microstructure and mechanical properties of steel-aluminium joints produced by metal arc joining. *Steel Res. Int.* **2008**, *79*, 530–535. [CrossRef]

38. Agudo Jácome, L.; Weber, S.; Leitner, A.; Arenholz, E.; Bruckner, J.; Hackl, H.; Pyzalla, A.R. Influence of filler composition on the microstructure and mechanical properties of steel-aluminum joints produced by metal arc joining. *Adv. Eng. Mater.* **2009**, *11*, 350–358. [CrossRef]

39. Zhang, H.T.; Feng, J.C.; He, P.; Hackl, H. Interfacial microstructure and mechanical properties of aluminium-zinc-coated steel joints made by a modified metal inert gas welding-brazing process. *Mater. Charact.* **2007**, *58*, 588–592. [CrossRef]

40. Zhang, H.T.; Feng, J.C.; He, P. Interfacial phenomena of cold metal transfer (CMT) welding of zinc coated steel and wrought aluminium. *Mater. Sci. Technol.* **2008**, *24*, 1346–1349. [CrossRef]

41. Madhavan, S.; Kamaraj, M.; Vijayaraghavan, L. Microstructure and mechanical properties of cold metal transfer welded aluminium/dual phase steel. *Sci. Technol. Weld. Joi.* **2016**, *21*, 194–200. [CrossRef]

42. Madhavan, S.; Kamaraj, M.; Vijayaraghavan, L.; Srinivasa Rao, K. Microstructure and mechanical properties of aluminium/steel dissimilar weldments: Effect of heat input. *Mater. Sci. Technol.* **2017**, *33*, 200–209. [CrossRef]

43. Cao, R.; Yu, G.; Chen, J.H.; Wang, P.-C. Cold metal transfer joining aluminum alloys-to-galvanized mild steel. *J. Mater. Process. Technol.* **2013**, *213*, 1753–1763. [CrossRef]

44. Ünel, E.; Taban, E. Properties and optimization of dissimilar aluminum steel CMT welds. *Weld. World* **2017**, *61*, 1–9. [CrossRef]

45. Silvayeh, Z.; Vallant, R.; Sommitsch, C.; Götzinger, B.; Karner, W. Experimental investigations on single-sided CMT welding of hybrid aluminum-steel blanks. In Proceedings of the 10th International Conference on Trends in Welding Research, Tokyo, Japan, 11–14 October 2016.

46. Silvayeh, Z.; Vallant, R.; Sommitsch, C.; Götzinger, B.; Karner, W.; Hartmann, M. Influence of filler alloy composition and process parameters on the intermetallic layer thickness in single-sided cold metal transfer welding of aluminum-steel blanks. *Metall. Mater. Trans. A* **2017**, *48*, 5376–5386. [CrossRef]

47. Bruckner, J. Cold Metal Transfer Has a Future Joining Steel to Aluminum. *Weld. J.* **2005**, *84*, 38–40.

48. Selvi, S.; Vishvaksenan, A.; Rajasekar, E. Cold metal transfer (CMT) technology—An overview. *Def. Technol.* **2018**, *14*, 28–44. [CrossRef]

49. Götzinger, B.; Karner, W.; Hartmann, M.; Silvayeh, Z. Optimised process for steel-aluminium welding. *Lightweight Des. Worldw.* **2017**, *10*, 42–46. [CrossRef]

50. Akdeniz, M.V.; Mekhrabov, A.O.; Yilmaz, T. The role of Si addition on the interfacial interaction in Fe-Al diffusion layer. *Scr. Metal. Mater.* **1994**, *31*, 1723–1728. [CrossRef]

51. Akdeniz, M.V.; Mekhrabov, A.O. The effect of substitutional impurities on the evolution of Fe-Al diffusion layer. *Acta Mater.* **1998**, *46*, 1185–1192. [CrossRef]

52. Hwang, S.-H.; Song, J.-H.; Kim, Y.-S. Effects of carbon content of carbon steel on its dissolution into a molten aluminum alloy. *Mater. Sci. Eng. A* **2005**, *390*, 437–443. [CrossRef]

53. Sasaki, T.; Yakou, T.; Mochiduki, K.; Ichinose, K. Effects of carbon contents in steels on alloy layer growth during hot-dip aluminum coating. *ISIJ Int.* **2005**, *45*, 1887–1892. [CrossRef]

54. Shibata, K.; Morozumi, S.; Koda, S. Formation of the alloy layer in the Fe-Al diffusion couple. *J. Japan Inst. Met.* **1966**, *30*, 382–388. [CrossRef]

55. Jindal, V.; Srivastava, V.C.; Das, A.; Ghosh, R.N. Reactive diffusion in the roll bonded iron-aluminum system. *Mater. Lett.* **2006**, *60*, 1758–1761. [CrossRef]

56. Kajihara, M. Quantitative Evaluation of Interdiffusion in Fe_2Al_5 during reactive diffusion in the binary Fe-Al system. *Mater. Trans.* **2006**, *47*, 1480–1484. [CrossRef]

57. Naoi, D.; Kajihara, M. Growth behavior of Fe_2Al_5 during reactive diffusion between Fe and Al at solid-state temperatures. *Mater. Sci. Eng. A* **2007**, *459*, 375–382. [CrossRef]

58. Zhe, M.; Dezellus, O.; Gardiola, B.; Braccini, M.; Viala, J.C. Chemical changes at the interface between low carbon steel and an Al-Si alloy during solution heat treatment. *J. Phase Equil. Diffus.* **2011**, *32*, 486–497. [CrossRef]

59. Xu, L.; Robson, J.D.; Wang, L.; Prangnell, P.B. The influence of grain structure on intermetallic compound layer growth rates in Fe-Al dissimilar welds. *Metall. Mater. Trans. A* **2018**, *49*, 515–526. [CrossRef]

60. Rong, J.; Kang, Z.; Chen, S.; Yang, D.; Huang, J.; Yang, J. Growth kinetics and thickness prediction of interfacial intermetallic compounds between solid steel and molten aluminum based on thermophysical simulation in a few seconds. *Mater. Charact.* **2017**, *132*, 413–421. [CrossRef]

61. Das, A.; Shome, M.; Goecke, S.-F.; De, A. Numerical modelling of gas metal arc joining of aluminium alloy and galvanised steels in lap joint configuration. *Sci. Technol. Weld. Joi.* **2016**, *21*, 303–309. [CrossRef]

62. Zhang, G.; Chen, M.J.; Shi, Y.; Huang, J.; Yang, F. Analysis and modeling of the growth of intermetallic compounds in aluminum-steel joints. *RSC Adv.* **2017**, *7*, 37797–37805. [CrossRef]

63. Silvayeh, Z.; Vallant, R.; Sommitsch, C.; Götzinger, B.; Karner, W.; Hartmann, M. Numerical estimation of the intermetallic layer thickness in aluminum-steel welding. In Proceedings of the European Congress on Advanced Materials and Processes (EUROMAT 2017), Thessaloniki, Greece, 17–22 September 2017.

64. Borrisutthekul, R.; Yachi, T.; Miyashita, Y.; Mutoh, Y. Suppression of intermetallic reaction layer formation by controlling heat flow in dissimilar joining of steel and aluminum alloy. *Mater. Sci. Eng. A* **2007**, *467*, 108–113. [CrossRef]

65. Mezrag, B.; Deschaux Beaume, F.; Rouquette, S.; Benachour, M. Indirect approaches for estimating the efficiency of the cold metal transfer welding process. *Sci. Technol. Weld. Join.* **2018**, *23*, 508–519. [CrossRef]

66. Haelsig, A.; Kusch, M.; Mayr, P. New findings on the efficiency of gas shielded arc welding. *Weld. World* **2012**, *56*, 98–104. [CrossRef]

67. Pépe, N.; Egerland, S.; Colegrove, P.A.; Yapp, D.; Leonhartsberger, A.; Scotti, A. Measuring the process efficiency of controlled gas metal arc welding processes. *Sci. Technol. Weld. Joi.* **2011**, *16*, 412–417. [CrossRef]

68. Goldak, J.A.; Chakravarti, A.; Bibby, M. A New Finite Element Model for Welding Heat Sources. *Metall. Trans. B* **1984**, *15*, 200–305. [CrossRef]

69. Goldak, J.A.; Akhiaghi, M. *Computational Welding Mechanics*; Springer: New York, NY, USA, 2005; pp. 30–33.

 *materials*

Article

EBSD Investigation of the Microtexture of Weld Metal and Base Metal in Laser Welded Al–Li Alloys

Li Cui [1,*], Zhibo Peng [1], Xiaokun Yuan [1], Dingyong He [1] and Li Chen [2]

[1] College of Materials Science and Engineering, Beijing University of Technology, Beijing 100124, China; pengzhibo@emails.bjut.edu.cn (Z.P.); yuanxiaokun@bjut.edu.cn (X.Y.); dyhe@bjut.edu.cn (D.H.)

[2] AVIC Manufacturing Technology Institute, Beijing 100024, China; ouchenxi@163.com

* Correspondence: cuili@bjut.edu.cn; Tel.: +86-10-6739-2523

Received: 29 October 2018; Accepted: 20 November 2018; Published: 23 November 2018

Abstract: Autogenous laser welding of 5A90 Al–Li alloy sheets in a butt-joint configuration was carried out in this study. The microstructure characteristics of the weld metal and base metal in the horizontal surface and the transverse section of the welded joints were examined quantitatively using electron back scattered diffraction (EBSD) technique. The results show that the weld metal in the horizontal surface and the transverse section exhibits similar grain structural features including the grain orientations, grain shapes, and grain sizes, whereas distinct differences in the texture intensity and misorientation distributions are observed. However, the base metal in the horizontal surface and the transverse section of the joints reveals the obvious different texture characteristics in terms of the grain orientation, grain morphology, predominate texture ingredients, distribution intensities of textures, and grain boundary misorientation distribution, resulting in the diversity of the microhardness in the base metal and the softening of the weld metal. However, the degree of the drop in the hardness of the weld metal is highly correlated to the microtexture developed in the base metal.

Keywords: Al–Li alloys; laser welding; weld metal; base metal; grain orientation; texture

---

1. Introduction

Lithium (Li) addition to aluminum alloys causes substantial reduction in the density accompanied by large increase in elastic modulus, appreciable improvement of specific strength and specific stiffness of the alloys, making Al–Li based alloys strong candidates used in high-performance lightweight aerospace structures [1,2]. These alloys in the welded form could be further lightening the structures with weight savings. Therefore, welding of Al–Li based alloys is a significant challenge to provide both weight superiorities and cost benefits. Until now many studies have been made on conventional arc welding of Al–Li based alloys using a wide variety of welding processes [3–8]. Serious mechanical property degradation and high deformation in arc welding of Al–Li based alloys have been reported [1,3–8]. The use of laser welding is particularly attractive for Al–Li based alloys due to the tight focus ability and high power density of the laser beam [1,9,10]. Due to the low heat input and the rapid cooling rate resulting from high travel speeds, the laser welded joint was characterized by a fine grained weld and a narrow heat affected zone (HAZ) [9,10] which makes the softening in HAZ negligible to the tensile strength of the joints [9]. Thus, they make the mechanical properties of the joints superior to those of the other arc welding processes with a lower power density [1,3,9–11].

5A90 Al–Li alloy provided by Southwest Aluminum Co., Ltd., Chongqing, China, has the advantages of excellent corrosion resistance and weldability [12]. In order to further understand 5A90 Al–Li alloys and expand its usage, laser welding of 5A90 Al–Li thin sheets has been developed to meet the needs of the medium strength applications in the aircraft and aerospace structures, including the vapor and plasmas characteristics, welding parameters, welding defects prevention, microstructures, and mechanical properties of the welded joints [13–16]. It has been shown that some distinct differences in the microstructures between the base metal and the weld metal [9,14–16] was found, and it is still important to further clarify the microstructural characteristics of the weld metal of 5A90 Al–Li alloys.

The microstructures of the materials are relative routinely characterized by the morphology and distribution of constituent phases. However, an elaborate and complete description of microstructure of a crystalline material must also include the knowledge about crystallographic orientation features and textures of the constituent grains [17]. Currently, there are increasing reports concerning the considerable deformation textures in the Al–Li based alloys, and crystallographic textures in the Al–Li alloys are quite significant to the properties via rendering them anisotropic [17,18]. Given that texture often causes anisotropic mechanical properties, its presence in the weld zones of the Al–Li alloys could be quite significant [19]. The information available on weld metal texture of 5A90 Al–Li alloys is, however, relatively scanty.

In general, the most common description of the texture for materials can be given in terms of various graphical plots via the grain orientation image mappings (OIM), pole figures (PF), misorientation angles and orientation distribution function (ODF) [17,20,21]. Although these approaches provide a useful description of the textures, the extracted texture information is insufficient. It is often desirable to determine the volume fractions of different texture components. Moreover, the difference in texture along the different directions (for example, sample directions including the rolling direction (RD), the transverse direction (TD), and the normal direction (ND)), which is quite essential to controlling of the welded joint performance, is still not very clear. Therefore, the major attempts in the current study are as follows: (1) to investigate the orientation bias of the weld metal grains along different sample directions; (2) to compare the orientation bias of the base metal grains along different sample directions. Summarizing the results of the above researches will provide quite beneficial information about the relationship between local orientation bias (of both grains and boundary planes) and local performance parameters, and will have broad meanings to the joining technique of the similar Al–Li alloys.

2. Materials and Methods

5A90 Al–Li alloy sheets of 3.0 mm thickness were used in this study. The chemical composition (wt%) of the 5A90 Al–Li alloy is shown in Table 1. Before welding, the specimens were chemically removed 0.2 mm thickness from each side using 10~20% NaOH solution (Beijing Chemical Works Co., Ltd., Beijing, China), and immersed in 20% nitric acid solution (Beijing Chemical Works Co., Ltd., Beijing, China), prior to polishing to minimize the presence of porosity. Full penetration I-butt joints were made using a Nd:YAG laser (GSI (Shanghai) Co., Ltd., Shanghai, United Kondom), with the nominal maximum laser power of 4.5 kW. The laser fixed to a six-axial welding robot with an emission wave length of 1.07 μm can deliver in continuous wave mode through an output fiber core diameter of 600 μm. A focusing lens with a focal length of 200 mm was used and the beam parameter product (BPP) of the laser beam at the focal point was 25 mm. The laser head was operated 15° leaning to the normal direction of the horizontal surface of the weld joint to prevent the fiber being burned. A 200 mm focal length lens was employed to focus the beam on the specimen surface. The laser power was 1.8 kW and the travel speed was 45 mm/s. An argon shielding gas at a flow rate of 20 L/min was used to shield the welding pool from the atmosphere and the back shielding gas was supplied by ultrahigh purity argon at a flow rate of 15 L/min during welding. Schematic diagram of laser welded joints of 5A90 Al–Li alloys was shown in Figure 1, where the horizontal surface refers to the RD-TD plane and the transversal section refers to the RD-ND plane.

Throughout the experiments, the welding operation was shielded by the trailing and back shielding gas supplied by ultrahigh purity argon at flow rates of 20 L/min and 15 L/min.

Table 1. Chemical composition of 5A90 Al–Li alloys (mass fraction/%).

Mg	Li	Zr	Fe	Si	Cu	Ti	Al
4.9–5.4	1.8–2.2	0.08–0.13	$\leq$0.12	$\leq$0.09	$\leq$0.05	$\leq$0.05	Bal

After welding, the visual checking of the weld surface and the X-ray inspection of the joints were performed. The weld ripples and weld width are uniform and the weld surface should be no visible porosity, hot cracking. Additionally, the porosity of the welds should be less than grade II. Consistent with the above two respects, the welded joints were considered as acceptable specimens selected for the further microtexture study by means of the EBSD technique. The locations of the EBSD samples selected from the base metal and weld metal of the welded joints are depicted in Figure 1, where the coordinate system consists of three (x,y,z) axes coinciding with the sample directions. As the next step, the samples for EBSD analysis were removed from the base metal and weld metal of both the horizontal surface and the transverse section, and were designated as sample HBM (abbreviated for horizontal base metal), HWM (abbreviated for horizontal weld metal), TBM (transversal base metal) and TWM (transversal weld metal), respectively. Afterwards, the four samples were mechanically polished and subsequently, electropolished by immersion in a 30% nitric acid in methanol (Beijing Chemical Works Co., Ltd., Beijing, China), solution cooled to $-25\ ^\circ$C at a voltage of 20 V for 30 s. These treatments yielded sample surfaces suitable for EBSD mapping.

The EBSD analysis was performed by using the high speed detector (EDAX Genesis 2000 system) (EDAX Co., Ltd., Salt Lake, USA), incorporated in a thermal field scanning electron microscope (SEM, JEOLJSM 6500F) (JEOL Co., Ltd., Tokyo, Japan), with an accelerating voltage of 15 kV. To ensure the accuracy of the EBSD measurements, the data were collected with a step size of 1 micron. The EBSD data were then transported into the TSL OIM Analysis 5.3 software (EDAX Co., Ltd., Salt Lake, USA), for further analysis. To describe the grain orientation and texture at different locations, the microstructures of the samples were indicated by the inverse pole figure maps and the image quality maps. For the selected samples, their corresponding orientation and texture were illustrated via the crystal orientation maps showing the spatial distributions and volume fractions for the ideal fcc rolling components. To describe the orientation bias of boundary planes at different locations, the misorientation across the observed boundaries was illustrated by the misorientation angle distribution functions (MDFs). Moreover, the grain boundary plane orientation distribution function (namely GBP-ODF, which describes the orientation bias of boundary planes, developed under the auspices of Carnegie Mellon University [22–24]), were used to study the grain boundary character distribution (GBCD) of the special Σ3 boundary.

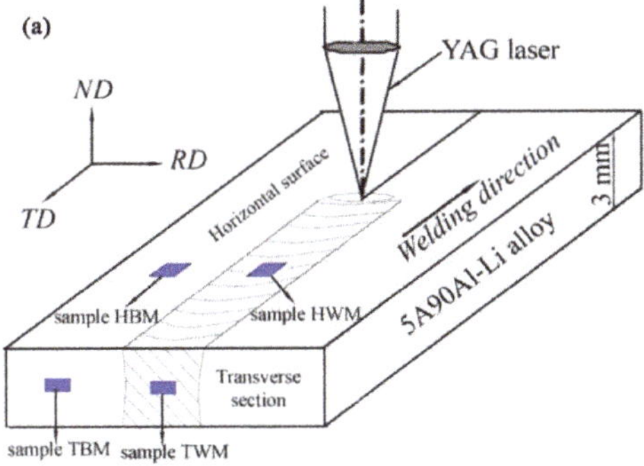

Figure 1. *Cont.*

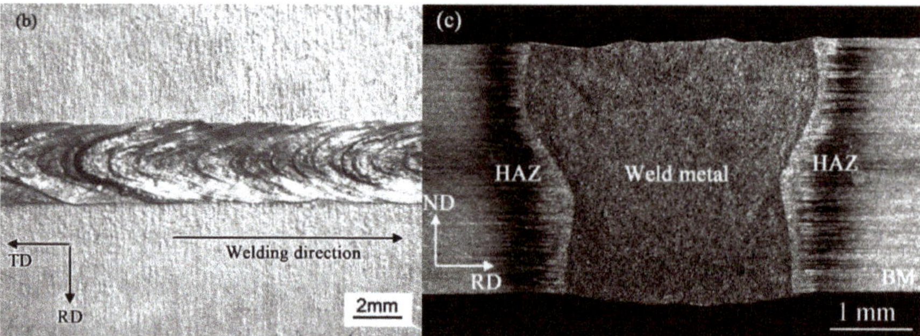

Figure 1. Laser welded 5A90 Al–Li alloys: (**a**) Schematic diagram of the locations of EBSD samples; (**b**) Weld appearance of the joint in the horizontal surface; and (**c**) cross-section of the joint in the transverse section.

3. Results

3.1. Grain Orientation, Grain Shape, and Grain Size

The microstructure of the four samples is depicted by inverse pole figure (IPF) map in Figure 2, where the grain color specifies the orientation according to the coloring indicated in the orientation legend for the cubic symmetry. The IPF maps of the sample HBM and TBM (Figure 2a,c) clearly illustrate that the base metal in the horizontal surface and the transverse section both have strong preferred orientations, however, the base metal has different preferred orientations on these two directions. Sample HBM predominately shows <101> orientations, while sample TBM strongly favors the <112> and <100> orientations. Moreover, the IPF maps of the sample HBM and TBM also reveal the difference in the grain shapes. The microstructure of the sample HBM is characterized by a coarse pancake shape, whereas the grains of the sample TBM present a lamellar structure along the rolling direction. In particular, several long and coarse "deformation bands" are observed in the sample TBM, indicating the microstructure of the base metal in the transverse section is not homogeneous after the rolling deformation. In short, the base metal in the horizontal surface and the transversal section exhibits obvious difference in grain orientation and grain morphology.

For the weld metal, the IPF maps of the sample HWM and sample TWM exhibit similar equiaxed grains (having a wide variety of colors corresponding to varied crystallographic orientations, see Figure 2b,d), indicating that the weld metal in the horizontal surface and the transversal section has nearly the same grains orientation features and grain shapes. This can also be supported by the fact that the grain sizes of the sample HWM and sample TWM are 31.0 μm and 36.8 μm, respectively. It can, therefore, be concluded that the weld metal in the horizontal surface and the transverse section exhibits similar structural features regarding the grain shape, grain orientation, and grain size parameters.

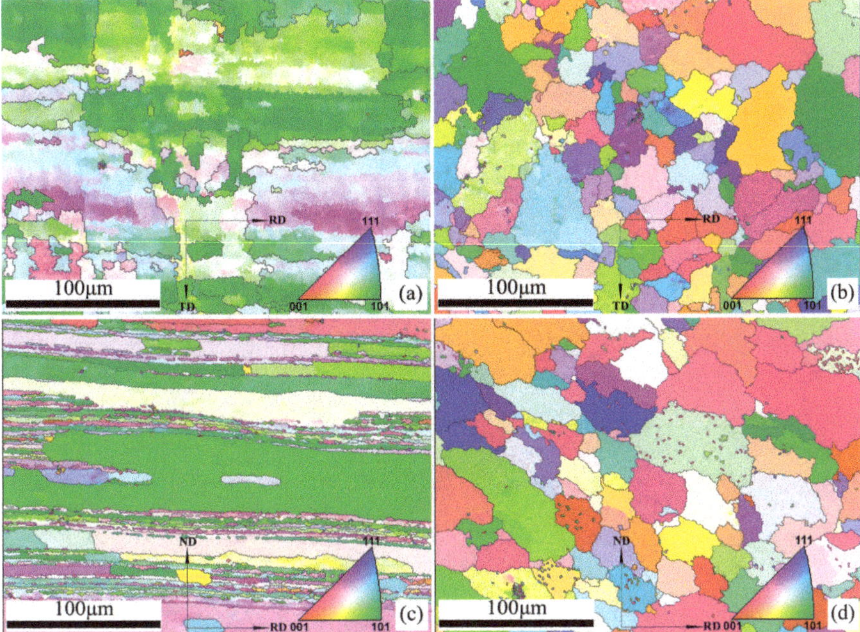

Figure 2. Inverse pole figure maps of base metal and weld metal in 5A90 Al–Li alloys along different sample directions with orientation legend for the cubic symmetry: (**a**) Sample HBM, (**b**) sample HWM, (**c**) sample TBM, and (**d**) sample TWM.

3.2. Spatial Distribution of the Textures

It is generally reported that the rolling texture of fcc symmetries mainly contains the Goss ({011} <100>), brass ({011} <211>), S ({123} <634>), and copper components ({112} <111>). To quantitatively determine the volume fractions of the typical texture components and to analyze the spatial distribution of the rolling textures, the crystal orientation maps of the four samples showing the spatial distribution of fcc rolling components are illustrated in Figure 3 (in which each grain color specifies a texture component), and the corresponding volume fractions of the rolling components are listed in Table 2. Together with the brass, copper, and Goss, the S1 ({241} <112>), S2 ({231} <124>), S3 ({231} <346>) component and Taylor component ({4411} <11118>) are also presented in the base metal and weld metal of the joints. As shown in Figure 3, the colors are orange for copper, green for S1, purple for S2, blue for S3, cyan for Taylor, yellow for brass, and red for Goss.

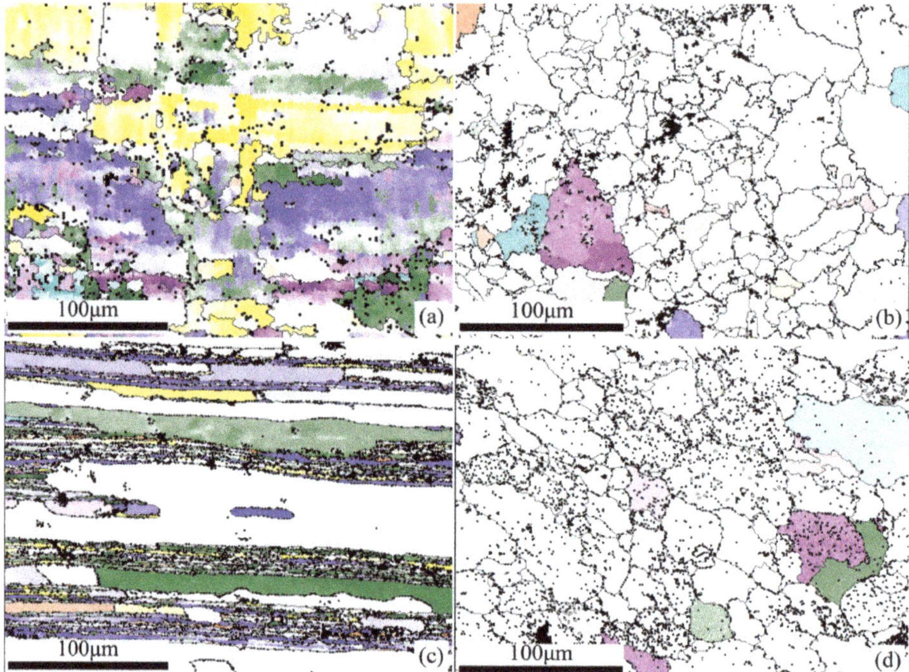

Figure 3. Spatial distributions of typical fcc rolling components of base metal and weld metal in 5A90 Al–Li alloys along different sample directions, with orange for copper, green for S1, purple for S2, blue for S3, cyan for Taylor, yellow for Brass, and red for Goss: (**a**) sample HBM; (**b**) sample HWM; (**c**) sample TBM; and (**d**) sample TWM.

Table 2. Volume fraction of the typical fcc rolling components (%).

Sample	Copper	S1	S2	S3	Taylor	Brass	Goss
HBM	0.3	12.6	4.9	22.0	2.4	27.2	0.0
HWM	1.6	2.4	2.3	2.1	2.0	1.3	0.6
TBM	1.6	13.7	2.5	13.7	0.8	6.0	0.0
TWM	0.4	1.5	3.0	0.8	3.2	0.2	0.3

It can be seen that the sample HBM (Figure 3a) exhibits significant Brass components and the overall intensity of the components is approximately 69.4%. This is obviously a consequence of the higher deformation degree during the rolling processing subjected to 5A90 Al–Li alloys [18]. In sample TBM, the rolling components show weak distribution intensities with a sum percentage about 38.3%, with the S components including S1, S2, and S3 as the predominate ingredients.

For the weld metal, the overall texture intensity of the sample HWM and the sample TWM is 12.3% and 9.4%, exhibiting a relatively weak texture in the weld metal (see Figure 3b,d and Table 1). In general, the overall texture of the weld metal is basically decided by the orientations of the grains [25] in the columnar zone [19] and the weld metal zone having the equiaxed grains is likely to have an almost random texture, while brass, copper and S components are normally encountered in the deformation texture of fcc materials [20,26]. The present work conforms this and the brass, copper, S and Goss components are in presence in the weld metal, which suggests some similar texture components in the weld metal were formed as base metal during the laser welding.

Previous reports [19,27] have suggested that the welded joint could develop several major textures and strong texture might exist at the base metal, HAZ and columnar grain zone of the weld metal, whereas the grain orientation at equiaxed grain zone in the center of weld metal was relatively random. The present work also confirms that the original base metal structure has been eliminated and replaced by a very fine equiaxed grain structure in the weld metal. All these phenomena remind us that during the solidification of aluminum alloys, the equiaxed grains formed in the weld metal are more prone to have random distributions with weaker intensities.

Generally speaking, the equiaxed grains are more likely be generated by the continuous dynamic recrystallization [26]. However, during the solidification of the welding pool, almost all equiaxed grains are those newly-nucleated and grown, hence the grain orientation is random [17]. Therefore, the equiaxed grain cannot be generated by the dynamic recrystallization occurred during laser welding. The formation of equiaxed grains in the weld metal could be ascribed to the heterogeneous nucleation mechanism aided by equilibrium Al_3Zr phase as well as the growth of pre-existing nuclei created by dendrite fragmentation, or by grain detachment resulted from Nd:YAG laser welding processes [16]. Hence, the random texture developed in the weld metal can be quite different from the normal random texture generated by the continuous dynamic recrystallization.

3.3. Texture Fiber Analysis

It is generally known that the rolling texture of fcc symmetries is near the location of so-called α-fiber and β-fiber textures in the Euler space [18,20]. The α-fiber mainly contains the Goss and the Brass orientations, while the β-fiber mainly contains brass, S and copper components.

Figure 4 shows the α-fiber and β-fiber texture analysis of the four samples (in reduced Euler spaces). It can be directly observed that both α-fiber and β-fiber textures are much stronger in base metal (sample HBM and sample TBM) than those of the weld metal (sample HWM and sample TWM). This is not surprising since the higher degree of deformation preferentially developed textures in base metal. Moreover, both α-fiber and β-fiber textures could be modified by different rolling rates, during which the major crystallographic features could be involved. Figure 4 clearly shows the difference in the fiber texture between sample HBM and TBM. On the α-fiber (Figure 4a), the intensity of sample HBM is obviously stronger than that in sample TBM when the ϕ_1 angel ranges from 20° to 40° (meaning that the brass texture is more prevalent), while on the β-fiber (Figure 4b), the intensity of sample TBM is stronger than that of sample HBM when the ϕ_2 angel ranges from 50° to 70° (meaning that the S texture is more prevalent). The outcome is consistent with the results in Table 1, and might indicate that S and Brass textures are preferentially developed with referring to the rolling direction in the base metal.

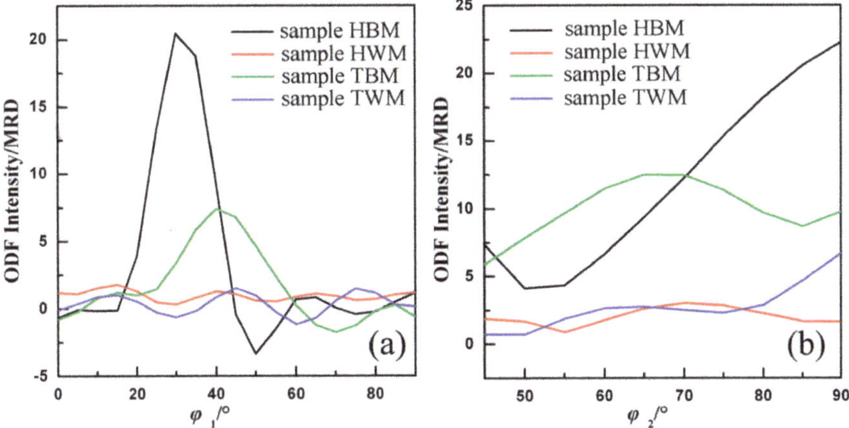

Figure 4. Texture fiber analysis of base metal and weld metal in 5A90 Al–Li alloys along the different sample directions: (**a**) α-fiber; and (**b**) β-fiber.

3.4. Orientation Bias of Boundary Planes

In the polycrystalline structure, the boundary can be defined either as a high-angle boundary when the misorientation angle across the boundary is higher than 15°, or a low-angle grain boundary when the misorientation angle across the boundary is between 2° and 15°. The image quality maps of the samples are displayed in Figure 5, where the high-angle and low-angle grain boundaries are highlighted in different colors.

It is evident that the sample HBM (Figure 5a) has prevalent low-angle boundaries with a fraction about 67%. As a contrast, the fraction of the low-angle boundaries is merely 30% for the sample TBM with an inhomogeneous lamellar structure (Figure 5c). Furthermore, the microstructure of the sample TBM shows several severely coarse "deformation bands" separated by thin lamellae. The lamellae grains contain high fractions of low-angle boundaries (up to 61%), while little low-angle boundaries and high-angle boundaries within the coarse "deformation bands" can be observed. Thus, a pronounce difference in the grain boundary misorientation of the base metal between the horizontal surface and the transversal section of the welded joint could be found.

For the weld metal, specific difference in the boundary misorientation distributions can be observed between sample HWM (Figure 5b) and sample TWM (Figure 5d). Here, the fraction of the high-angle boundaries for samples HWM and TWM are 67.6% and 53.8%, respectively, that is to say, the weld metal in the horizontal surface has much higher fraction of the high-angle boundaries. A comparison of these results with those of the base metal demonstrates the clear shift in the boundary misorientation distributions from low to high angles. Another difference is, the clustering of the low-angle boundaries is often found within grains, whereas the high-angle boundaries are more prone to occur along grain contours, or, between grain pairs. Hence, there might have a distinct difference in the grain boundary misorientation of the weld metal along different directions.

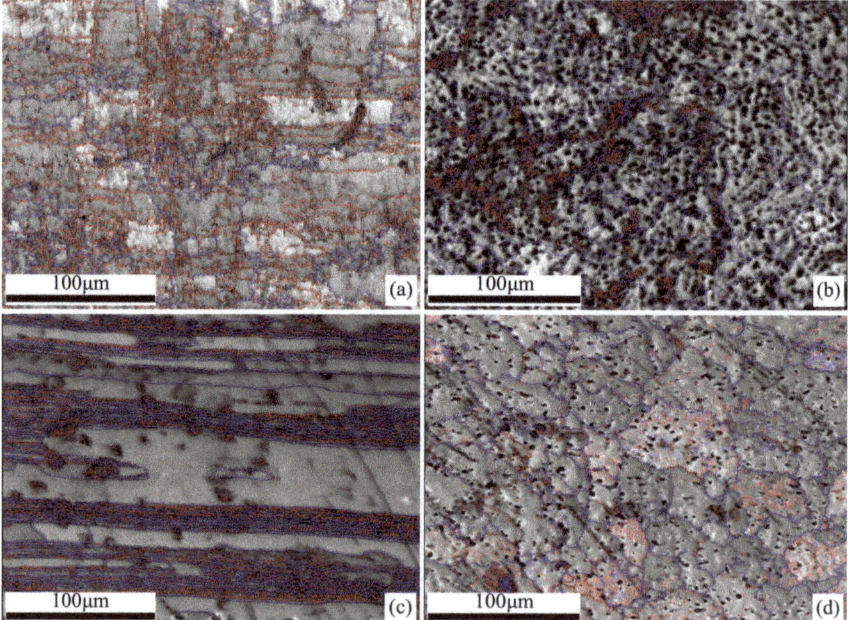

Figure 5. Grain boundary traces imposed on image quality map of base metal and weld metal in 5A90 Al–Li alloys along different sample directions: (**a**) Sample HBM, (**b**) sample HWM, (**c**) sample TBM, and (**d**) sample TWM, with high angle (>15°) grain boundaries shown as blue and low angle (2~15°) boundaries as red.

A frequency distribution of misorientation angles for sample HWM and sample TWM with a superimposed random McKenzie distribution [28] is shown in Figure 6. The average misorientation angles deduced from the histograms are 30.4°, and 24.9° form sample HWM and sample TWM, respectively. Figure 6 overall illustrates that the misorientation distribution of sample HWM and sample TWM is much different from random distribution. In addition, the relative frequency of intermediate (30° to 50°) for the sample HWM is higher than that of the sample TWM. Meanwhile, the relative frequency of low-angle boundaries for sample HWM is lower than that of the sample TWM. The above results on the whole illustrate the heterogeneity of misorientation of the weld metal on different directions.

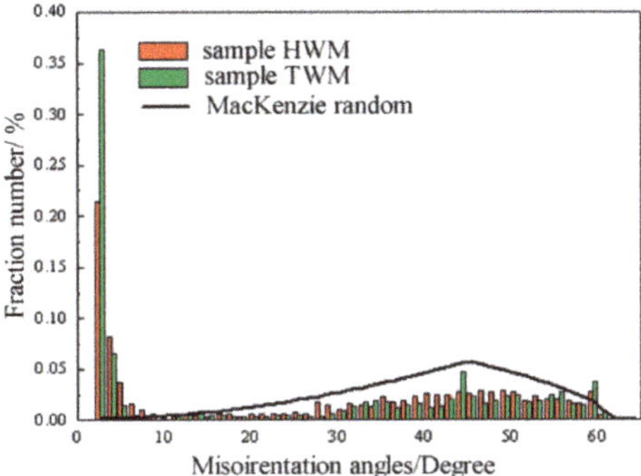

Figure 6. Misorientation angles distributions of the laser weld metal of 5A90 Al–Li alloys along different directions.

The GBP-ODF requires large amount of boundary quantity for a complete analysis [22] and in the current work, only sample HBM and sample TBM can meet this standard. The GBP-ODFs of Σ3 boundaries in the sample HBM and sample TBM were calculated and the outcomes are presented in Figure 7. In the sample HBM, the Σ3 boundary shows a tilt boundary feature [29], with distribution intensity in units of 63.5 multiples of a random distribution (MRD); while in the sample TBM, the Σ3 boundary shows a twist boundary character [29], with distribution intensity of 97.3 MRD. Since the Σ3 boundary can be regarded as a 60° rotation along the <111> axis [20], the above results show that the Σ3 boundary exhibit simple geometries. Nevertheless, Σ3 tilt (in sample HBM) and Σ3 twist (in sample TBM) may correspond to specific structures and, consequently, to special physical properties of boundaries. The indexed twist and tilt Σ3 boundaries in the cubic case clearly illustrated that the misorientation distribution in the base metal is not homogeneous along different directions, which could cause the diversity of mechanical properties along different directions in the weld metal. The microhardness across the welded joint from left to right including the base metal, HAZ and weld metal for sample HWM and sample TWM were measured, and the results are shown in Figure 8a,b, respectively. For the base metal, the average hardness value of the base metal is 128 HV in sample HWM, while the average hardness value of the base metal in sample TWM is 117 HV. It indicates that the indexed twist and tilt Σ3 boundaries in the base metal along different directions have resulted in the diversity of the microhardness due to the homogeneous microstructure along different directions. For the weld metal, the microhardness along different directions is affected strongly by its own base metal. As observed in Figure 8, the hardness value measured in the weld metal is ~102 HV for sample HWM, whereas the hardness value measured in the weld metal is ~92 HV for sample TWM. Thus, it can be seen that the hardness value of weld metal is evidently lower than that of the base metal, meaning the softening of the weld metal. Although both sample HWM and sample TWM reveal the softening of the weld metal, the degree of the drop in the hardness of the weld metal is highly correlated to the microtexture developed in the base metal.

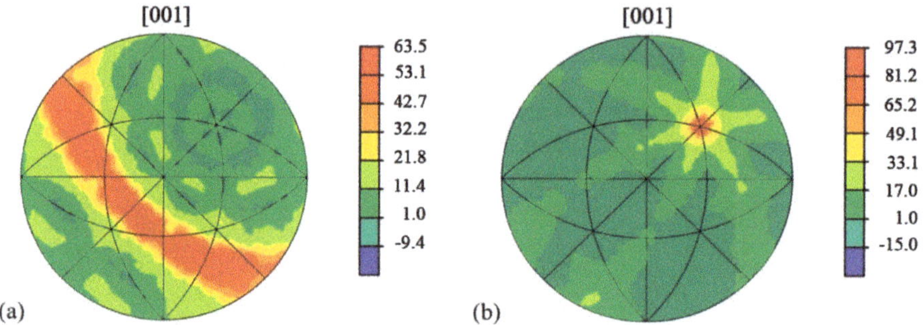

Figure 7. Grain boundary plane orientation distribution function along different sample directions in the base metal of 5A90 Al–Li alloys, showing orientation texture of boundary planes plotted along the [001] direction: (**a**) sample HBM, and (**b**) sample TBM. Texture intensities are given in unit of MRD.

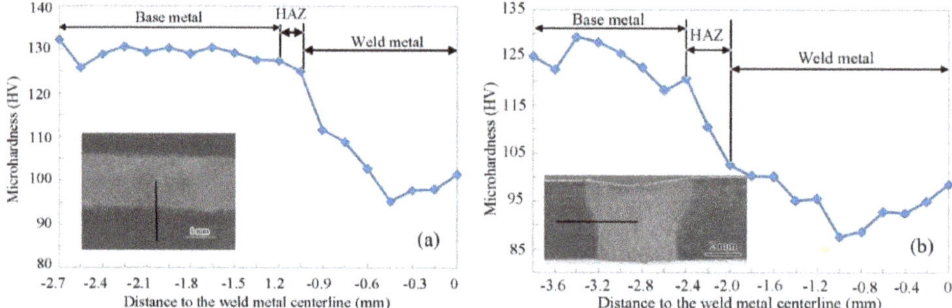

Figure 8. Hardness profiles along different sample directions in the welded 5A90 Al–Li alloys: (**a**) sample HBM, and (**b**) sample TBM.

4. Conclusions

Based on the above results and discussion, the following conclusions can be made from this work:

(1) For the base metal, there is an obvious difference in the grain morphology and orientation in the horizontal surface and the transversal section. However, the weld metal in the horizontal surface and the transverse section exhibits similar structural features regarding the grain shape, grain orientation, and grain size parameters.

(2) For the weld metal, there is an obvious difference in the texture intensity in the horizontal surface and the transversal section, despite the weld metals exhibit the similar grain shapes, grain orientations and grain size. Moreover, the texture intensity are much weaker compared to those of the base metal. Particularly, the brass, copper, S and Goss components observed in the base metal are also presented in the weld metal.

(3) For the boundary plane misorientation, the low-angle boundaries are most predominant in the base metal in the horizontal surface. The large fraction of high-angle boundaries of the weld metal in the horizontal is higher than that of the transversal section of the welded joint. The misorientation distribution of the weld metal is much different from random distribution.

(4) The overall GBP-ODF of Σ3 boundary is not homogeneous in the base metal, resulting in the diversity of the microhardness in the base metal. In addition, the hardness value of weld metal is evidently lower than that of the base metal, meaning the softening of the weld metal. Although both sample HWM and sample TWM reveal the softening of the weld metal, the degree of the drop in the hardness of the weld metal is highly correlated to the microtexture developed in the base metal.

Author Contributions: L.C. and D.H. conceived and designed the experiment. X.Y. and Z.P. performed the experiments; Z.P. and X.Y. analyzed the data; D.H. and L.C. contributed materials and analysis tools; and L.C. wrote the paper.

Funding: This research was funded by the National Natural Science Foundation of China (grant no. 51475006, no. 51471007) and the Key Program of Science and Technology Projects of Beijing Municipal Commission of Education (grant no. KZ201610005004).

Conflicts of Interest: The authors declare no conflict of interest.

References

1. Shi, Y.W.; Zhong, F.; Li, X.Y.; Gong, S.L.; Chen, L. Effect of laser beam welding on tear toughness of a 1420 aluminum alloy thin sheet. *Mater. Sci. Eng. A* **2007**, *465*, 153–159. [CrossRef]

2. Pickens, J.R. Recent developments in the weldability of lithium-containing aluminium alloys. *J. Mater. Sci.* **1990**, *25*, 3035–3407. [CrossRef]

3. Kostrivas, A.; Lippold, J.C. Weldability of Li-bearing aluminium alloys. *Int. Mater. Rev.* **1999**, *44*, 217–237. [CrossRef]

4. Padmanabham, G.; Pandey, S.; Schaper, M. Pulsed gas metal arc welding of Al–Cu–Li alloy. *Sci. Technol. Weld. Join.* **2005**, *10*, 67–75. [CrossRef]

5. Peng, Y.; Fu, Z.Y.; Wang, W.M.; Zhang, J.Y.; Wang, Y.C.; Wang, H.; Zhang, Q.J. Phase transformation at the interface during joining of an Al–Mg–Li alloy by pulsed current heating. *Scr. Mater.* **2008**, *58*, 49–52. [CrossRef]

6. Kostrivas, A.D.; Lippold, J.C. Simulating weld-fusion boundary microstructures in aluminum alloys. *JOM* **2004**, *56*, 65–72. [CrossRef]

7. Lin, D.C.; Wang, G.X.; Srivastan, T.S. A mechanism for the formation of equiaxed grains in welds of aluminum-lithium alloy. *Mater. Sci. Eng. A* **2003**, *351*, 304–309. [CrossRef]

8. Chaturvedi, M.C.; Chen, D.L. Effect of specimen orientation and welding on the fracture and fatigue properties of 2195 Al–Li alloy. *Mater. Sci. Eng. A* **2004**, *387–389*, 465–469. [CrossRef]

9. Xiao, R.S.; Zhang, X.Y. Problems and issues in laser beam welding of aluminum-lithium alloys. *J. Manuf. Process.* **2014**, *16*, 166–175. [CrossRef]

10. Zhao, H.; White, D.R.; Debroy, T. Current issues and problems in laser welding of automotive aluminum alloys. *Int. Mater. Rev.* **1999**, *44*, 238–265. [CrossRef]

11. Molian, A.P.; Srivatsan, T.S. Weldability of aluminum lithium alloy 2090 using laser welding. *J. Mater.Sci.* **1990**, *25*, 3347–3358. [CrossRef]

12. Zhang, P.; Ye, L.Y.; Zhang, X.M.; Gu, G.; Jiang, H.C.; Wu, Y.L. Grain structure and microtexture evolution during superplastic deformation of 5A90 Al–Li alloy. *Trans. Nonferrous Met. Soc. China* **2014**, *24*, 2088–2095. [CrossRef]

13. Duan, A.Q.; Chen, L.; Guo, L. Characteristics of vapor/plasmas during YAG laser welding of 5A90 Al–Li alloy. *Rare Met. Mater. Eng.* **2009**, *38*, 160–164.

14. Chen, L.; Xu, W.W.; Gong, S.L. The effect of weld size on joint mechanical properties of 5A90Al–Li alloy. *Adv. Mater. Res.* **2011**, *146–147*, 1831–1838. [CrossRef]

15. Xu, F.; Chen, L.; Gong, S.L.; Li, X.Y.; Yang, J. Microstructure and mechanical properties of Al–Li alloy by laser welding with filler wire. *Rare Met. Mater. Eng.* **2011**, *40*, 1775–1779.

16. Cui, L.; Li, X.Y.; He, D.Y.; Chen, L.; Gong, S.L. Effect of Nd:YAG laser welding on microstructure and hardness of an Al–Li based alloy. *Mater. Charact.* **2012**, *71*, 95–102. [CrossRef]

17. Mishin, O.V.; Gertsman, V.Y.; Gottstein, G. Distributions of orientations and misorientations in hot-rolled copper. *Mater. Charact.* **1997**, *38*, 39–48. [CrossRef]

18. Chen, Z.Y.; Cai, H.N.; Chang, Y.Z.; Zhang, X.M.; Liu, C.M. Texture evolution of polycrystalline aluminum during rolling deformation. *Acta Metall. Sin.* **2008**, *44*, 1316–1321.

19. Hector, J.L.; Chen, Y.L.; Agarwal, S.; Briant, C.L. Texture characterization of autogenous Nd:YAG laser welds in AA5182-O and AA6111-T4 aluminum alloys. *Metall. Mater. Trans. A* **2004**, *35A*, 3032–3038. [CrossRef]

20. Sidor, J.J.; Petrov, R.H.; Kestens, L.A.I. Microstructural and texture changes in severely deformed aluminum alloys. *Mater. Charact.* **2011**, *62*, 228–236. [CrossRef]

21. Cui, L.; Li, X.Y.; He, D.Y.; Chen, L.; Gong, S.L. Study on microtexture of laser welded 5A90 aluminium-lithium alloys using electron backscattered diffraction. *Sci. Technol. Weld. Join.* **2013**, *18*, 204–209. [CrossRef]

22. Saylor, D.M.; Dasher, B.S.; Adams, B.L.; Rohrer, G.S. Measuring the five-parameter grain-boundary distribution from observations of planar sections. *Metall. Mater. Trans. A* **2004**, *35*, 1981–1989. [CrossRef]

23. Rohrer, G.S.; Randle, V.; Kim, C.S.; Hu, Y. Changes in the five-parameter grain boundary character distribution in alpha-brass brought about by iterative thermomechanical processing. *Acta Mater.* **2006**, *54*, 4489–4502. [CrossRef]

24. Watanabe, T. Grain boundary engineering: Historical perspective and future prospects. *J. Mater. Sci.* **2011**, *46*, 4095–4115. [CrossRef]

25. Randle, V. Applications of electron backscatter diffraction to materials science: Status in 2009. *J. Mater. Sci.* **2009**, *44*, 4211–4218. [CrossRef]

26. Sathiaraj, G.D.; Ahmed, M.Z.; Bhattacharjee, P.P. Microstructure and texture of heavily cold-rolled and annealed fcc equiatomic medium to high entropy alloys. *J. Alloy. Compd.* **2016**, *664*, 109–119. [CrossRef]

27. Yu, L.; Nakata, K.; Yamamoto, N.; Liao, J. Texture and its effect on mechanical properties in fiber laser weld of a fine-grained Mg alloy. *Mater. Lett.* **2009**, *63*, 870–872. [CrossRef]

28. Mackenzie, J.K. The distribution of rotation axes in a random aggregate of cubic crystals. *Acta Metall.* **1964**, *12*, 223–225. [CrossRef]

29. Morawiec, A. Low-Sigma twist and tilt grain boundaries in cubic materials. *J. Appl. Crystallogr.* **2011**, *44*, 1152–1156. [CrossRef]

Article

Effect of Arc Pressure on the Digging Process in Variable Polarity Plasma Arc Welding of A5052P Aluminum Alloy

Bin Xu [1,2], Shinichi Tashiro [2], Fan Jiang [1,3,*], Shujun Chen [1] and Manabu Tanaka [2]

[1]   Engineering Research Center of Advanced Manufacturing Technology for Automotive Components, Ministry of Education, Beijing University of Technology, Beijing 100124, China; cougarxbin@163.com (B.X.); sjchen@bjut.edu.cn (S.C.)
[2]   Joint and Welding Research Institute, Osaka University, Osaka 5670047, Japan; tashiro@jwri.osaka-u.ac.jp (S.T.); tanaka@jwri.osaka-u.ac.jp (M.T.)
[3]   Beijing Engineering Researching Center of laser Technology, Beijing 100124, China
*   Correspondence: jiangfan@bjut.edu.cn; Tel.: +86-137-2008-7645

Received: 11 March 2019; Accepted: 29 March 2019; Published: 1 April 2019

Abstract: The keyhole digging process associated with variable polarity plasma arc (VPPA) welding remains unclear, resulting in poor control of welding stability. The VPPA pressure directly determines the dynamics of the keyhole and weld pool in the digging process. Here, through a high speed camera, high frequency pulsed diode laser light source and X-ray transmission imaging system, we reveal the potential physical phenomenon of a keyhole weld pool. The keyhole depth changes periodically corresponding to the polarity conversion period if the current is same in the electrode negative (EN) phase and electrode positive (EP) phase. There exist three distinct regimes of keyhole and weld pool behavior in the whole digging process, due to the arc pressure attenuation and energy accumulation effect. The pressure in the EP phase is smaller than that of the EN phase, causing the fluctuation of the weld pool free surface. Based on the influence mechanism of energy and momentum transaction, the arc pressure output is balanced by separately adjusting the current in each polarity. Finally, the keyhole fluctuation during the digging process is successfully reduced and welding stability is well controlled.

Keywords: VPPA welding; keyhole; digging process; plasma arc pressure; electrode energy balance; X-ray transmission system

---

1. Introduction

VPPA (Variable Polarity Plasma Arc) keyhole welding is an ideal method to achieve joints made of middle thick aluminum with high quality and efficiency [1]. There are two polarities in one current cycle: an EN (Electrode Negative) phase and an EP (Electrode Positive) phase. The cycle can achieve both deep penetration and remove the oxide layer on the surface of the base metal [2,3]. The keyhole digging process is very important, because the welding easily fails without proper keyhole generation [4]. However, the evolution of the keyhole and weld pool has not been clearly understood, especially in the digging process of VPPA welding. Therefore, it is necessary to explore the physical phenomenon and the mechanism of the digging process for achieving stable welding.

As a unique characteristic of plasma arc welding, keyhole behavior determines the process stability of plasma arc welding. Keyhole detection research has been carried out in view of DC-PAW (Direct Current Plasma Arc Welding) for steel. Liu et al. [5–7] directly captured the keyhole exit image using a CCD (Charge Coupled Device) camera. The keyhole exit deviation distance and keyhole size parameters were put forward to evaluate the thermal state, which provides the feasibility for accurately controlling the keyhole stability. Zhang et al. [8] developed a charge sensor to monitor the plasma cloud dynamics, corresponding to the keyhole status. Metcalfe's investigation shows that it is possible to indicate whether the keyhole is open or not yet formed by monitoring the efflux plasma. Its fluctuations can reflect the keyhole stability [9]. Regarding to VPPA welding of aluminum, Zheng et al. [10] focused on the front image sensing of the keyhole. An algorithm for extracting the keyhole's geometrical size was proposed to establish the linkages with weld formation. Wu et al. [11] developed a vision sensor system to acquire the keyhole images. A novel hybrid approach was used to recognize the keyhole status. The prediction is realized through the detection, which is a step forward for accurate control of the keyhole. In addition to directly observing the keyhole with a CCD camera, the unique signal characteristics of VPPA welding were used to measure the keyhole state. Saad et al. [12] achieved identification between three models of VPPA keyhole welding (no-keyhole, keyhole and cutting) using acoustic signal measurement. Wu et al. [13] also investigated the relationship between the keyhole geometry and acoustic signatures with a dual-sensor system. An extreme learning machine model was built for predicting keyhole geometry. However, these methods are mainly used to indirectly obtain the keyhole information. The keyhole boundary inside the weld pool in real time is difficult to measure with the above methods. An X-ray transmission observation system was successfully adopted to investigate the keyhole and weld pool dynamics [14–16], which highly contributed to the understanding of keyhole evolution. Anh et al. [17] adopted stereo synchronous imaging of tracer particles with two sets of X-ray transmission systems to clarify the weld pool formation process in DC-PAW.

The above research mainly focused on the keyhole weld pool evolution in a welding quasi-steady state rather than the keyhole digging process. Zheng et al. [18] pointed out that a smooth transition from start-up segment to main body segment was very important for welding stability. By optimizing the waveforms of current and plasma gas flow rate, a relative proper keyhole generation segment was obtained. A synchronous increase of gas flow rate and current contributes to the smooth transition. Chen et al. [19] added a pre-cleaning segment before the parameters increased for easier penetration. The numerical simulation was carried out to reveal the mechanism of the keyhole digging process of VPPA welding. The keyhole weld pool is still prone to be unstable when the welding condition (such as welding position) changes [20–22]. Due to the thermal-physical properties of aluminum alloys, welding defects, such as porosity and cutting, are easily generated if the welding process is unstable [23]. Therefore, it becomes very important to understand the plasma arc pressure output and analyze the instability mechanism of the keyhole weld pool. Han et al. [24] measured the plasma arc pressure and analyzed the effect of the arc shape on arc pressure in VPPA. It found that the arc pressure in the EP phase is smaller than in the EN phase when the arc current of different polarities are the same. The existence of a double arc in the EP phase makes the plasma arc pressure reduce further. Jiang et al. [25] measured the VPPA pressure using both pressure transducer and U-tube barometer methods, while the effects of welding parameters were analyzed. The influence of EP on the pressure output is minimal because its time ratio is much less than that of the EN phase. The increase of plasma gas flow rate cools the arc further, resulting in greater constraint of the arc, thus the arc pressure obviously increases. At present, the arc pressure change due to the polarity transaction process is not well understood. Also, the influence of pressure on keyhole weld pool evolution in VPPA welding digging process has not been studied.

Here, we have observed the fluctuation of molten pool surface in the digging process by high speed camera with a high frequency pulsed diode laser light source system. The keyhole boundary in real time was also obtained by an imaging system of an X-ray transmission. In order to analyze the factors influencing the keyhole stability, the plasma arc pressure is measured by the pressure transducer. Combining the energy and momentum balance between electrodes and arc, the physical mechanism of plasma arc pressure was obtained, based on which we optimized the pressure output. Finally, the optimized parameters were verified by the weld formation and weld pool free surface fluctuation situation.

2. Materials and Methods

2.1. Measurement of Weld Pool Surface Deformation

A5052P aluminum alloy is adopted as the work piece and the chemical composition is in Table 1. The size of aluminum plate is 150 mm × 100 mm × 5 mm.

Table 1. Chemical composition of A5052P aluminum alloy (wt. %).

Si	Fe	Cu	Mn	Mg	Cr	Zn	Al
<0.25	<0.40	<0.10	<0.10	2.2–2.8	0.15–0.35	<0.10	balance

The setup for observing weld pool free surface is set as shown in Figure 1. The plasma arc torch includes tungsten electrode, plasma gas nozzle and shielding gas nozzle. The parameters are listed in Table 2. The plasma gas and shielding gas both are argon gas.

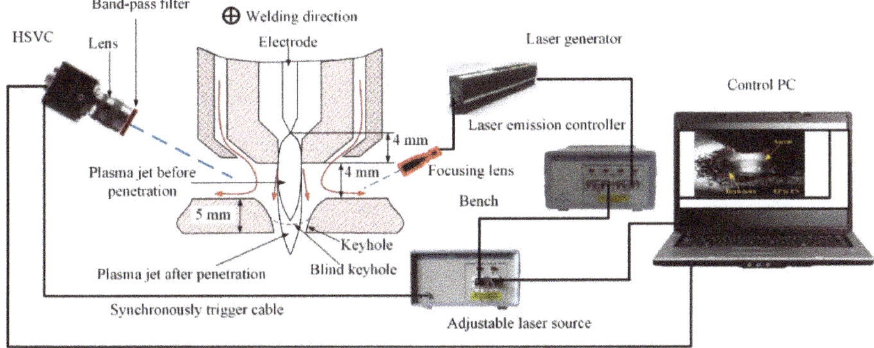

Figure 1. Measurement system of weld pool free surface in digging process of VPPA welding.

Table 2. Parameters of VPPA welding in digging process.

Process Parameter	Value	Unit
Tungsten diameter	4.8	mm
Nozzle diameter	3.2	mm
Plasma gas flow rate	2.0	L/min
Shielding gas flow rate	10	L/min
Current of EP	150	A
Current of EN	150	A
Duration of EP	4	ms
Duration of EN	21	ms
The standoff	4	mm
Tungsten setback	4	mm

In order to clearly observe the weld pool free surface, a telephoto micro lens (AF Micro Nikkor 200 mm, Nikon, Tokyo, Japan) with a 640 nm band-pass filter is used to filter out the arc light, a high-frequency pulsed diode laser light source system (Cavilux HF system, Cavitar, Tampere, Finland) to illuminate the weld pool, a high speed video camera (HSVC) (Memrecam-Q1V, Nac Image Technology, Tokyo, Japan) to record the images. The frame rate of HSVC is 2000 fps. The HSVC sends out an electrical pulse signal to the laser light generator at the start moment of shutter opening, ensuring each picture is clear and reducing the effect of laser heat on the weld pool.

Figure 2 shows the rectangular wave current of VPPA. It is in the EN phase when the current is positive. The duration of the EN phase is longer. The tungsten electrode is connected to the negative pole of power supply while the workpiece is connected to the positive pole. Correspondingly, the current of the EP phase is negative. The duration is shorter. The current flows through the following path: tungsten electrode-workpiece-power supply-tungsten electrode.

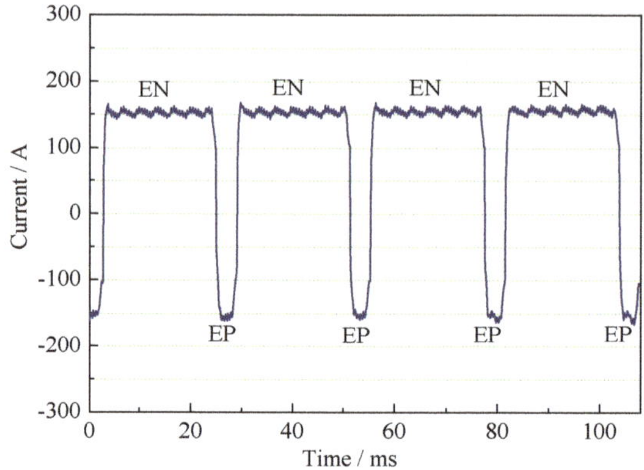

Figure 2. Current waveform of VPPA.

2.2. Measurement of the Keyhole Boundary during Digging Process of VPPA Welding

Figure 3 is the X-ray transmission system used to observe the keyhole boundary in digging process [26]. It consists of one set of X-ray power sources, HSVC and image intensifiers, welding power source and control personal computer (PC). The X-ray power source maximum outputs a tube current of 1.0 mA and a tube voltage of 230.0 kV. In this measurement, X-ray power source is used at 700 μA tube current and180 kV tube voltage. The aluminum plate and parameters are same with that of Section 2.1. As a result, it is difficult for the X-ray to penetrate the entire base metal horizontally. In order to get the relative clear image, the axis line between X-ray source and HSVC is set at 30° with the flat direction to make it easy for X-ray transaction. The X-ray transmission images are recorded by the HSVC at the frame rate of 2000 fps. The camera has an image resolution setting of 800 × 600 pixels to capture a real dimension of 22 mm × 20 mm, as shown in Figure 4. It should be pointed out that due to the angle setting, there is a blind vision region in the observed image. The blind vision region is ignored because of its small size is related to the whole thickness of base metal.

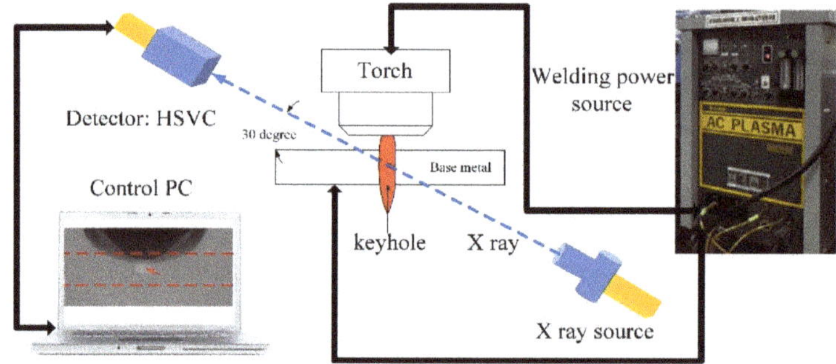

Figure 3. X-ray observation system of the keyhole boundary in digging process of VPPA welding.

2.3. Measurement of Variable Polarity Plasma Arc Pressure

To understand the plasma arc pressure, and thus to analyze its influence on the digging process, the measurement platform is established as shown in Figure 5. Two water tanks are used for cooling plasma torch and the copper, respectively. Arc pressure is measured by the diffusible silicon pressure transducer (Beijing HuaKong xingye technology development Co., Ltd., Beijing, China). The range is from 0 to 5 kPa. A hall sensor is used to get the current synchronously. The measuring frequency is 10,000 Hz. The plane diagram of pressure transducer device is shown in Figure 6. The copper with 6 routes of cooling water is outside. A square tungsten plate is in the center. A small hole with 1 mm diameter is in the center of tungsten plate. Thermally conductive silicone is smeared between the pressure transducer and the cooling copper, which is used to quickly transfer the heat of pressure sensor to the cool copper. The base metal moves with the speed of 1 mm/s for measuring the radial distribution of pressure. The parameters are same with that in Section 2.1. The EP current changes from 150 A to 200 A.

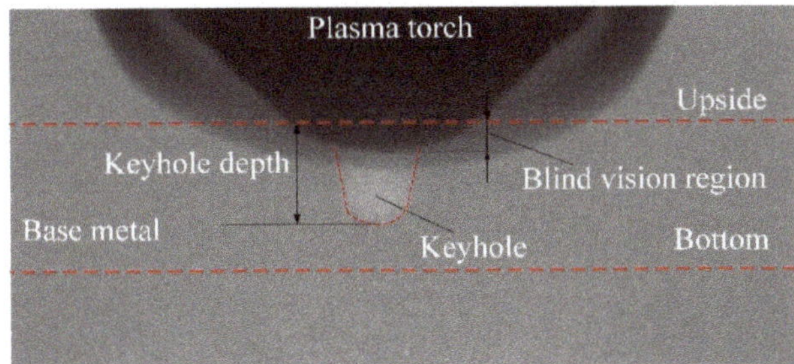

Figure 4. Diagram of the observation results of the small hole boundary.

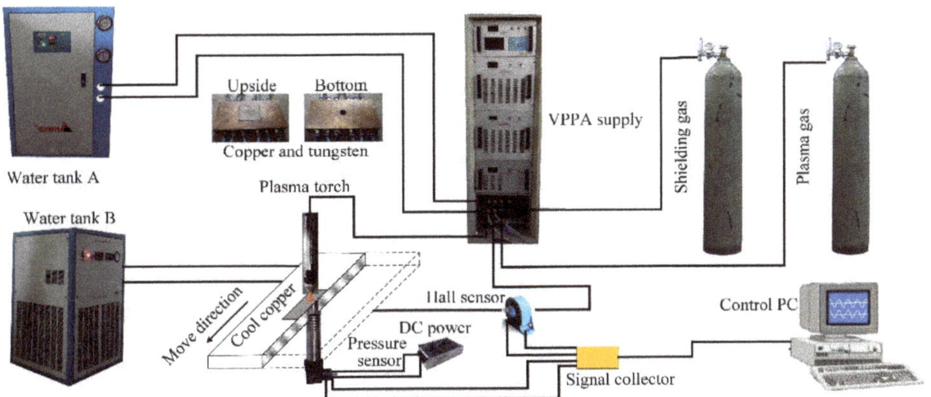

Figure 5. Experimental system of measuring variable polarity plasma arc pressure.

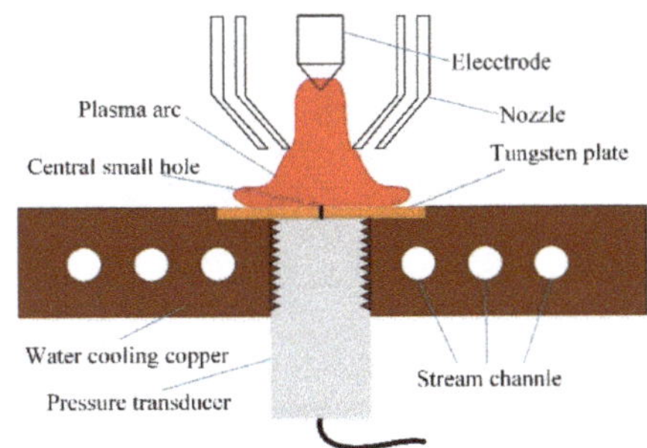

Figure 6. Diagram of measurement structure cross section.

3. Experimental Results and Discussion

3.1. The Evolution of Keyhole and Weld Pool in Digging Process of VPPA Welding

In the digging process of VPPA welding, the variation of free surface of weld pool in one current period is analyzed, as shown in Figure 7. The time interval for each picture is 1 ms. The TSM means the time to starting moment. The start moment is the fifth second after arc ignition. One period is separated into four stages: EN phase, EN to EP phase (the polarity switches from EN to EP), EP phase and EP to EN phase (the polarity switches from EP to EN). The disturbance of the weld pool free surface can be obtained from the above four stages. As Figure 7a shows, the weld pool surface in center region concaves while the weld pool edge protrudes up in the EN phase. It does not change when the polarity switches from EN to EP as shown in Figure 7b. After the polarity changes to the EP phase, the weld pool surface in the center rises to the upside. The edge region descends as shown from Figure 7c to Figure 7e. The surface continue to rise up in the EP to EN phase and the start segment of the EN phase is shown from Figure 7f to Figure 7h. Subsequently, the free surface begins to deform rapidly from the center extending outward. The state remains basically stable after TSM-9 ms. Then it moves on to the next cycle and repeats the above phenomenon. The free surface moves up within the red line while it moves down outside the red line.

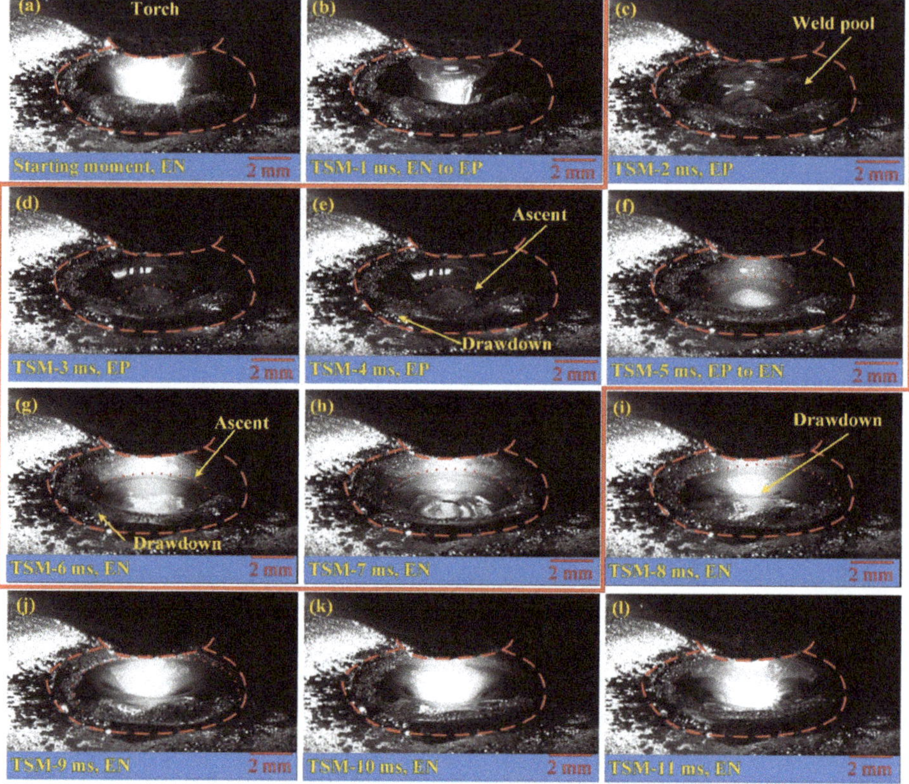

Figure 7. The evolution of weld pool free surface in one current cycle. (**a**) Starting moment (EN); (**b**) TSM-1 ms (EN to EP); (**c**) TSM-2 ms (EP); (**d**) TSM-3 ms (EP); (**e**) TSM-4 ms (EP); (**f**) TSM-5 ms (EP to EN); (**g**) TSM-61 ms (EN); (**h**) TSM-7 ms (EN); (**i**) TSM-8 ms (EN); (**j**) TSM-9 ms (EN); (**k**) TSM-10 ms (EN); (**l**) TSM-11 ms (EN).

Figure 8 shows the keyhole dynamics measured by the X-ray image system, from which the keyhole boundary inside the weld pool can be obtained. Results in one current cycle (about 8 s from arc ignition) are selected to analyze the keyhole status in different polarity. The schematic illustration is drawn based on the X-ray image. The keyhole size including depth and width obviously reduces from the start of the EP phase to the end. The keyhole size gradually increases from the start of the EN phase as shown from 0 to 3 ms. It remains basically stable when the keyhole increases to a certain size as shown from 4 ms to 7 ms. Through the above observation of the weld pool free surface and keyhole boundary, it is found that there are periodic fluctuations in the state of weld pool during the digging process in VPPA welding for aluminum alloy, which contrasts with the previous study [27,28].

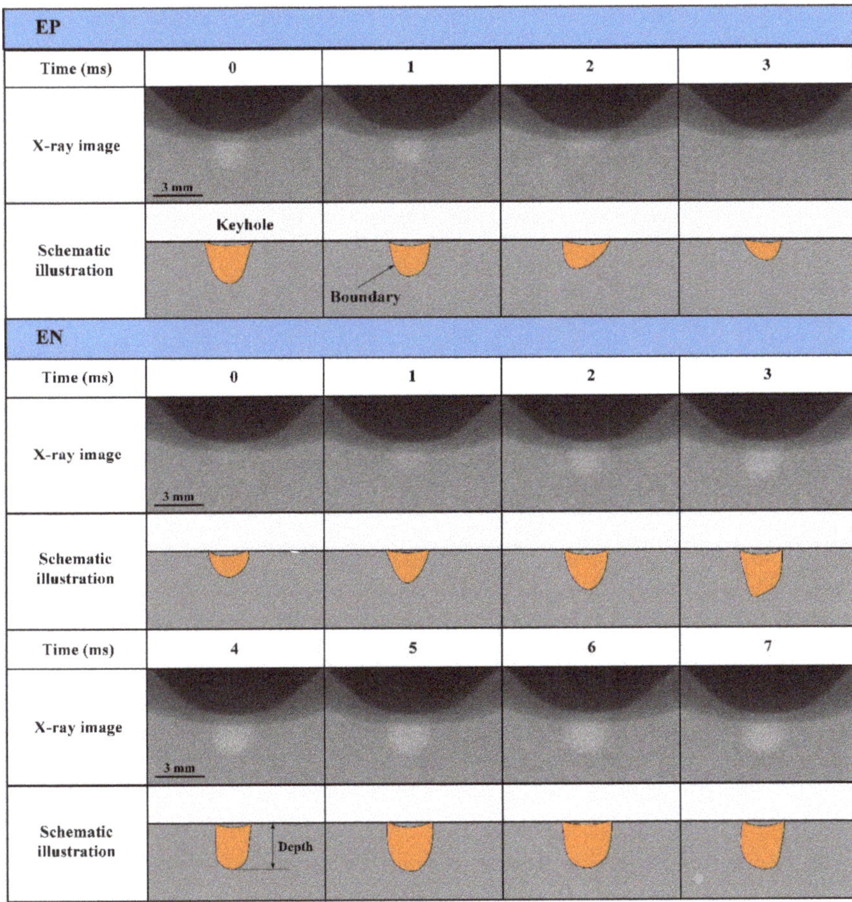

Figure 8. Dynamic characteristics of keyhole during the digging process shown by X-ray imaging technology.

The keyhole boundary is obtained by image edge extracting technology based on the X-ray results, as shown in Figure 9. The variation of keyhole depth with time can be quantitatively analyzed by the keyhole boundary results.

The keyhole depth with time in the whole digging process is shown in Figure 10. It includes the results of the EP and EN phases, measured depending on the above keyhole boundary. The keyhole digging process can be divided into three distinct regimes of behavior, which are the stages of RPF, VF and BP. The RPF means regular periodic fluctuation, VF means violent fluctuation and BP is blasting penetration, respectively. In the RPF stage, the keyhole depth in the EN phase gradually increases. The keyhole of the EP phase in the RPF stage is in the blind vision region. Combing with the observation results of weld pool free surface, it is considered that the keyhole depth of the EP phase in the RPF stage is close to zero. In the VF stage, the plasma arc in EN phase continually digs the keyhole and the depth increases by about 2.5 mm. Then it gradually increases with a little fluctuation. In this stage, the keyhole of the EP phase appears in the view region and the depth also increases with fluctuation. At the end of the VF stage, the depth of the EN phase is 4.15 mm and that of the EP phase is 2.78 mm. After that it goes into the BP stage and the keyhole is quickly fully established.

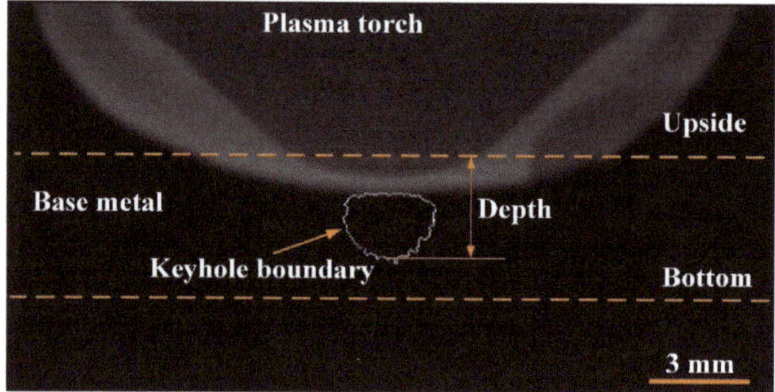

Figure 9. The keyhole boundary indicated by image edge extraction.

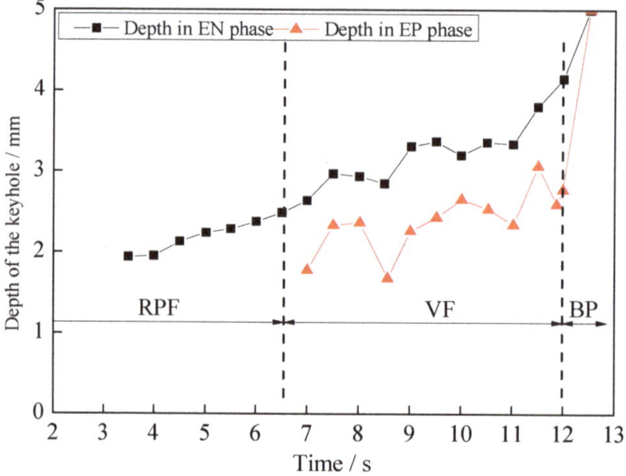

Figure 10. The keyhole depth of a different polarity in the whole digging process.

The diagram of keyhole and weld pool during digging process is shown in Figure 11. In this figure, EP-S means the start moment of the EP phase. The EP-E means the end of the EP phase. EN-S and EN-E have equivalent meanings for the EN phase. In the stage of RPF, the fusion depth is smaller than the unmelt height, as shown in Figure 11b. The heat of the weld pool is easily transferred out because the unmelt region is large, resulting in a relatively slow melting speed and small size weld pool. Moreover, the plasma arc keeps a strong action on the weld pool if the intensity does not reduce much when the keyhole depth is small. Therefore, in this stage, the response of the keyhole weld pool to the arc status is rapid. Then the plasma arc intensity attenuates seriously with the keyhole depth increase, causing the keyhole weld pool fluctuation, as shown in the VF stage.

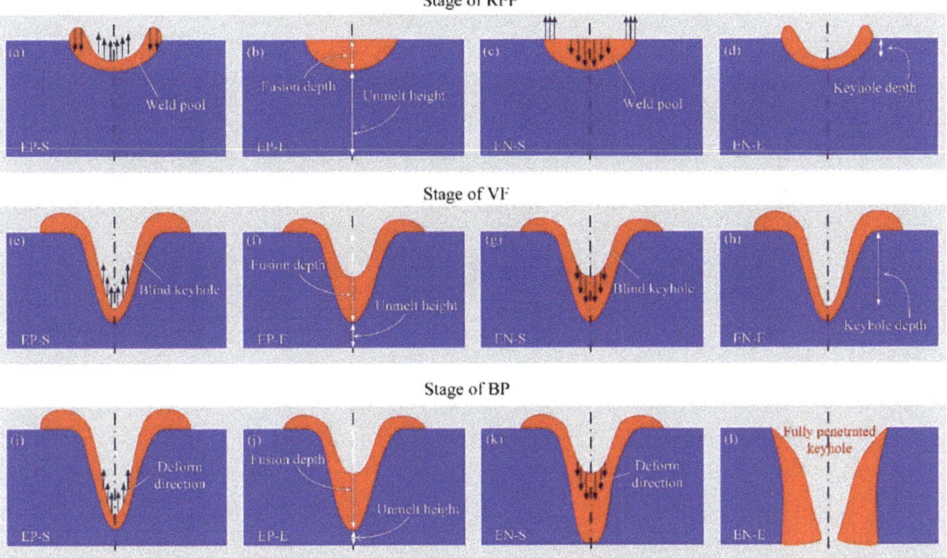

Figure 11. Diagram of keyhole and weld pool in the three stages during digging process. (**a**) start of EP in RPF stage; (**b**) end of EP in RPF stage; (**c**) start of EN in RPF stage; (**d**) end of EN in RPF stage; (**e**) start of EP in VF stage; (**f**) end of EP in VF stage; (**g**) start of EN in VF stage; (**h**) end of EN in VF stage; (**i**) start of EP in BP stage; (**j**) end of EP in BP stage; (**k**) start of EN in BP stage; (**l**) end of EN in BP stage.

In addition, the unmelt height becomes small with the increase of the melting depth, leading to heat loss at the weld pool bottom becoming more difficult. Heat gradually accumulates in the unmelt region. The melting speed gets faster. The keyhole digging speed also increases. Predictably, with the keyhole depth increase, the thermal accumulation also increases. The keyhole digging process and thermal accumulation are mutually reinforcing, resulting in the blasting type penetration in the last stage. Through the above analysis, the keyhole instability during digging process mainly occurs in the first two stages. Adjusting the output of plasma arc pressure to stabilize the keyhole state of EN and EP phases is crucial to realizing stable penetration.

3.2. The VPPA Pressure and its Influence Mechanism on Keyhole and Weld Pool Evolution

Figure 12 is the radial distribution of VPPA pressure as the function of elapsed time from the beginning of the EP phase. The currents in EP and EN phase both are 150 A. The pressure is the Gaussian distribution in the radial direction. The radial distribution width of arc pressure has little difference in the EP and EN phases. However, there is an obvious difference between two polarities in the plasma arc center. The pressure of the EP phase is clearly lower than that of the EN phase. Combining with pressure evolution in arc center region shown in Figure 13, we can understand the pressure difference between EN and EP more clearly. In the EN phase, the pressure can keep at a stable value about 3.3 kPa. After the polarity switches from EN to EP, the pressure firstly drops quickly, then gradually decreases. The minimum pressure appears at the moment of polarity switching from EP to EN. Then it increases to the stable value of EN phase, which takes about 6 ms. Therefore, we can conclude that the difference between the two polarities is mainly due to the arc pressure decreasing process and the rising process in the start segment of each phase.

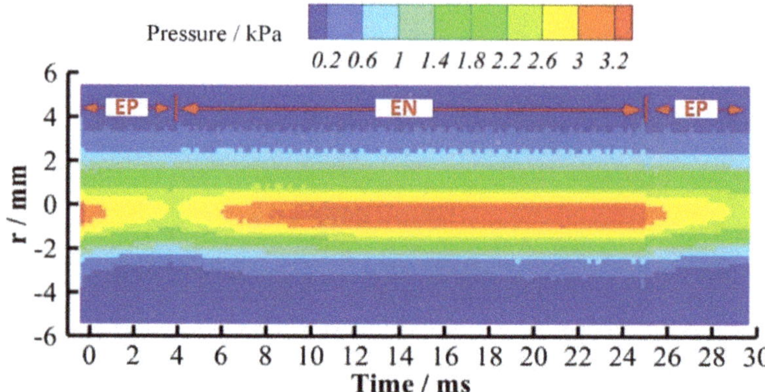

Figure 12. Radial distribution of plasma arc pressure as a function of elapsed time from the beginning of EP phase with the current of 150 A.

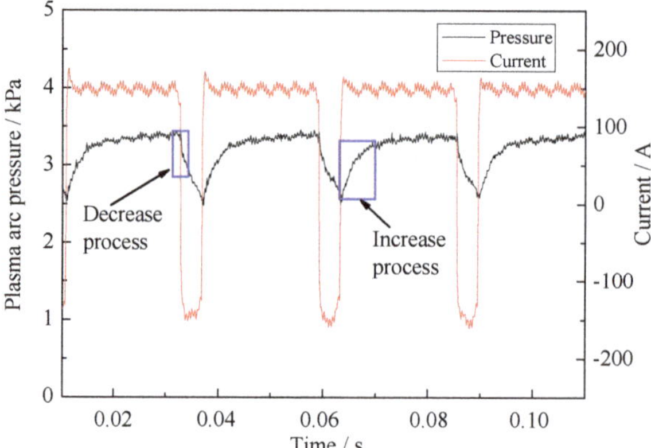

Figure 13. The evolution of plasma arc pressure and current in the center region of arc with 150 A in EN and EP phase.

In order to understand the evolution mechanism of the VPPA pressure and adjust the pressure output reasonably, the energy and momentum balance between electrodes and plasma arc in different polarities is analyzed in Figure 14. The balance items in the start moment of the EP phase are listed out in Figure 14a, based on which the variation tendencies of the tungsten electrode temperature field and the plasma arc shape are also described, marked by the pink solid line. In the start of the EP phase, the tungsten becomes positive polarity and begins to absorb electrons with the high temperature that comes from the arc column, leading to the temperature of tungsten gradually increasing. Therefore, the melting region becomes larger as shown in Figure 14b. Tanaka et al. [29] synchronously measured the electrode temperature and work function of the tungsten electrode. They found that the work function of the tungsten electrode with the temperature above melting point is obviously lower than that of a solid electrode. Therefore, due to the low work function, the region of current outflow increases. Current density decreases with the increase of the melting region. The plasma arc pressure gradually decreases because of the changing of the current density.

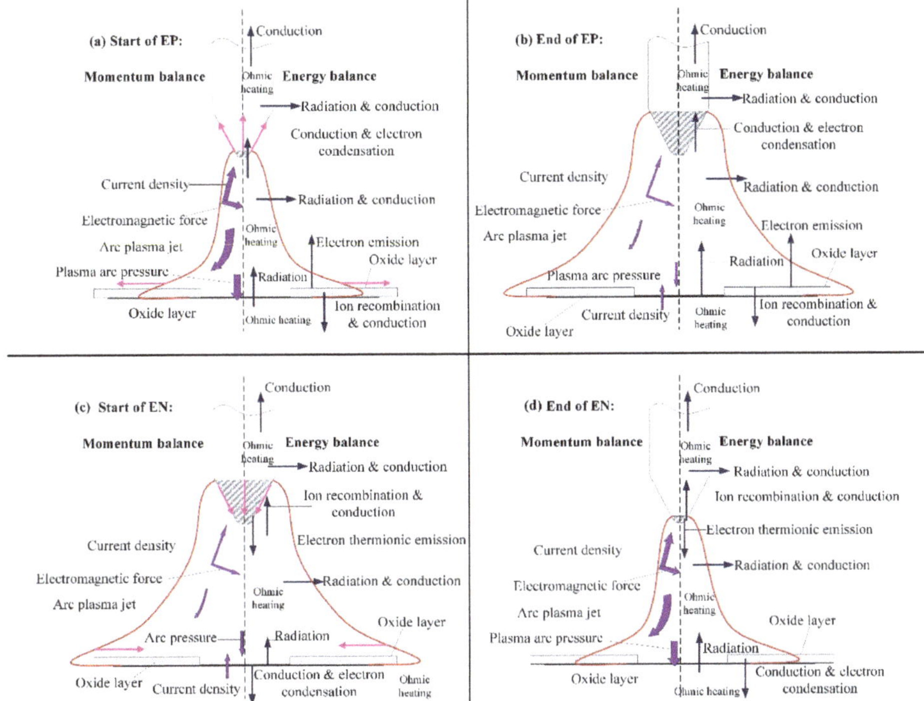

Figure 14. The energy and momentum balance of plasma arc in different polarities. (**a**) start of EP pahse; (**b**) end of EP phase; (**c**) start of EN phase; (**e**) end of EN phase.

Another reason for the pressure reduction in the EP phase is the different physical process on the interaction of the plasma arc and the base metal in EN and EP phases. Tashiro et al. [30] pointed out that the cathode spot tended to be produced on the oxide layer. The current flowing between the arc and base metal is conducted mainly through the cathode spots. The gradual cleaning of the oxide film from the center to the circumference of the arc also leads to the expansion of the arc shape, causing the current density to decrease further. As shown in Figure 14c,d, the opposite process occurs in the EN phase, which makes the pressure of the EN phase higher.

Moreover, Basins et al. [31] pointed out that the arc pressure can be calculated by the following equation.

$$P_{arc} = \frac{\mu_0 I^2}{4\pi^2 r^2} \tag{1}$$

where P_{arc} is arc pressure, μ_0 is the space permeability, I is the current, r is the radius of plasma arc. Therefore, the difference of pressure between EN phase and EP phase can be got.

$$P_{arc}^{EN} - P_{arc}^{EP} = \frac{\mu_0}{4\pi^2 r_{EN}^2}\left(I_{EN}^2 - \frac{I_{EP}^2}{\varphi^2}\right) \tag{2}$$

Expressed in terms of current density, Equation (2) gives

$$P_{arc}^{EN} - P_{arc}^{EP} = \frac{\mu_0}{4}\left(j_{EN}^2 r_{EN}^2 - j_{EP}^2 r_{EP}^2\right) \tag{3}$$

where P_{arc}^{EN} is the plasma arc pressure of the EN phase, P_{arc}^{EP} is the plasma arc pressure of the EP phase, r_{EN} is the radius of the EN plasma arc, r_{EP} is the radius of the EP plasma arc, I_{EN} is the current of the

EN phase, I_{EP} is the current of the EP phase, φ is the radius ratio of the EP arc to the EN arc, j_{EN} is the current density of the EN phase, j_{EP} is the current density of the EP phase.

Through the analysis of momentum and energy balance in the interface between electrodes and plasma arc, we can know that the tungsten temperature increase and cleaning of oxide layer on the surface of base metal during EP phase both result in the decrease of current density. Thus, the arc pressure depends not only on the square of the current but also on the square of arc radius based on the Equations (2) and (3). However, Lin et al. [32] pointed out that welding parameters have little effect on the arc pressure distribution radius. Therefore, the best method to balance the pressure of the EP and EN phases is to adjust the current of different phases to change the arc current density separately.

3.3. The Optimization of Plasma Arc Pressure and the Molten POOL Stability

In this section, the current of the EP phase is adjusted to balance the pressure output. Figure 15 is the evolution of plasma arc pressure in the arc center with a different EP current. The current in the EN phase is fixed to 150 A. The EP current changes from 160 A to 200 A at intervals of 20 A. When the EP current is 160 A as shown in Figure 15a, the pressure of the EP phase is lower than that of the EN phase. The average pressure difference between two phases is around 0.25 kPa. When the EP current increases to 180 A as shown in Figure 16b, the pressure of the EP phase is still lower than that of the EN phase. The pressure between two phases gets closer and the average difference is 0.15 kPa. When the EP current is 200 A, the pressure of the EP phase already becomes a little bigger than that of the EN phase.

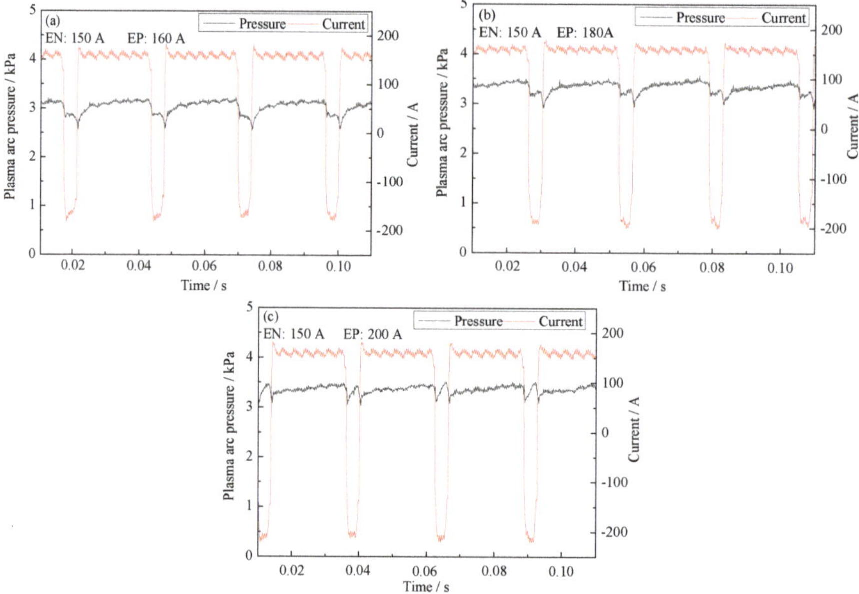

Figure 15. The plasma arc pressure and current waveform with different current of EP phase. (**a**) 150 A in EN phase and 160 A in EP phase; (**b**) 150 A in EN phase and 180 A in EP phase; (**c**) 150 A in EN phase and 200 A in EP phase.

The variation of plasma arc pressure with the change in current can be clearly presented by the distribution contour shown in Figure 16. By comparison, it is found that there is no obvious expansion of the arc pressure distribution radius with the changing current. The difference of arc pressure between two polarities is obvious when the EP current is 160 A as shown in Figure 16a. When continuously increasing the current of the EP phase to 180 A and 200 A, the arc pressure of the EP and EN phases tend to be consistent, as shown in Figure 16b,c. Therefore, it can be inferred that the balance of pressure output between the EP and EN phases is achievable by separately adjusting the current in two polarities. Thus, the keyhole weld pool instability during the digging process of VPPA welding can be weakened.

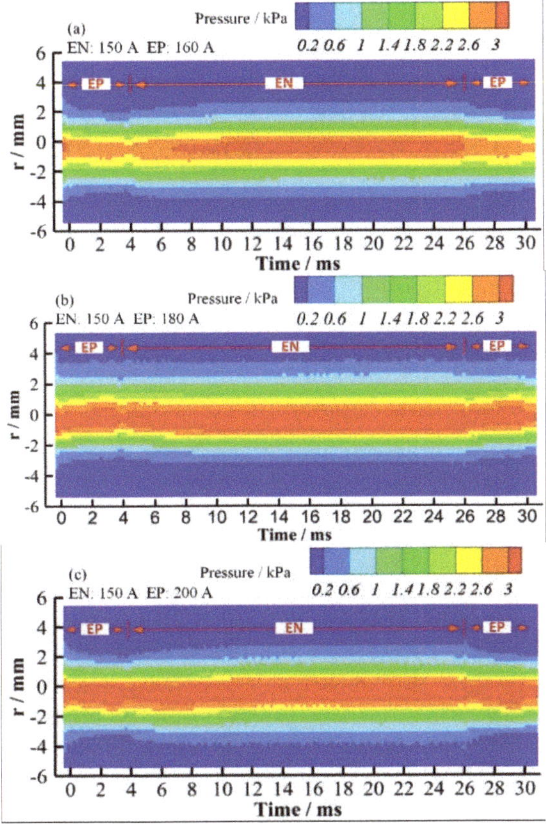

Figure 16. Radial distribution of plasma arc pressure as a function of elapsed time from the beginning of the EP phase with different EP currents. (**a**) 150 A in EN phase and 160 A in EP phase; (**b**) 150 A in EN phase and 180 A in EP phase; (**c**) 150 A in EN phase and 200 A in EP phase.

Figure 17 shows the distribution point and fitting line of the average pressure in EP and EN phases with the change of the EP current. The pressure of the EP phase gradually increases and there's a small fluctuation in the EN pressure. According to the fitting equation, the current of EP phase is 196 A if the pressure of two polarities is equal to each other, when the current of EN phase is 150 A.

The weld pool free surface with different EP currents is shown in Figure 18. The influence of balanced pressure output on the weld pool fluctuation is analyzed by comparing the different case. The time 0 in the figures with EN results in around 5 s from arc ignition. Four weld pool statuses in one current cycle are presented: EN, EN to EP, EP, EP to EN, respectively. When the current of EN phase is 150 A and EP current is 160 A as shown from Figure 18a to Figure 18d, the difference of the weld pool free surface between two polarities is obvious. Due to the decrease of plasma arc pressure in the EP phase, the weld pool free surface moves up and close to the upper surface of the base metal. When the current of EP phase is 180 A which is 30 A bigger than that of the EN phase, the difference of free surface deformation in different polarities becomes very small, as shown in Figure 18e–h. When the EP current increases to 200 A, there is also a small weld pool fluctuation between two polarities, which is shown in Figure 17i–l. Therefore, the keyhole and weld pool of the digging process can be stabilized by separately adjusting the current to balance the pressure output.

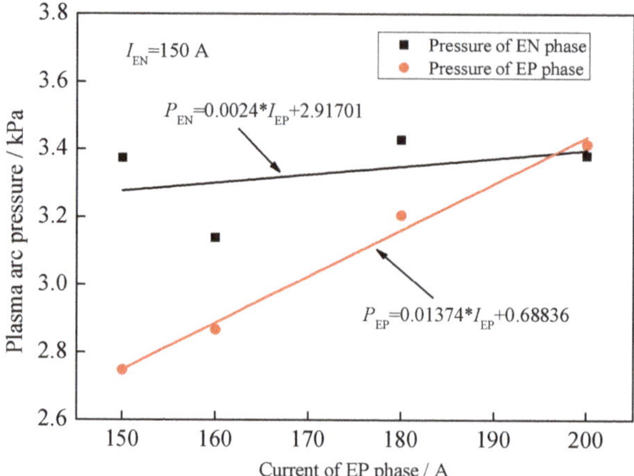

Figure 17. Average plasma arc pressure at different EP currents.

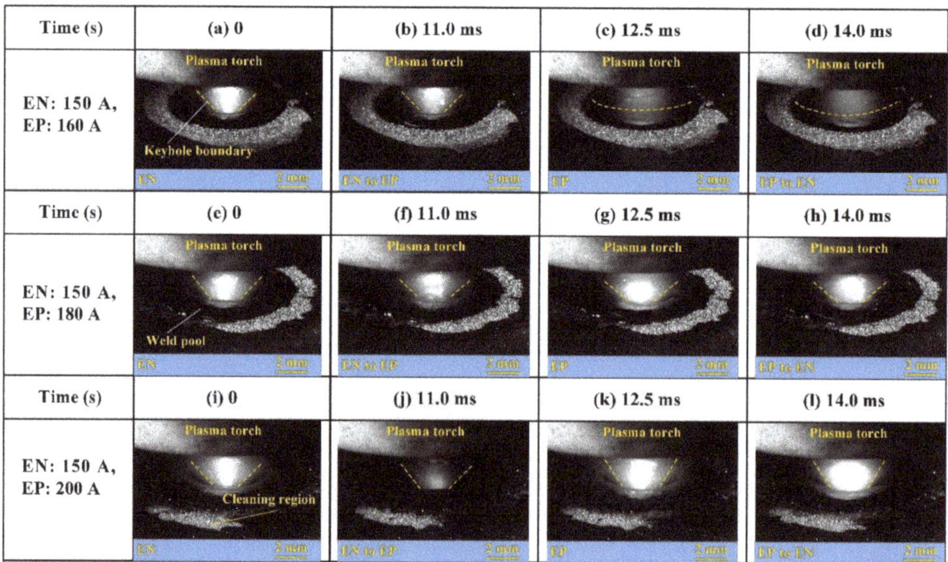

Figure 18. The evolution of weld pool upside surface in one cycle with different EP current: (**a–d**) EP = 160 A; (**e–h**) EP = 180 A; (**i–l**) EP = 200 A.

Through a large amount of welding experiments, it is found that the success rate of forming weld bead is low if the keyhole weld pool fluctuates. It is easy to form cutting. As Figure 19 shows, when the current of EN and EP phase is 150 A, all the liquid metal falls to the workpiece bottom in the keyhole generation segment, which makes it difficult to close the keyhole, resulting in cutting. After balancing the pressure output (EN: 150 A, EP: 180 A), the keyhole is more easily closed by avoiding a liquid metal overall drop due to the weak fluctuation. Thus, the weld bead is easier to form.

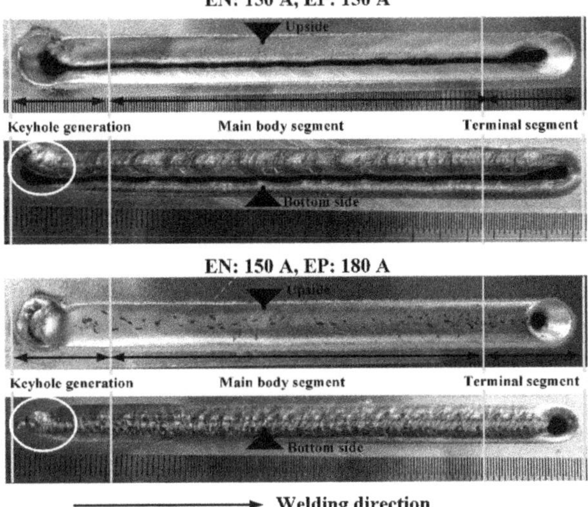

Figure 19. The weld formation with different EP current.

4. Conclusions

This work mainly investigates the stability of keyhole digging process in VPPA welding. Using comprehensive experimental measurement of weld pool free surface, keyhole boundary and plasma arc pressure, the evolution of keyhole weld pool and the influence mechanism of plasma arc pressure are obtained. The specific conclusions are as follows.

(1) The keyhole and weld pool fluctuate with the periodic change of plasma arc state during the digging process of VPPA welding, rather than continuously increasing the keyhole depth over time. By observing the keyhole boundary of digging process in real time, it is found that the keyhole depth increases rapidly in the EN phase, but decreases gradually in the EP phase.

(2) The three stages of keyhole digging process are found, which are regular periodic fluctuation, violent fluctuation and blasting penetration, respectively. The fluctuation of the second stage is caused by the attenuation of plasma arc pressure due to the increase of the keyhole depth. The positive feedback between the thermal accumulation and the continuous increase in the depth of the keyhole leads to the rapid penetration in the third stage.

(3) In order to accurately balance the plasma arc pressure output, the influence mechanism of energy and momentum balance on the pressure is analyzed. The difference between the plasma arc pressure of the EP and EN phases can be effectively reduced if the EP current is 30 A to 50 A larger than that of EN. The pressure balance output can reduce the keyhole weld pool fluctuation in the digging process and improve the welding success rate.

Author Contributions: B.X. performed all experiments and wrote this manuscript. S.C. and F.J. provided the basic idea. S.T. and M.T. participated in the discussion on the results and guided the writing of the article.

Funding: This work is Supported by National Natural Science Foundation of China (No.51875004), Beijing Natural Science Foundation (3172004), the China Scholarship Council (CSC) for one-year study in University of Alberta for Fan Jiang, the China Scholarship Council (CSC) for one-year study in Osaka University for Bin Xu, International Research Cooperation Seed Fund of Beijing University of Technology (No. 2018B16), the open project of Beijing Engineering Researching Center of Laser Technology (BG0046-2018-09) and start-up funding from Beijing University of Technology.

Conflicts of Interest: The authors declare no conflict of interest.

References

1. Jenney, C.L.; O'Brien, A. *Welding Handbook*; American Welding Society: Miami, FL, USA, 2001; pp. 303–336.
2. Cho, J.; Jung-Jae, L.; Seung-Hwan, B. Heat input analysis of variable polarity arc welding of aluminum. *Int. J. Adv. Manuf. Technol.* **2015**, *81*, 1273–1280. [CrossRef]
3. Fuerschbach, P.W. Cathodic cleaning and heat input in variable polarity plasma arc welding of aluminum. *Weld. J.* **1998**, *77*, 76s–85s.
4. Liu, Z.M.; Cui, S.; Luo, Z.; Zhang, C.; Wang, Z.; Zhang, Y. Plasma arc welding: Process variants and its recent developments of sensing, controlling and modeling. *J. Manuf. Process.* **2016**, *23*, 315–327. [CrossRef]
5. Liu, Z.; Wu, C.S.; Gao, J. Vision-based observation of keyhole geometry in plasma arc welding. *Int. J. Therm. Sci.* **2013**, *63*, 38–45. [CrossRef]
6. Liu, Z.M. Keyhole Behaviors Influence Weld Defects in Plasma Arc Welding Process. *Weld. J.* **2015**, *94*, 281–290.
7. Liu, Z.M.; Wu, C.S.; Cui, S.L.; Luo, Z. Correlation of keyhole exit deviation distance and weld pool thermo-state in plasma arc welding process. *Int. J. Heat Mass Transf.* **2017**, *104*, 310–317. [CrossRef]
8. Zhang, Y.M.; Zhang, S.B.; Liu, Y.C. A plasma cloud charge sensor for pulse keyhole process control. *Meas. Sci. Technol.* **2001**, *12*, 1365–1370. [CrossRef]
9. Metcalfe, J.C.; Quigley, M.B.C. Keyhole Stability in Plasma Arc Welding. *J. Eng. Manuf.* **2000**, *214*, 401–404.
10. Zheng, B.; Wang, H.J.; Wang, Q.L. Front image sensing of the keyhole puddle in the variable polarity PAW of aluminum alloys. *J. Mater. Process. Technol.* **1998**, *83*, 286–298. [CrossRef]
11. Wu, D.; Chen, H.; Huang, Y.; He, Y.; Hu, M.; Chen, S. Monitoring of weld joint penetration during variable polarity plasma arc welding based on the keyhole characteristics and PSO-ANFIS. *J. Mater. Process. Technol.* **2017**, *239*, 113–124. [CrossRef]
12. Saad, E.; Huijun, W.; Radovan, K. Classification of molten pool modes in variable polarity plasma arc welding based on acoustic signature. *J. Mater. Process. Technol.* **2006**, *174*, 127–136. [CrossRef]
13. Wu, D.; Chen, H.; He, Y.; Song, S.; Lin, T.; Chen, S. A prediction model for keyhole geometry and acoustic signatures during variable polarity plasma arc welding based on extreme learning machine. *Sens. Rev.* **2016**, *36*, 257–266. [CrossRef]
14. Cunningham, R.; Zhao, C.; Parab, N.; Kantzos, C.; Pauza, J.; Fezzaa, K.; Sun, T.; Rollett, A.D. Keyhole threshold and morphology in laser melting revealed by ultrahigh speed x-ray imaging. *Science* **2019**, *363*. [CrossRef]
15. Leung, C.L.A.; Marussi, S.; Atwood, R.C.; Towrie, M.; Withers, P.J.; Lee, P.D. In situ X-ray imaging of defect and molten pool dynamics in laser additive manufacturing. *Nat. Commun.* **2018**, *9*, 1–9. [CrossRef]
16. Miyagi, M.; Wang, H.; Yoshida, R.; Kawahito, Y.; Kawakami, H.; Shoubu, T. Effect of alloy element on weld pool dynamics in laser welding of aluminum alloys. *Sci. Rep.* **2018**, *8*, 1–10. [CrossRef] [PubMed]
17. Nguyen Van, A.; Tashiro, S.; Van Hanh, B.; Tanaka, M. Experimental investigation on the weld pool formation process in plasma keyhole arc welding. *J. Phys. D. Appl. Phys.* **2018**, *51*, 015204.
18. Zheng, B.; Wang, Q.I.; Kovacevic, R. Parameters optimization for the generation of a keyhole weld pool during the start-up segment in variable-polarity plasma arc welding of aluminum alloys. *Proc. Inst. Mech. Eng. Part B J. Eng. Manuf.* **2000**, *214*, 393–400. [CrossRef]
19. Chen, S.; Bin, X.; Fan, J. Blasting type penetrating characteristic in variable polarity plasma arc welding of aluminum alloy of type 5A06. *Int. J. Heat Mass Transf.* **2018**, *118*, 1293–1306. [CrossRef]
20. Chen, S.; Yan, Z.Y.; Jiang, F.; Zhang, W. Gravity effects on horizontal variable polarity plasma arc welding. *J. Mater. Process. Technol.* **2018**, *255*, 831–840. [CrossRef]
21. Yan, Z.Y.; Chen, S.; Jiang, F.; Huang, N.; Zhang, S.L. Material flow in variable polarity plasma arc keyhole welding of aluminum alloy. *J. Manuf. Process.* **2018**, *36*, 480–486. [CrossRef]
22. Xu, B.; Chen, S.; Jiang, F.; Phan, H.; Tashiro, S.; Tanaka, M. The influence mechanism of variable polarity plasma arc pressure on flat keyhole welding stability. *J. Manuf. Process.* **2019**, *37*, 519–528. [CrossRef]
23. Wang, Z.; Oliveira, J.; Zeng, Z.; Bu, X.; Peng, B.; Shao, X. Laser beam oscillating welding of 5A06 aluminum alloys: Microstructure, porosity and mechanical properties. *Opt. Laser Technol.* **2019**, *111*, 58–65. [CrossRef]
24. Han, Y.; Yaohui, L.; Shujun, C.; Shuyan, Y. Influence of variable polarity plasma arc shape on arc force. *Trans. China Weld. Inst.* **2005**, *26*, 49–52.

25. Jiang, Y.; Binshi, X.; Yaohui, L.; Cunlong, L.; Ming, L. Experimental Analysis on the Variable Polarity Plasma Arc Pressure. *Chin. J. Mech. Eng.* **2011**, *24*, 607. [CrossRef]
26. Morisada, Y.; Fujii, H.; Kawahito, Y.; Nakata, K.; Tanaka, M. Three-dimensional visualization of material flow during friction stir welding by two pairs of X-ray transmission systems. *Scr. Mater.* **2011**, *65*, 1085–1088. [CrossRef]
27. Pan, J.J.; Yang, L.J.; Hu, S.S.; Chen, S.J. Numerical analysis of keyhole formation and collapse in variable polarity plasma arc welding. *Int. J. Heat Mass Transf.* **2017**, *109*, 1218–1228. [CrossRef]
28. Wang, H.X.; Wei, Y.H.; Yang, C.L. Numerical calculation of variable polarity keyhole plasma arc welding process for aluminum alloys based on finite difference method. *Comput. Mater. Sci.* **2007**, *40*, 213–225. [CrossRef]
29. Tanaka, M.; Ushio, M.; Ikeuchi, M.; Kagebayashi, Y. In situ measurements of electrode work functions in free-burning arcs during operation at atmospheric pressure. *J. Phys. D. Appl. Phys.* **2005**, *38*, 29–35. [CrossRef]
30. Tashiro, S.; Minoru, M.; Manabu, T. Numerical analysis of AC tungsten inert gas welding of aluminum plate in consideration of oxide layer cleaning. *Thin Solid Films* **2011**, *519*, 7025–7029. [CrossRef]
31. Basins, M. Pressures produced by gas tungsten arcs. *Metall. Trans.* **1986**, *17B*, 601–607.
32. Lin, M.; Eagar, T. Influence of arc pressure on weld pool geometry. *Weld. J.* **1985**, 163–169.

 *materials*

Article

Weibull Statistical Analysis of Strength Fluctuation for Failure Prediction and Structural Durability of Friction Stir Welded Al–Cu Dissimilar Joints Correlated to Metallurgical Bonded Characteristics

Chung-Wei Yang * and Shiau-Jiun Jiang

Department of Materials Science and Engineering, National Formosa University, No. 64, Wunhua Road, Huwei, Yunlin 63201, Taiwan; 10152114@gm.nfu.edu.tw
* Correspondence: cwyang@nfu.edu.tw; Tel.: +886-5-6315478

Received: 14 December 2018; Accepted: 3 January 2019; Published: 9 January 2019

Abstract: In this paper, dissimilar Al–Cu joints of AA1050H/C1100-Cu, AA6061-T6/C1100-Cu, and AA1050H/C2600-brass are successfully welded by a friction stir welding (FSW) process. The aim of the present study is not only to examine the tensile strength, but also to investigate the reliability, durability, and failure behaviors of joints as correlated with the metallurgical bonded microstructures of varied Al–Cu joints. Experimental evidence confirms that good welding quality for an FSW Al–Cu dissimilar joint is obtained when pure Cu and brass plates are positioned at the advancing side. Cross-sectional microstructures reveal that the AA6061-T6/C1100-Cu joint exhibits an extensive metallurgical bonded region with significant onion rings in the welding zone, whereas the AA1050H/C2600-brass joint generally displays a clear mechanical kissing bonded boundary at the joint interface. Al_2Cu, Al_4Cu_9, and γ-Cu_5Zn_8 are major intermetallic compounds (IMCs) that are formed within the metallurgical bonded welding zone. The Weibull model provides a statistical method for assessing the failure mechanism of FSW Al–Cu joints. Better welding reliability and tensile properties with ductile dimpled ruptures are obtained for the Al–Cu joints with a typical metallurgical bonded zone. However, a mechanical kissing bonded interface and thick interfacial IMCs result in the deterioration of tensile strength with a brittle fracture and a rapid increase in the failure probability of Al–Cu joints.

Keywords: friction stir welding; dissimilar joints; metallurgical bonding microstructure; the Weibull model; failure strength; engineering reliability

1. Introduction

The welding of dissimilar metals is becoming an important subject for industrial applications nowadays due to their technical and beneficial advantages [1–5]. Aluminum (Al) and copper (Cu) are two common engineering metals with favorable mechanical strength, ductility, and good corrosion resistance. Moreover, Al–Cu dissimilar joints have been widely used in engineering structural components, electronic packaging, and the electric power industries, and are of interest in electrical connections because of their excellent electrical and thermal conductivities. In contrast with these advantages, the joining of dissimilar Al and Cu alloys is generally difficult by a conventional fusion welding process due to the differences in their physical, thermal, and mechanical characteristics. Thus, the development of a promising welding technique for joining dissimilar Al and Cu alloys has been made by a number of researches [6–10].

Friction stir welding (FSW) is a solid-state joining process [11,12] that was first invented by The Welding Institute (TWI) of the United Kingdom (UK) [13], and can be considered an important development in joining dissimilar metals [11]. The FSW process is considered energy efficient and environmentally friendly because no toxic fumes are produced during the welding process. FSW can be commonly used to join nonferrous light and plastic metals such as Al, Mg and Ti alloys with other dissimilar metal alloys that are hard to weld by conventional fusion welding [2,5,14–18]. Recently, FSW has also been recognized as an effective technique to overcome the welding problems of Al–Cu dissimilar joints [19–27]. Some studies indicated that not only the welding quality, but also mechanical properties of Al–Cu dissimilar joints are significantly influenced by controlling the FSW parameters [21–25] and microstructural features, especially for the intermetallic compounds (IMCs) layer formed at the bonding interface within the welding zone of Al–Cu dissimilar joints [26–31]. The brittleness of the IMCs layer usually results in easier cracks propagation and failures at the joint interface [32,33]. However, the enhanced mechanical properties of Al–Cu joints can also be achieved by controlling the particle size and distribution of the IMCs within the welding zone of FSW joints [26,28,32].

Since the failure of FSW-joined structural components depends on the applied stress to approach a critical weakest strength within the welding zone, the variability of the failure strength and the durability of FSW joints are fairly correlated with the welding qualities. Therefore, it is worthwhile to investigate the correlation of microstructural features to the data fluctuation of mechanical strength and welding reliability of FSW dissimilar joints in detail. The failure prediction can be effectively achieved through a statistical reliability engineering method [34], and the Weibull model [34,35] of survival analysis has been developed as a popular and powerful engineering design method for various structural materials and joint performance [36–40]. The advantage of Weibull statistical analysis is that it provides reasonably accurate failure analysis and failure predictions with a small number of samples. Solutions can be acquired at earlier indications of problems, and fewer samples also enable cost-effective component testing.

Therefore, in order to clarify the influence of metallurgical factors on the failure strength and the durability of FSW Al–Cu dissimilar joints, the aim of the present study is to examine the microstructural features of the welding zone and evaluatethe mechanical strength of dissimilar joints under tensile tests. In addition, a statistical analysis of the Weibull model with the examination of fracture surface morphologies will be applied to investigate the welding reliability, joints durability, and failure mechanism of FSW Al–Cu dissimilar joints.

2. Materials and Methods

The base metals that were used in this study were three-mm thick commercial pure copper (99.9% purity, annealed, JIS C1100), brass (JIS C2600), AA1050H, and AA6061-T6 aluminum rolled sheets. The chemical compositions of raw materials that are given in Table 1 were determined by inductively coupled plasma-atomic emission spectrometry (ICP-AES). The base metals were machined into rectangular specimens with dimensions of 80 (*l*) × 30 (*w*) × 3 (*t*) mm for the FSW process. The surfaces of FSW-joined specimens were ground with 2000-grit SiC papers, and then ultrasonically cleaned with acetone prior to welding. The friction stir welded AA1050H/C1100, AA6061-T6/C1100, and AA1050H/C2600 dissimilar joints were carried out in this study, and these joints were labeled as A1/C1, A6/C1, and A1/C2 specimens in the following, respectively.

Table 1. Chemical compositions of the used Al and Cu base metals (in wt.%).

Cu Base Metals	Cu	Zn	Pb	Fe	Si	Mg	Al
C1100 (Pure Cu)	Bal.	0.03	0.02	0.01	0.01	-	0.02
C2600 (brass)	69.3	Bal.	0.01	0.03	0.15	0.03	0.02

Al Base Metals	Al	Mg	Si	Fe	Mn	Cr	Cu	Zn	Ti
AA1050H	Bal.	0.05	0.15	0.38	0.05	-	0.05	0.01	0.02
AA6061-T6	Bal.	1.03	0.66	0.35	0.11	0.14	0.22	0.03	0.02

Figure 1a schematically illustrates the FSW process of Al–Cu joints. The welding tool that was used in this study was made of AISI H-13 tool steel with a shoulder that was 15 mm in diameter and a stirring probe with a four-mm diameter and two-mm depth, as shown in Figure 1b. Pure Cu and brass plates were always positioned at the advancing side (AS). The tool was in the center of the Al–Cu joints, and it was not shifted to the Al or Cu side. According to our preliminary trial experiments of several Al–Cu joints and the macrograph/micrograph inspections for the welds of various parameters, optimum welding parameters were obtained for the welds with an optimal tensile strength and defect-free microstructures in the welding zone at the stable part of the seam for the present study. Based on our preliminary trials and the literature review [22], a low transverse speed and high tool rotational speed are usually required for obtaining defect-free joints. The tool rotational speed was set at 3000 rpm. The rotating tool was tilted 1.5° opposite to the welding direction, and the stirring probe moved along the butt line of the Al–Cu joint specimens at a constant traverse speed of about 60 mm min^{-1}. The downward push pressure was controlled at about 45 MPa. During the FSW process, the downward push pressure was maintained for an appropriate time to generate sufficient frictional heat. The frictional heat softened the Al–Cu joint specimens, and the stirring probe caused material plastic flow in both circumferential and axial directions. The welded direction (WD), normal direction (ND), and transverse direction (TD) of the FSW Al–Cu dissimilar joints were also defined in Figure 1a. After the friction stir welding process, the upper and lower surfaces of welded specimens were carefully ground with 2000-grit SiC abrasive paper to eliminate the defects and stress concentrators located on the surface of the seam.

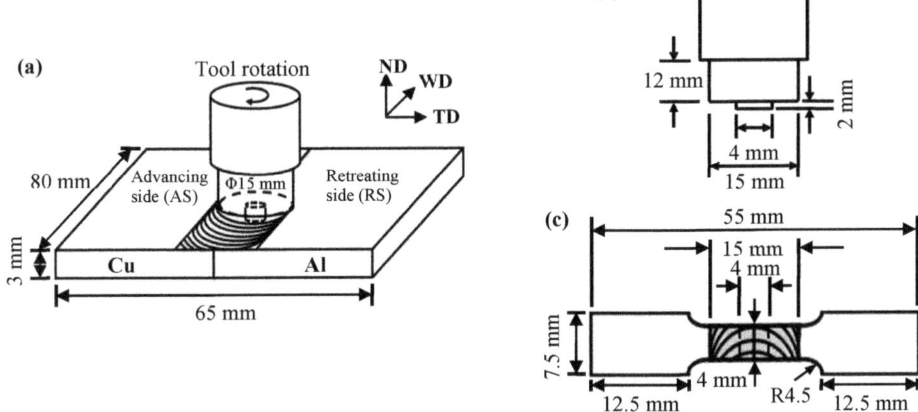

Figure 1. Schematic illustrations of the (**a**) friction stir welding (FSW) process of Al–Cu dissimilar joints, (**b**) the welding tool geometries used for FSW, and (**c**) the dimensions of the tensile specimen prepared from welded samples.

The FSW joined A1/C1, A6/C1, and A1/C2 dissimilar joints were cross-sectioned along the TD. For the study of the welding microstructures, the cross-sections (on the WD plane) taken at the stable part of the seam were ground and polished with a diamond polishing agent. The specimens were etched in a Keller's solution for the Al alloy side, and the Cu alloy side was etched with a solution of five grams of FeCl$_3$, 50 mL of HCl, and 100 mL of H$_2$O. The phase composition within the welding zone (WZ) of the Al–Cu dissimilar joints were identified by an X-ray diffractometer (XRD, Bruker D8A25, Bruker Corp., Karlsruhe, Germany), using CuKα radiation at 40 kV and 40 mA with a scan speed of 3° (2θ) min^{-1} (step size, 0.02°). Microstructures of the A1/C1, A6/C1, and A1/C2 dissimilar joints were observed by the backscattering electron image (BEI) taken with a scanning electron microscopy (SEM, JEOL/JSM-6360, JEOL Ltd., Tokyo, Japan). SEM equipped with an energy-dispersive X-ray spectroscopy (EDS) and electron probe micro-analyzer (EPMA, JEOL JXA-8530F, JEOL Ltd., Tokyo, Japan) were used to identify the elemental compositions and distribution of compounds formed within the WZ of FSW Al–Cu dissimilar joints for the investigation of metallurgical interactions.

The Micro-Vickers hardness test (HV) was applied to evaluate the variations of microhardness after the FSW process. The Micro-Vickers hardness test across the cross-sections of the A1/C1, A6/C1, and A1/C2 joints was applied using a Vickers indenter (Future-Tech Corp., Kawasaki, Japan) with a 50-g load for 10 s of dwell time. Each measured microhardness datum was the average of at least three tests.The tensile strength of the FSW A1/C1, A6/C1, and A1/C2 dissimilar joints was measured according to the standard tension testing of ASTM E8M-11. Uniaxial tensile tests of the FSW Al–Cu joints were carried out in the directions perpendicular to the WD. Figure 1c shows the dimensions of the tensile specimens prepared from welded samples. The specimens were tested at room temperature with an initial strain rate of 8×10^{-4} s^{-1} per mm. The tensile strength measurements of each condition (A1/C1, A6/C1, and A1/C2 joints) were performed on 20 test specimens ($n = 20$) for the statistical significance of following Weibull statistical analysis. The fracture surfaces and sub-surfaces of the FSW Al–Cu joints were further examined using SEM/BEI with EDS mapping to analyze the fracture morphologies and behaviors.

3. Results

3.1. Microstructures and Microhardness Variation of FSW Al–Cu Dissimilar Joints

Figure 2 represents the backscattering electron images (SEM/BEI) of cross-sectional micrographs within the WZ of various FSW Al–Cu dissimilar joints for illustration. The cross-sectional images are taken at the stable part of the seam. The light gray area is the Cu matrix, and the dark gray area is the Al matrix. It can be seen that sound A1/C1, A6/C1, and A1/C2 dissimilar joints are successfully achieved by the FSW process, while the pure Cu (C1100) and brass (C2600) plates are always positioned at the advancing side (AS). The good welding quality of all of the Al–Cu joints is obtained in the present study without the typical cavity defects that tend to occur within the WZ region.

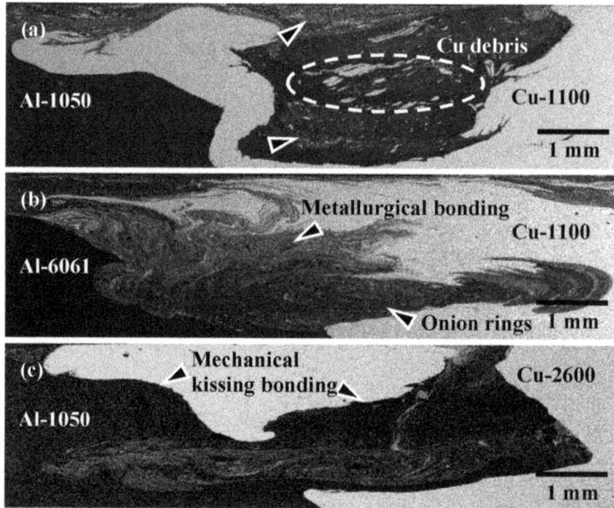

Figure 2. Scanning electron microscopy (SEM)/backscattering electron image (BEI) cross-sectional micrographs (on the welded direction (WD) plane) of the FSW (**a**) A1/C1, (**b**) A6/C1, and (**c**) A1/C2 joints (both of the C1 and C2 base metals were positioned at the advancing side).

Comparing Figure 2a with Figure 2b, it is apparent that the plastic flow of the Al and Cu base metals within the WZ region is quite different between the A1/C1 and A6/C1 joints. Figure 2a shows the cross-sectional microstructure of the A1/C1 joint for illustration. A large amount of pure Cu matrix and several dispersed large irregular Cu debris (as those encircled in Figure 2a) are mixed with the Al matrix in the WZ of the A1/C1 joint. A clear A1/C1 joint boundary represents that just a mechanical kissing bond is formed between AA1050H-aluminum and C1100-pure Cu at the welding interface of A1/C1 specimens after the FSW process. In addition, a small amount of Al–Cu reacting mixture, which is the result of the intense material plastic flow during FSW, is observed at the top and near the bottom of the WZ (indicated by the triangular marks). The mixture can be considered the metallurgical bonded zone of AA1050H-aluminum and C1100-pure Cu base metals. Unlike the A1/C1 joint, no typical dispersed Cu debris or small Cu particles are observed within the WZ of the A6/C1 joint, as shown in Figure 2b. It is worth noting that a significant wide range of material plastic flow is observed between the Al and Cu matrixes with an obvious onion rings microstructure (indicated by the triangular marks) in the WZ of the A6/C1 joint. It can be recognized that the A6/C1 joint displays much better welding quality, with a larger area fraction of its metallurgical bonded zone and no mechanical kissing bond between the AA6061-T6-aluminum and C1100-pure Cu base metals, compared with the A1/C1 joint. Figure 2c shows the cross-sectional microstructure of the A1/C2 joint for illustration. Comparing Figure 2c with Figure 2b, an Al–Cu metallurgical bonded zone, which is the result of the material plastic flow of AA1050H-aluminum and C2600-brass during FSW, is formed within the WZ of the A1/C2 joint (Figure 2c). However, the Al–Cu metallurgical bonding area fraction of the A1/C2 joint is apparently less than that of the A6/C1 joint. Moreover, a clear joint boundary between AA1050H-aluminum and C2600-brass base metals is observed in the WZ region. Unlike the metallurgical bonding effect within the Al–Cu reacting mixture during FSW, the clear joint boundary observed in the A1/C2 jointis thought of as a mechanical kissing bonded interface, as indicated by the triangular marks in Figure 2c.

Figure 3a,b show the SEM/BEI images and EPMA analysis results for the element distribution maps of Al and Cu within the WZ of the A1/C1 and A6/C1 joints, respectively. As shown in Figure 3a, an apparent joint boundary exists at the welding interface of the AA1050H-aluminum (the dark gray region) and C1100-pure Cu (the light gray region) base metals. In addition, some bright small particles, which are mainly composed of the Cu element, are detected within the AA1050H-aluminum matrix (see Figure 3a). These regions can be recognized as the reacting mixtures resulted from the intense material plastic flow of Al and Cu base metals during FSW, as indicated by the triangular marks in Figure 2a. It can be reasonably deduced that some Al–Cu intermetallic compounds (IMCs) are formed within the Al–Cu reacting mixture region, and the phase composition of Al–Cu IMCs will be identified by following X-ray diffraction analysis. Figure 3b shows the SEM/BEI images and EPMA analysis results in the WZ of the A6/C1 joint. The dark gray and light gray regions correspond to the AA6061-T6-aluminum and C1100-pure Cu base metals, respectively. As shown in Figure 3b, the FSW A6/C1 joint displays a significant material plastic flow and reacting mixtures within the WZ compared with the A1/C1 joint. Therefore, comparing Figure 3b with Figure 2b demonstrates that the metallurgical bonded zone consists of severe stirring plastic flow between AA6061-T6-aluminum and C1100-pure Cu base metals with a large amount of reacting mixtures of dispersed Al–Cu IMCs particles. The distribution of Al–Cu IMCs particles and microstructural morphologies of the A6/C1 joint are quite different from that of the A1/C1 joint. Figure 3c shows the SEM/BEI image and a line scanning analysis result with the distribution of Al and Cu elements in the WZ of the FSW A1/C2 joint. The dwell time via the SEM/BEI EDS mapping to collect the data for the Al and Cu elements is about 10 minutes. The dark gray and light gray regions correspond to the AA1050H-aluminum and C2600-brass base metals, respectively. However, unlike the reacting mixtures represented in Figure 3a,b, an obvious interfacial layer with mixing Al and Cu elements is observed between the base metals, and it can be recognized as the Al–Cu IMCs thick layer that was formed on the interface of AA1050H-aluminum and C2600-brass during the FSW process. Generally, the FSW dissimilar joints are failed at the WZ or along the interface between the base metals after the mechanical tests. Therefore, the tensile strength, measuring data fluctuation, failure behaviors, and welding reliability of the FSW A1/C1, A6/C1, and A1/C2 joints will be strongly affected by the metallurgical bonding effect, the particle distribution, and the morphologies of reacted Al–Cu IMCs within the WZ. As a result of the above-mentioned observation and analysis of microstructures, the correlation between the microstructural features, tensile strength, and reliability of these FSW Al–Cu dissimilar joints will be discussed in the following sections.

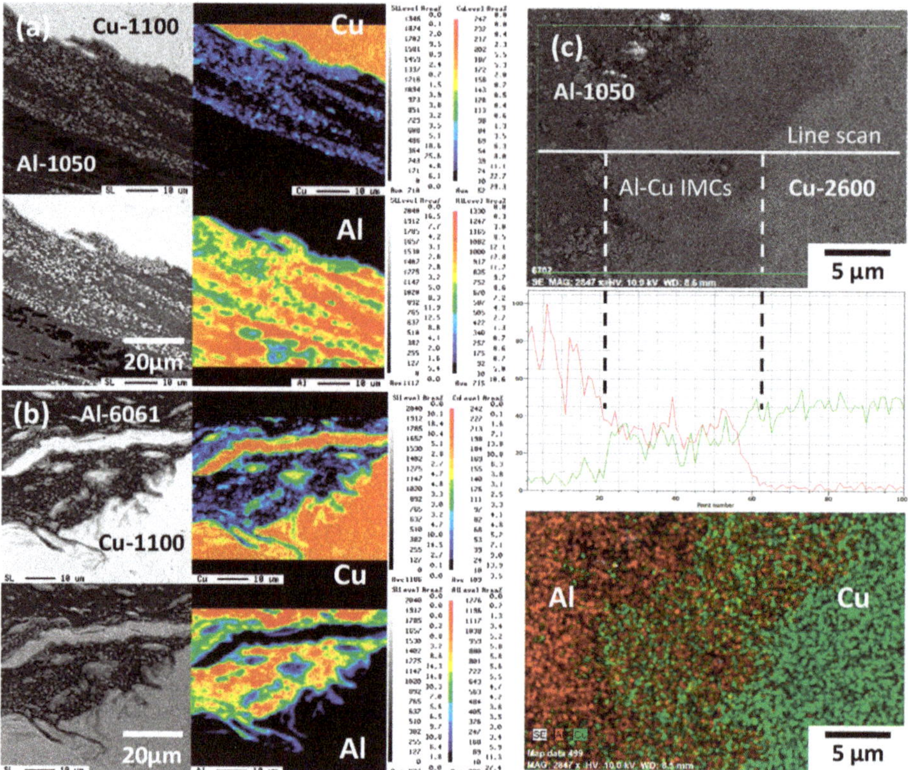

Figure 3. The electron probe micro-analyzer (EPMA) analysis for Al and Cu elements distribution within the WZ of (**a**) the FSW A1/C1 joint, and (**b**) the FSW A6/C1 joint. (**c**) The SEM/BEI line-scanning and Al–Cu elements mapping at the Al/Cu interface of the FSW A1/C2 joint.

The X-ray diffraction patterns obtained from the WZ region of various FSW dissimilar joints are given in Figure 4. Figure 4a,b show the XRD patterns of the A1/C1 and A6/C1 joints, respectively. In addition to the strong diffraction peaks of major α-Al (main peaks detected at $2\theta = 38.47°$, $44.74°$, $65.13°$, and $78.23°$, JCPDS 04-0787) and Cu (main peaks detected at $2\theta = 43.30°$, $50.43°$, and $74.13°$, JCPDS 04-0836) base metals, some relatively weak peaks are also detected within the WZ region of these Al–Cu dissimilar joints. According to the Al–Cu equilibrium phase diagram, several Al–Cu IMCs, such as AlCu (η_2), Al_2Cu (θ), Al_2Cu_3 (ε), Al_3Cu_4 (ζ_2), and Al_4Cu_9 (γ) can be found in the Al–Cu binary alloy system. However, it is recognized that the Al-rich Al_2Cu phase and Cu-rich Al_4Cu_9 phase are the two major IMCs formed during the Al–Cu metallurgical reaction [23,26,41]. Therefore, these weak diffraction peaks obtained in the WZ region of A1/C1 joints are then identified as the IMCs of Al_2Cu (main peaks detected at $2\theta = 37.87°$, $42.59°$, $47.33°$, and $47.81°$, JCPDS 25-0012) and Al_4Cu_9 (the diffraction peak detected at 2θ = 44.12°, JCPDS 24-0003). Comparing Figure 4b with Figure 4a, we can see that the diffraction peaks of the Al_2Cu and Al_4Cu_9 IMCs for the A6/C1 joint are significantly sharper and much more obvious than those of the A1/C1 joint. The differencein peak intensity means that more Al_2Cu and Al_4Cu_9 IMCs are obtained within the WZ region of the A6/C1 joint. Referring to the SEM/BEI images and the element distribution maps obtained by EPMA, the XRD analysis result of the FSW A6/C1 joint (see Figure 4b cf. Figures 2b and 3b) corresponds to the microstructural feature, which displays a more widely Al–Cu reacting mixture distribution with a larger area fraction of the metallurgical bonded zone than the FSW A1/C1 joint (see Figure 4a cf. Figures 2a and 3a). Figure 4c is the XRD pattern of the FSW A1/C2 dissimilar joint, which included strong diffraction peaks of α-Al and $Cu_{0.64}Zn_{0.36}$ (α-brass, main peaks detected at 2θ = 42.33°, 49.28°, and 72.25°, JCPDS 50-1333) base metals. The result illustrates that the WZ region of the A1/C2 joint also consists mainly of Al and Cu (i.e., α-brass) base metals with a fair amount of Al_2Cu and Al_4Cu_9 IMCs. In addition, a γ-Cu_5Zn_8 compound (diffraction peaks at 2θ = 43.30° and 62.88°, JCPDS25-1228) is also observed in the A1/C2 joint, as indicated by the triangular marks in Figure 4c. As a result, it can be recognized that the γ-Cu_5Zn_8 compound is another minor reaction product accompanied by the formation of Al_2Cu and Al_4Cu_9 IMCs for the A1/C2 joint during FSW. Since less Al–Cu reacting mixture and a clear mechanical kissing bonded boundary of the A1/C2 joint are observed (see Figure 2c), it implies that the weldability of A1/C2 joint is worse than that of the A1/C1 and A6/C1 joints.

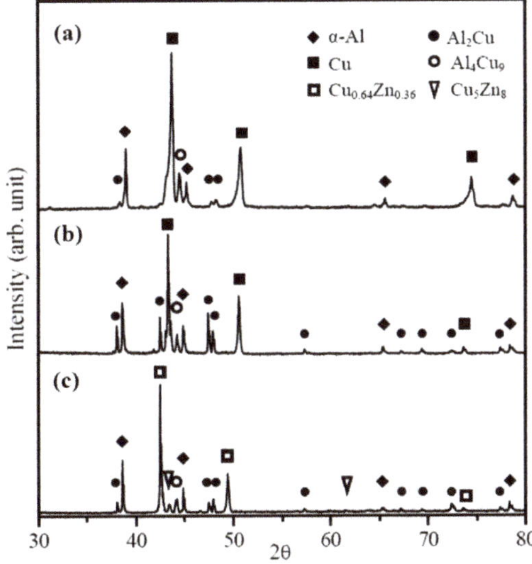

Figure 4. X-ray diffraction patterns of the friction stir welded (**a**) A1/C1, (**b**) A6/C1, and (**c**) A1/C2 dissimilar joints within the welding zone (WZ).

Figure 5 displays the distribution of micro-Vickers hardness (HV) profiles measured within the WZ regions of FSW A1/C1, A6/C1, and A1/C2 dissimilar joints. The hardness data recorded on the transverse cross-sections of FSW-joined specimens, and the Vickers indenter testing positions are located at 1.5 mm depth from the surface. The micro-Vickers hardness test reveals the average values of AA1050H, AA6061-T6 aluminum alloys, commercial pure copper (C1100), and brass (C2600) base metals to be about HV32.6 ± 4.7, HV61.2 ± 5.3, HV62.1 ± 3.0, and HV103.2 ± 4.5, respectively. According to the profiles represented in Figure 5, the hardness of FSW Al–Cu dissimilar joints significantly increases in the WZ region relative to both Al and Cu base metals. The data fluctuation at the transverse direction is the resulted of the heterogeneous microstructure of the WZ. Referring to the cross-sectional microstructures as mentioned in Figure 2, the distribution range of the increased hardness profiles (in the position of about $\pm$ 4 mm from the center, as shown in Figure 5) is almost the same as that of the Al–Cu metallurgical bonded region with obvious metallic plastic flow. Therefore, it can be recognized that the increased hardness profiles of the A6/C1 and A1/C2 joints can be attributed to the apparent Al–Cu reacting mixture within the WZ region (see Figure 5 cf. Figure 2b,c). In addition, the high hardness values of the A6/C1 and A1/C2 joints are also the result of the significant strain-hardening effect of intense materials plastic deformation as well as the particle strengthening effect with the formation and uniform redistribution of Al_2Cu, Al_4Cu_9, and a fair amount of γ-Cu_5Zn_8 IMCs (such as those observed by the XRD analysis in Figure 4) during the FSW process. However, compared with the A6/C1 and A1/C2 joints, the hardness profile for the A1/C1 joint is lower as a result of an insufficient metallurgical bonding reaction and a lesser amount of Al–Cu reacting IMCs mixture (see Figure 2a cf. Figure 4).

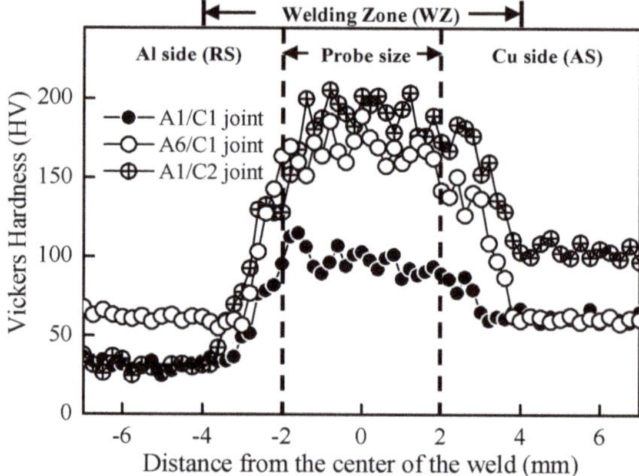

Figure 5. The variation of microhardness of various FSW Al–Cu dissimilar joints. The indentations are made with a spacing of 0.2 mm on the WD plane (AS: advancing side; RS: retreating side).

3.2. Tensile Failure Strength of FSW Al–Cu Dissimilar Joints

The tensile strength of the FSW dissimilar Al–Cu joint is evaluated as a criterion to the joint performance. Table 2 lists a comparison of the tensile strength and elongation of base metals and the FSW A1/C1, A6/C1, and A1/C2 dissimilar joints. Since a sound FSW A6/C2 joint without welding cavity defects in the WZ region is not successfully obtained, therefore, the tensile properties of A6/C2 joints are too low to be used in this study. In addition, the data fluctuation of A6/C2 joints is dispersed to show a valid statistical significance for the following reliability analysis in this study. It can be seen that a maximum tensile strength of about 212.7 MPa and the highest elongation of about 22.3% are obtained for the FSW A6/C1 joint. Figure 6 shows the representative fractured specimens of FSW A1/C1, A6/C1, and A1/C2 dissimilar joints after tensile tests. The Al–Cu dissimilar joints failed with distinguished fracture characteristics, and the fracture location can be clearly observed in these photos. The fracture of FSW A1/C1 joints is usually located at the WZ region, as shown in Figure 6a. Since the average tensile strength of A1/C1 joints is about 85% of the AA1050H aluminum alloy (121.8 $\pm$ 2.3 MPa) according to earlier measurements for base metals, the fracture characteristics that are displayed in Figure 6a and the decrease in the tensile strength of the A1/C1 joints should be the result of the inhomogeneous welding microstructure, which features a mechanical kissing bond at the interface between AA1050H-aluminum and C1100-pure Cu (see Figure 2a cf. Figure 6a).

Comparing Figure 6b with Figure 6a, it is worth noting that the fracture region of the FSW A6/C1 joints is distant from the WZ and nearly located at the heat-affected zone (HAZ) with an obvious necking deformation of the AA6061-T6 Al base metal. As a result, both the tensile strength and the elongation of the A6/C1 joints are significantly higher than that of the A1/C1 joints (see Table 2), and the fracture characteristic without failure at the WZ of the FSW A6/C1 joints, which is represented in Figure 6b, can be regarded as an illustration of better joint quality. Referring to the SEM/BEI cross-sectional images and EPMA analysis results shown in Figures 2b and 3b, it is noted that the much better tensile strength of the A6/C1 joints is achieved from their good joint quality with a larger area fraction of the metallurgical bonded zone and a firmly material metallurgical joining between the AA6061-T6 Al alloy and the C1100 commercial pure copper through the FSW process. The intense material plastic flow with obvious onion rings (see Figure 2b) and uniformly dispersive refined hard IMCs particles in the Al–Cu reacting mixture enhance the metallurgical bonding effect within the WZ. Meanwhile, the movement of dislocations will be effectively impeded by the tangle of high dislocation density and the refined hard IMCs particles during plastic deformation. Therefore, it can be recognized that the FSW A6/C1 dissimilar joint with increased hardness profiles and much higher tensile strength is obtained due to the strong metallurgical bonding effect.

Table 2. Tensile properties of base metals and various FSW Al–Cu dissimilar joints.

Samples	TensileStrength (MPa)	Elongation (%)
AA1050H *	121.8 ± 2.3	15.2±1.7
AA6061-T6 *	293.1 ± 2.6	12.8± 2.5
C1100 Cu *	227.9 ± 1.8	30.6 ± 1.8
C2600 Brass *	365.2 ± 1.5	27.7 ± 2.3
A1/C1 Joint [†]	108.6 ± 9.1	10.5 ± 3.3
A6/C1 Joint [†]	212.7 ± 8.5	22.3 ± 2.8
A1/C2 Joint [†]	53.2 ± 5.9	5.7 ± 0.9

* Each value was the average of a least three tests; [†] Each value was the average of 20 tests (n = 20).

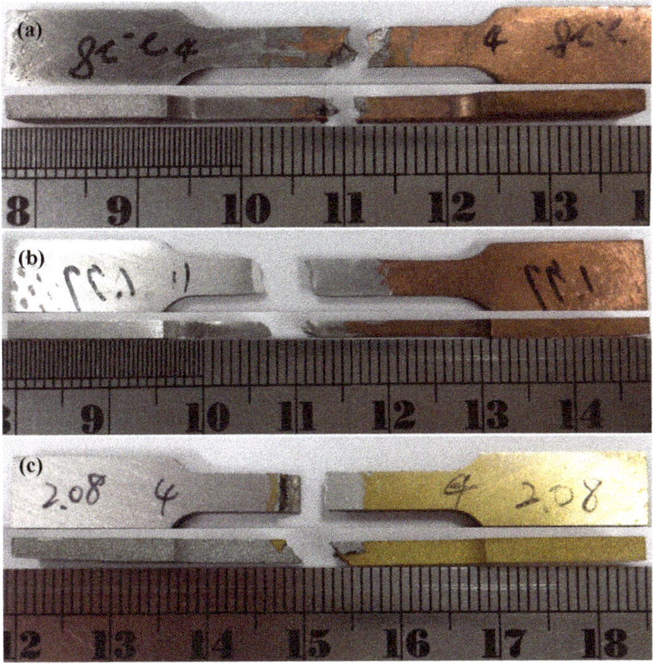

Figure 6. Macrographs of fractured FSW (**a**) A1/C1, (**b**) A6/C1, and (**c**) A1/C2 dissimilar joints after tensile tests.

Among these Al–Cu dissimilar joints, the FSW A1/C2 joint apparently exhibits the worst tensile strength and displays much lower elongation (see Table 2), with an obvious brittle fracture at the WZ region along the joining interface of AA1050H-aluminum and C2600-brass base metals, as illustrated in Figure 6c. The notable degradation in the tensile strength of the A1/C2 joints should be closely correlated with the structural inhomogeneity. Referring to Figures 2c and 3c cf. Figure 6c, a large area fraction of the mechanical kissing bond with a continuous thick IMCs layer was significantly formed at the Al/brass interface within the WZ of the A1/C2 joints. It demonstrates that the reduction of effective metallurgical bonding and the formation of a continuous thick IMCs layer yielded a weak interfacial A1/C2 joint. As a result, the detrimental decohesive ruptures notably occurred at the weak mechanical kissing bonded boundary of AA1050H-aluminum and C2600-brass before other structural defects caused the failures of the A1/C2 joint throughout the WZ.

According to the above-mentioned examinations, FSW A1/C1, A6/C1, and A1/C2 dissimilar joints display quite different welding microstructures, interfacial joining characteristics, hardness profiles, and phase compositions within the WZ regions. Since the tensile strength of FSW Al–Cu dissimilar joints should be closely affected by the variation of welding qualities, the statistical analysis by the Weibull model on the data variance of tensile tests made it possible to assess the microstructural features in determining the failure behaviors and reliability of FSW Al–Cu dissimilar joints. The Weibull statistical analysis detail on the tensile mechanical properties and failures related to the welding qualities of FSW Al-Cu joints will be discussed in the following section.

4. Discussion

4.1. The Weibull Statistical Analysis on the Failure Probability of Al–Cu Joints

The failure of an engineering component, as defined by reliability engineering, can be recognized as "the event, or inoperable state, in which any item or part of an item does not, or would not, perform as previously specified" [42]. The advantage of Weibull statistical analysis for reliability engineering is to effectively evaluate failure probability and provide reasonable failure predictions for engineering components from the stages of design and manufacturing processes with extremely small amount of testing samples. In this study, the cumulative failure probability function (Equation (1)) determined by the three-parameter Weibull model is applied to simulate the data variability of the measured tensile strength of FSW Al–Cu dissimilar joints. The cumulative failure probability $F(\sigma_i)$ of the testing samples is estimated using Bernard's median rank [43], and the reliability function $R(\sigma_i)$ with a relation of $R(\sigma_i) = 1 - F(\sigma_i)$ is determined as the survival probability. The cumulative failure probability is controlled by parameters m, σ_c, and σ_0. The Weibull modulus (m) is a measure of data variability. The characteristic life (σ_c) corresponds to the tensile strength at which the cumulative failure probability is equal to 63.2%. The minimum strength (σ_0) is so-called the failure-free strength, which means that the failure probability of the FSW Al–Cu dissimilar joints below this tensile strength is zero.

$$F(\sigma_i) = \int_{\sigma=0}^{\sigma=\sigma_i} f(\sigma)d\sigma = 1 - \exp\left[-\left(\frac{\sigma_i - \sigma_0}{\sigma_c}\right)^m\right] \tag{1}$$

Fitting the measured tensile strength data into the cumulative failure probability function of Equation (1), and subsequently the failure probability density function ($f(\sigma_i)$) of FSW A1/C1, A6/C1, and A1/C2 dissimilar joints are calculated and plotted in Figure 7a. Figure 7b illustrates the Weibull plot, which is a natural logarithmic (ln) graph for the cumulative failure probability at each corresponding σ_i of various FSW Al–Cu dissimilar joints. The horizontal scale of the Weibull plot is a measure of tensile strength, and the vertical scale is the cumulative percentage failed. The Weibull modulus, which is particularly significant and provides a clue to the physics of the failure, is graphically evaluated from the slope of Weibull plots by the least squares fitting method at a maximum coefficient of determination (R^2). According to the definition of the Weibull model, the critical coefficient of determination (CCD) for 20 failures ($n = 20$) should be higher than 0.95, and then the Weibull plot can be considered a good fit for the three-parameter Weibull cumulative distribution function [44]. Therefore, it is recognized that a good linear relationship is obtained for the experimental data in Figure 7b, and the Weibull model is valid to describe the failure behaviors of the FSW Al–Cu joints. The Weibull statistical analysis results are listed in Table 3.

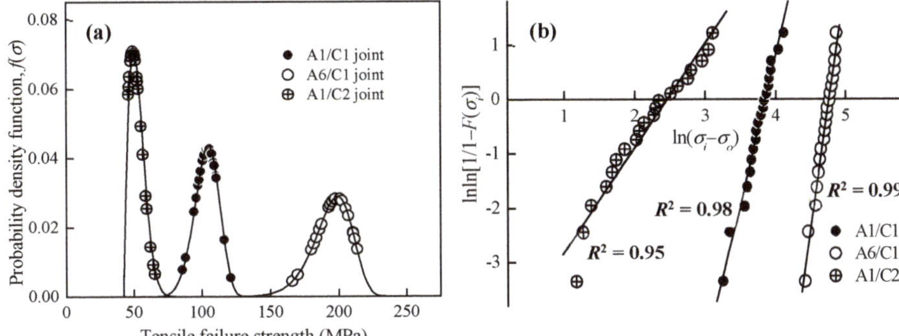

Figure 7. (a) The failure probability density function $f(\sigma_i)$ curves, and (b) the Weibull distribution plots of various FSW Al–Cu dissimilar joints. $F(\sigma_i)$ is the cumulative failure probability for a corresponding tensile strength (σ_i), and the slope represents the Weibull modulus (m) calculated by the least squares fitting method at a maximum coefficient of determination (R^2).

Table 3. Statistical analysis results * of the Weibull model for the tensile strength of various FSW A1–Cu dissimilar joints.

Samples	Weibull Modulus, m	Characteristics Strength (MPa), σ_c	Minimum Strength (MPa), σ_0	Coefficient of Determination, R^2
A1/C1 Joint	5.4	46.9	59.7	0.98
A6/C1 Joint	9.2	117.5	83.0	0.99
A1/C2 Joint	1.7	11.7	41.5	0.95

* Data were calculated from Weibull plots (Figure 7b).

4.2. Microstructural Variations Affect Data Fluctuation and Failure Behaviors

Referring to the aforementioned microstructural observations and tensile testing results, a fair amount of the data fluctuation for the FSWA1/C1, A6/C1, and A1/C2 dissimilar joints shown in Figure 7a should be significantly related to the joint inhomogeneity with a different metallurgical bonding area fractionin the WZ region. Since the Weibull model is commonly adopted to forecast the reliability and failure behaviors of engineering components, the hazard function ($\lambda(\sigma_i)$) listed in Equation (2) at each corresponding tensile strength is further defined as the ratio of the failure probability density function and reliability functionin the present study for the assessment of failure behaviors. Figure 8a shows the reliability function ($R(\sigma_i)$) curves of the Al–Cu joints. These curves start from the minimum strength (σ_0), and the reliability of joints is decreased with increasing tensile loading. The failure rate curves calculated from the hazard function ($\lambda(\sigma_i)$) of Al–Cu joints are shown in Figure 8b.

$$\lambda(\sigma_i) = \frac{f(\sigma_i)}{R(\sigma_i)} = \frac{m}{\sigma_c{}^m}(\sigma_i - \sigma_0)^{m-1} \tag{2}$$

The Weibull modulus (m) is a dimensionless value, and it is a main factor for the Weibull model to determine which Al–Cu joint displays better engineering reliability. It is noted that the Weibull modulus represents the variability of experimental data, which becomes larger as the degree of tensile strength fluctuation decreases and the reliability of the joints increases. As a result of Table 3 and Figure 8b, the Weibull statistical analysis demonstrates that all of the A1/C1, A6/C1, and A1/C2 welding conditions are reliable FSW dissimilar joints with a wear-out failure model ($m > 1$) of the increasing failure rate (IFR) behavior. However, the A1/C1 joint ($m = 5.4$) and the A6/C1 joint ($m = 9.2$) significantly show a larger Weibull modulus compared with the A1/C2 joint ($m = 1.7$). The A1/C2 joints with a lower Weibull modulus generally characterize an early failure behavior because of the much higher initial failure probability of the tensile specimens compared with the A1/C1 and A6/C1 joints, as shown in Figure 7a. Therefore, referring to the microstructural observations illustrated in Figures 2 and 3, it is demonstrated that the FSW Al-Cu dissimilar joints with a large amount of Al–Cu reacting mixtures of uniformly dispersed Al–Cu IMCs particles can resist a higher tensile failure load and also display better engineering reliability because of the larger Weibull modulus. The FSW A6/C1 dissimilar joint with a successful metallurgical bonded WZ region (see Figures 2a and 3a cf. Figure 6b) represent better tensile strength and reliability than a mainly mechanical kissing bonded A1/C2 joint interface (see Figure 2c cf. Figure 6c). Moreover, the reliability of FSW A1/C2 joints is rapidly decreased through just slightly increasing the tensile loading to be larger than the minimum strength, as shown in Figure 8a. Since the minimum strength can be recognized as the safety value of an engineering component, the existence of minimum strength σ_0 is needed in order to evaluate the critical reliable tensile strength of varied FSW Al–Cu dissimilar joints. Therefore, FSW Al–Cu dissimilar joints with a larger Weibull modulus can properly be selected for engineering application, as this may be an indicator of lower technique sensitivity and less reliability decrease (see Figure 8a) while the applied tensile loading exceeds the minimum strength.

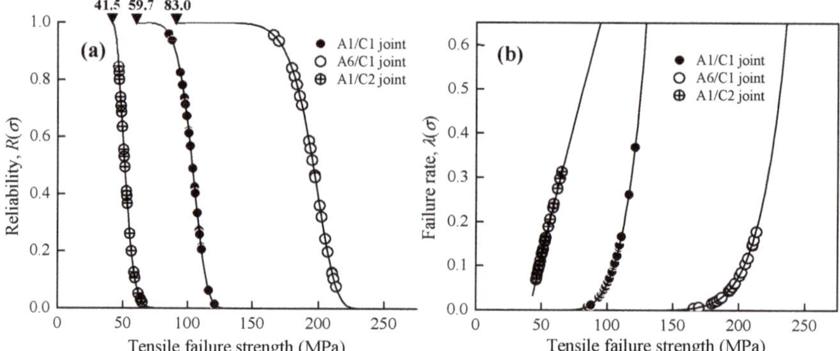

Figure 8. (**a**) The reliability function curves $R(\sigma_i)$ and (**b**) the failure rate curves from the hazard function $\lambda(\sigma_i)$ of various FSW Al–Cu dissimilar joints. These curves start from the minimum strength (σ_0), which is the safety tensile strength for FSW Al–Cu dissimilar joints.

Properly friction stir welded joints should have high tensile strength and display ductility, whereas the joints with low tensile strength will fail in a brittle fracture at the WZ region. In order to realize the fracture mechanism with the above-mentioned Weibull statistical analysis results, Figures 9–11 show the representative SEM fracture morphologies of the FSW A1/C1, A6/C1, and A1/C2 dissimilar joints, respectively. Figure 9a gives the BEI fracture sub-surface of A1/C1 joints to identify the Al and Cu base metals and Al–Cu IMCs more clearly. A fair amount of reacting mixtures (the light gray region indicated by the triangular mark) composed of Al_2Cu and Al_4Cu_9 IMCs (identified by XRD analysis, see Figure 4a) are observed within the WZ region of the A1/C1 joint. In addition, cracks propagation is observed at the IMCs/Cu interface, and the internal cracks are perpendicular to the tensile direction, as those encircled in Figure 9a. Figure 9b shows the BEI tensile fracture surface of the A1/C1 joints. Figure 9 c,d represent the EDS mapping analysis region denoted by the triangular mark in Figure 9b, and the composition of this region is about 54.8 Al and 45.2 Cu (in atomic %) by semi-quantitative SEM/EDS analysis. It can be seen that a brittle fracture appears, and the fracture can be recognized as a result of the interfacial cracks propagation along the Al–Cu IMCs within the WZ region (see Figure 9b cf. Figures 2a and 6a).

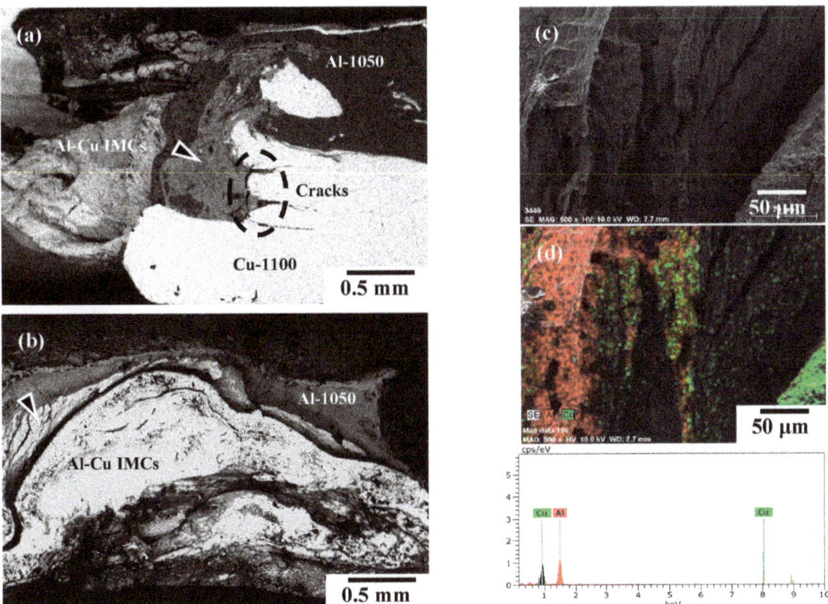

Figure 9. (**a**) Failure sub-surface, (**b**) fracture surface, (**c**) representative fracture morphology denoted by the arrow in (**b**), and (**d**) Al, Cu elements energy-dispersive X-ray spectroscopy (EDS) mapping of the A1/C1 joint.

Figure 10a shows the SEM tensile fracture surface of the A6/C1 joints for illustration. As seen from the fracture morphologies of the A6/C1 joint, a ductile failure behavior appears with necking and dimpled ruptures (see Figure 10a cf. Figure 6b) at the AA6061 Al base metal. Therefore, it is recognized that A6/C1 joints obviously display better elongation (as listed in Table 2) than A1/C1 and A1/C2 joints. Figure 10b displays the EDS mapping analysis results of Al and Cu elements (with chemical compositions of 31.2 Al and 68.8 Cu, in atomic %) obtained from the WZ region of tensile failed A6/C1 joints. Referring to the above-mentioned microstructural observations, significant reacting mixtures of dispersed Al–Cu IMCs particles represent that intense material plastic flow effectively induces a larger area fraction of the metallurgical bonded zone (see Figures 2b and 3b cf. Figure 4b) within the WZ region of the A6/C1 joint during the FSW process. As a result, Figure 10b illustrates a typical WZ region of good welding quality with a textureof elongated joining materials flow along the tensile direction without an apparent cracking effect after tensile tests. Since the Al–Cu IMCs can be used as reinforcing phases through the distribution of particles [27,45], as shown in the experimental results mentioned in Table 2, Figures 8a and 10b demonstrate that a large fraction of metallurgical bonding of obvious onion rings with dispersed Al–Cu IMCs particles provides a successful firm welding structure, higher tensile strength, better welding quality, and joint reliability of the FSW A6/C1 joint compared with the A1/C1 joint.

Figure 11a,b shows the BEI fracture sub-surface and fracture surface of the A1/C2 joints, respectively. Figure 11c,d represent the EDS mapping analysis result of the rectangular region denoted in Figure 11b, and the composition in this region is about 51.7 Al and 48.3 Cu (in atomic %) by SEM/EDS analysis. Referring to Figure 6c, the fracture of the A1/C2 joints is almost located at the A1050H-aluminum and C2600-brass joint interface. It is reported that FSW dissimilar joints with an excessively generated thick interfacial IMCs layer generally display poor mechanical properties because of the brittleness of IMCs and quite easier crack propagation [27,32]. Therefore, comparing Figure 11a with Figure 6c, it can be found that many of the cracks that were significantly initiated at the AA1050H/C2600brass joint interface (cracks are also observed on the fracture surface of Figure 11c), and the failure of joints occurred while the cracks penetrated through Al–Cu IMCs (as denoted by the triangular marks in Figure 11a). This phenomenon demonstrates that a thick continuous interfacial IMC results in a significant deterioration of tensile strength and the presence of much lower Weibull modulus ($m = 1.7$) with a rapid increase in the failure probability (see Table 3 and Figure 7a) for the A1/C2 joint. Therefore, FSW A1/C2 joints obviously display the lowest tensile strength with a brittle fracture and an unfavorable welding reliability in this study (see Figure 11a,b cf. Figure 8a).

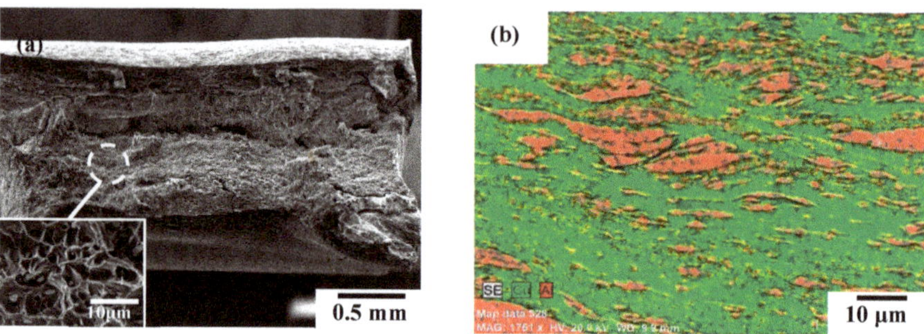

Figure 10. (**a**) Fracture surface and (**b**) EDS mapping of the Al and Cu elements at the WZ region of the tensile failed A6/C1 joint.

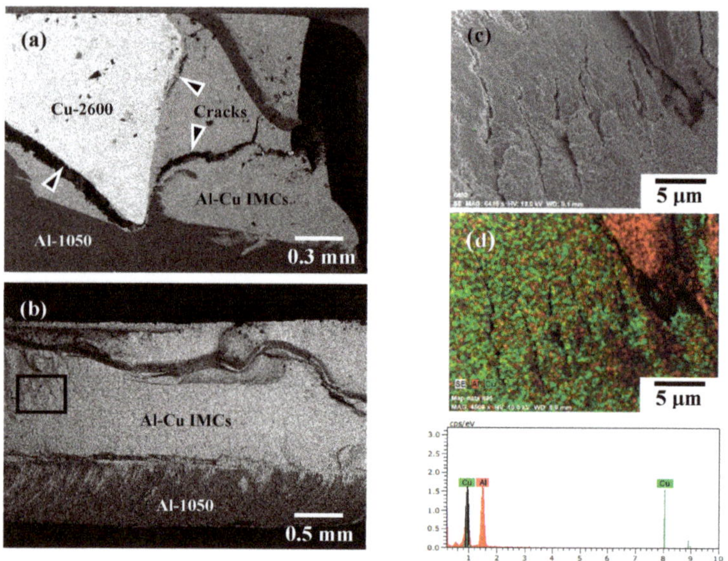

Figure 11. (**a**) Failure sub-surface, (**b**) fracture surface, (**c**) representative fracture morphology of the rectangular area in (**b**), and (**d**) EDS mapping of the Al and Cu elements of the A1/C2 joint.

5. Conclusions

The welding qualitiesand tensile mechanical properties related to the metallurgical bonded microstructural characteristicsof FSW Al–Cu dissimilar joints are identified and discussed in this study. In addition, the relationship between welding quality and tensile strength, as well as the joint reliability, is investigated by the Weibull statistical analysis. The conclusions are drawn based on the above results and discussions:

(1) Dissimilar Al–Cu joints of AA1050H/C1100-Cu (A1/C1), AA6061-T6/C1100-Cu (A6/C1), and AA1050H/C2600-brass (A1/C2) couples are successfully joined without typical cavity defects in the welding zone (WZ) by the friction stir welding process (FSW).

(2) Al_2Cu and Al_4Cu_9 are the major intermetallic compounds (IMCs) formed in the metallurgical bonded welding zone of FSW Al–Cu dissimilar joints, and γ-Cu_5Zn_8 is another reacted IMC observed in the WZ of the AA1050H/C2600-brass joint.

(3) The microhardness of FSW Al–Cu joints in the WZ is increased as a result of the formation of Al–Cu IMCs and intense plastic deformation during FSW.

(4) The AA6061-T6/C1100-Cu joint exhibits a significant metallurgical bonded zone with onion rings in the WZ region, whereas the AA1050H/C2600-brass joint usually displays a mechanical kissing bonded boundary at the Al–Cu joining interface.

(5) Through the powerful statistical analysis of the Weibull model, FSW Al–Cu dissimilar joints, which display a wear-out failure model, can be recognized as reliable joints for further engineering applications.

(6) Better welding reliability and a higher tensile strength with ductile dimpled ruptures can be obtained for those FSW Al–Cu joints with IMCs particles uniformly dispersed in a large area fraction of the metallurgical bonded WZ region.

(7) FSW Al–Cu joints with a mechanical kissing bonded boundary and a thick continuous interfacial IMC layer results in a rapid increase in the failure probability and the deterioration of tensile strength with a brittle fracture at the WZ region of the joints.

Author Contributions: This study presented here was carried out in collaboration between all authors. C.-W.Y. conceived and designed the experiments; S.-J.J. performed the experiments; C.-W.Y. analyzed the data and wrote the manuscript. All authors reviewed the manuscript, and the authors hope that this manuscript can make its due contribution to the successful application of the high-performance FSW Al-Cu dissimilar joints.

Funding: The Ministry of Science and Technology (MOST) is the government agency dedicated to scientific and technological development. Its three main missions are promoting nationwide science & technology development, supporting academic research, and developing science parks.

Acknowledgments: This study was financially supported by the Ministry of Science and Technology, Taiwan (Contract No. MOST 107-2221-E-150-005) for which we are grateful.

Conflicts of Interest: The authors declare that they have no conflict of interest.

References

1. Wang, H.T.; Wang, G.Z.; Xuan, F.Z.; Liu, C.J.; Tu, S.T. Local mechanical properties of a dissimilar metal welded joint in nuclear power systems. *Mater. Sci. Eng. A* **2013**, *568*, 108–117. [CrossRef]

2. Feistauer, E.E.; Bergmann, L.A.; Barreto, L.S.; Dos Santos, J.F. Mechanical behaviour of dissimilar friction stir welded tailor welded blanks in Al-Mg alloys for Marine applications. *Mater. Des.* **2014**, *59*, 323–332. [CrossRef]

3. Ren, D.; Liu, L. Interface microstructure and mechanical properties of arc spot welding Mg-steel dissimilar joint with Cu interlayer. *Mater. Des.* **2014**, *59*, 369–376. [CrossRef]

4. Zhang, C.Q.; Robson, J.D.; Ciuca, O.; Prangnell, P.B. Microstructural characterization and mechanical properties of high power ultrasonic spot welded aluminum alloy AA6111-Ti6Al4V dissimilar joints. *Mater.Charact.* **2014**, *97*, 83–91. [CrossRef]

5. Jagadeesha, C.B. Dissimilar friction stir welding between aluminum alloy and magnesium alloy at a low rotational speed. *Mater. Sci. Eng. A* **2014**, *616*, 55–62. [CrossRef]

6. Zhou, X.; Zhang, G.; Shi, Y.; Zhu, M.; Yang, F. Microstructures and mechanical behavior of aluminum-copper lap joints. *Mater. Sci. Eng. A* **2017**, *705*, 105–113. [CrossRef]

7. Li, H.; Chen, W.; Dong, L.; Shi, Y.; Jiu, J.; Fu, Y.Q. Interfacial bonding mechanism and annealing effect on Cu-Al joint produced by solid-liquid compound casting. *J. Mater. Process. Technol.* **2018**, *252*, 795–803. [CrossRef]

8. Li, H.; Cao, B.; Yang, J.W.; Liu, J. Modeling of resistance heat assisted ultrasonic welding of Cu-Al joint. *J. Mater. Process. Technol.* **2018**, *256*, 121–130. [CrossRef]

9. Jarwitz, M.; Fetzer, F.; Weber, R.; Graf, T. Weld seam geometry and electrical resistance of laser-welded aluminum-copper dissimilar joints produced with spatial beam oscillation. *Metals* **2018**, *8*, 510. [CrossRef]

10. Wang, X.G.; Li, X.G.; Wang, C.G. Influence of diffusion brazing parameters on microstructure and properties of Cu/Al joints. *J. Manuf. Process.* **2018**, *35*, 343–350. [CrossRef]

11. Mishra, R.S.; Ma, Z.Y. Friction stir welding and processing. *Mater. Sci. Eng. R* **2005**, *50*, 1–78. [CrossRef]

12. DebRoy, T.; Bhadeshia, H.K.D.H. Friction stir welding of dissimilar alloys—A perspective. *Sci. Technol. Weld. Join.* **2010**, *15*, 266–270. [CrossRef]

13. Thomas, W.M.; Nicholas, E.D.; Needham, J.C.; Murch, M.G.; Templesmith, P.; Dawes, C.J. Friction Welding. Patent Application No. 9125978.8, 6 December 1991.

14. Anouma, M.; Nakata, K. Dissimilar metal joining of ZK60 magnesium alloy and titanium by friction stir welding. *Mater. Sci. Eng. B* **2012**, *177*, 543–548. [CrossRef]

15. Zhao, Y.; Lu, Z.; Yan, K.; Huang, L. Microstructural characterizations and mechanical properties in under water friction stir welding of aluminum and magnesium dissimilar alloys. *Mater. Des.* **2015**, *65*, 675–681. [CrossRef]

16. Tamjidy, M.; Baharudin, B.H.T.; Paslar, S.; Matori, K.A.; Sulaiman, S.; Fadaeifard, F. Multi-objective optimization of friction stir welding process parameters of AA6061-T6 and AA7075-T6 using a biogeography based optimization algorithm. *Materials* **2017**, *10*, 533. [CrossRef] [PubMed]

17. Song, G.; Li, T.; Yu, J.; Liu, L. A review of bonding immiscible Mg/steel dissimilar metals. *Materials* **2018**, *11*, 2515. [CrossRef] [PubMed]

18. Dong, J.; Zhang, D.; Zhang, W.; Zhang, W.; Qiu, C. Microstructure evolution during dissimilar friction stir welding of AA7003-T4 and AA6060-T4. *Materials* **2018**, *11*, 342. [CrossRef] [PubMed]

19. Eslami, N.; Harms, A.; Deringer, J.; Fricke, A.; Böhm, S. Dissimilar friction stir butt welding of aluminum and copper with cross-section adjustment for current-carrying components. *Metals* **2018**, *8*, 661. [CrossRef]

20. Esmaeili, A.; Givi, M.K.B.; Rajani, H.R.Z. A metallurgical and mechanical study on dissimilar friction stir welding of aluminum 1050 to brass (CuZn30). *Mater. Sci. Eng. A* **2011**, *528*, 7093–7102. [CrossRef]

21. Galvão, I.; Loureiro, A.; Verdera, D.; Gesto, D.; Rodrigues, D.M. Influence of tool offsetting on the structure and morphology of dissimilar aluminum to copper friction-stir welds. *Metall. Mater. Trans. A* **2012**, *43*, 5096–5105. [CrossRef]

22. Bisadi, H.; Tavakoli, A.; Sangsaraki, M.T.; Sangsarki, K.T. The influences of rotational and welding speeds on microstructures and mechanical properties of friction stir welded Al5083 and commercially pure copper sheets lap joints. *Mater. Des.* **2013**, *43*, 80–88. [CrossRef]

23. Sahu, P.K.; Pal, S.; Pal, S.K.; Jain, R. Influence of plate position, tool offset and tool rotational speed on mechanical properties and microstructures of dissimilar Al/Cu friction stir welding joints. *J. Mater. Process. Technol.* **2016**, *235*, 55–67. [CrossRef]

24. Shojaei Zoeram, A.; Mousavi Anijdan, S.H.; Jafarian, H.R.; Bhattacharjee, T. Welding parameters analysis and microstructural evolution of dissimilar joints in Al/Bronze processed by friction stir welding and their effect on engineering tensile behavior. *Mater. Sci. Eng. A* **2017**, *687*, 288–297. [CrossRef]

25. Regensburg, A.; Schürer, R.; Weigl, M.; Bergmann, J.P. Influence of pin length and electrochemical platings on the mechanical strength and macroscopic defect formation in stationary shoulder friction stir welding of aluminum to copper. *Metals* **2018**, *8*, 85. [CrossRef]

26. Zhang, W.; Shen, Y.; Yan, Y.; Guo, R. Dissimilar friction stir welding of 6061 Al to T2 pure Cu adopting tooth-shaped joint configuration: Microstructure and mechanical properties. *Mater. Sci. Eng. A* **2017**, *690*, 355–364. [CrossRef]

27. Zhou, L.; Li, G.H.; Zhang, R.X.; Zhou, W.L.; He, W.X.; Huang, Y.X.; Song, X.G. Microstructure evolution and mechanical properties of friction stir spot welded dissimilar aluminum-copper joint. *J. Alloys Compd.* **2019**, *775*, 372–382. [CrossRef]

28. Ouyang, J.H.; Yarrapareddy, E.; Kovacevic, R. Microstructural evolution in the friction stir welded 6061 aluminum alloy (T6-temper condition) to copper. *J. Mater. Process. Technol.* **2006**, *172*, 110–122. [CrossRef]

29. Xue, P.; Xiao, B.L.; Ni, D.R.; Ma, Z.Y. Enhanced mechanical properties of friction stir welded dissimilar Al-Cu joint by intermetallic compounds. *Mater. Sci. Eng. A* **2010**, *527*, 5723–5727. [CrossRef]

30. Liu, H.J.; Shen, J.J.; Zhou, L.; Zhao, Y.Q.; Liu, C.; Kuang, L.Y. Microstructural characterization and mechanical properties of friction stir welded joints of aluminum alloy to copper. *Sci. Technol. Weld. Join.* **2011**, *16*, 92–99. [CrossRef]

31. Galvão, I.; Oliveira, J.C.; Loureiro, A.; Rodrigues, D.M. Formation and distribution of brittle structures in friction stir welding of aluminum and copper: Influence of process parameters. *Sci. Technol. Weld. Join.* **2011**, *16*, 681–689. [CrossRef]

32. Galvão, I.; Oliveira, J.C.; Loureiro, A.; Rodrigues, D.M. Formation and distribution of brittle structures in friction stir welding of aluminum and copper: Influence of shoulder geometry. *Intermetallics* **2012**, *22*, 122–128. [CrossRef]

33. Chen, C.Y.; Chen, H.L.; Huang, W.S. Influence of interfacial structure development on the fracture mechanism and bond strength of aluminum/copper bimetal plate. *Mater. Trans.* **2006**, *47*, 1232–1239. [CrossRef]

34. O'Connor, P.; Kleyner, A. *Practical Reliability Engineering*, 5th ed.; John Wiley and Sons, Ltd.: Hoboken, NJ, USA, 2012; pp. 78–140, ISBN 978-0-470-97981-5.

35. Weibull, W. A statistical distribution function of wide applicability. *J. Appl. Mech.* **1951**, *18*, 293–297.

36. Yang, C.W. Development of hydrothermally synthesized hydroxyapatite coatings on metallic substrates and Weibull's reliability analysis of its strength. *Int. J. Appl. Ceram. Technol.* **2015**, *12*, 282–293. [CrossRef]

37. Effertz, P.S.; Infante, V.; Quintino, L.; Suhuddin, U.; Hanke, S.; Dos Santos, J.F. Fatigue life assessment of friction spot welded 7075-T76 aluminum alloy using Weibull distribution. *Int. J. Fatigue* **2016**, *87*, 381–390. [CrossRef]

38. Blacha, L.; Karolczuk, A. Validation of the weakest link approach and the proposed Weibull based probability distribution of failure for fatigue design of steel welded joints. *Eng. Fail. Anal.* **2016**, *67*, 46–62. [CrossRef]

39. Pérot, N.; Bousquet, N. Functional Weibull-based models of steel fracture toughness for structural risk analysis: Estimation and selection. *Reliab. Eng. Syst. Saf.* **2017**, *165*, 355–367. [CrossRef]

40. Xiong, J.; Wang, S.; Li, X.; Yang, Z.; Zhang, J.; Yan, C.; Tang, A. Mechanical behavior and Weibull statistics based failure analysis of vanadium flow battery stacks. *J. Power Sources* **2019**, *412*, 272–281. [CrossRef]

41. Jiang, H.G.; Dai, J.Y.; Tong, H.Y.; Ding, B.Z.; Song, Q.H.; Hu, Z.Q. Interfacial reactions on annealing Cu/Al multilayer thin films. *J. Appl. Phys.* **1993**, *74*, 6165–6169. [CrossRef]

42. Dodson, B.; Nolan, D. *Reliability Engineering Handbook (Quality and Reliability)*, 1st ed.; Taylor & Francis; CRC Press Inc.: Boca Raton, FL, USA, 2002; pp. 33–63, ISBN 978-0-824-70364-6.

43. Faucher, B.; Tyson, W.R. On the determination of Weibull parameters. *J. Mater. Sci. Lett.* **1988**, *7*, 1199–1203. [CrossRef]

44. Abernethy, R.B. *The New Weibull Handbook: Reliability and Statistical Analysis for Predicting Life, Safety, Survivability, Risk, Cost and Warranty Claims*, 4th ed.; Gulf Professional Publishing Books: North Palm Beach, FL, USA, 2000; pp. 3-1–3-9, ISBN 0-9653062-1-6.

45. Hsu, C.J.; Kao, P.W.; Ho, N.J. Ultrafine-grained Al-Al$_2$Cu composite produced in situ by friction stir processing. *Scr. Mater.* **2005**, *53*, 341–345. [CrossRef]

Article

Effect of Heat Treatment on the Microstructure and Properties of a Ti$_3$Al Linear Friction Welding Joint

Xiaohong Li [1,2], Jianchao He [2], Tiancang Zhang [2], Jun Tao [2], Ju Li [2] and Yanhua Zhang [1,*]

1   School of Mechanical Engineering and Automation, Beihang University, Beijing 100191, China; lixhamti06@163.com
2   Aeronautical Key Laboratory for Welding and Joining Technologies, AVIC Manufacturing Technology Institute, Beijing 100024, China; hjch1985@gmail.com (J.H.); bjsztc@126.com (T.Z.); taojun40@126.com (J.T.); hfutlju@163.com (J.L.)
*   Correspondence: zhangyh@buaa.edu.cn

Received: 3 March 2019; Accepted: 8 April 2019; Published: 10 April 2019

Abstract: Heat treatment at different temperatures was carried out on a Ti$_3$Al linear friction welding joint. The characteristics and evolution of the microstructure in the weld zone (WZ) and the thermo-mechanically affected zone (TMAZ) of the Ti$_3$Al LFW joint were analyzed. Combined with the heat treatment after welding, the effect of the heat treatment temperature on the joint was discussed. The test results indicated that the linear friction welding (LFW) process can accomplish a reliable connection between Ti$_3$Al alloys and the joint can avoid defects such as microcracks and voids. The weld zone of the as-welded Ti$_3$Al alloy joint was mainly composed of metastable β phase, while the TMAZ was mainly composed of deformed α_2 phase and metastable β phase. After being heat treated at different temperatures, the WZ of the Ti$_3$Al LFW joint exhibited a significantly different microstructure. After heat treatment at 700 °C, dot-like structures precipitated and the joint microhardness increased significantly. Subsequently, the joint microhardness decreases with the increase in temperature. Under heat treatment at temperatures above 850 °C, the formed structure was acicular α_2 phase and the joint microhardness after heat treatment was lower than that of the as-welded joint.

Keywords: intermetallic compound; linear friction welding; microstructure; heat treatment

1. Introduction

As a type of Ti-Al series intermetallic compound, Ti$_3$Al-based alloys have better high-temperature performance, oxidation resistance, creep resistance, and higher service temperature than ordinary Ti alloys. Compared with the Ni-based alloys, the density of Ti$_3$Al-based alloys is lower and about half that of Ni-based alloys, which is beneficial for the reduction in equipment weight. As a high-temperature structural material that can fill the temperature gap between the serving temperatures of Ti alloys and Ni-based superalloys, Ti$_3$Al has a good application prospect in the aerospace field, such as to substitute for structural materials with lower strength, to reduce the weight of engines of various vehicles as well as the vehicle's own weight, and to enhance the specific thrust and efficiency of engines. Owing to the above advantages, Ti$_3$Al-based alloys have broad application prospects, since they can meet the urgent needs of future aerospace component structures for light structural materials with high specific strength, high specific modulus, and excellent overall performance [1–4]. In the application of Ti$_3$Al, it is important to solve the jointing problem between the same or different materials that will directly affect its application [5,6].

The main problem in the welding of Ti$_3$Al alloys is that the joint has low room-temperature plasticity and is very sensitive to cracks during solidification. The transition from β phase to α phase is one of the main factors affecting the welding of Ti$_3$Al. During the welding process of Ti$_3$Al alloys, the cooling rate effect on the joint performance plays a decisive role, i.e., as the cooling rate increases, the β phase is cooled rapidly, and the main phase in the microstructure of the welded joint is metastable β phase [7]. The β phase is relatively soft and has a lower fracture toughness, and it is thermodynamically unstable, thus, the transformation from β phase to α_2 phase will occur in high temperature applications. At a slightly lower cooling rate, a very hard α_2 (martensite) phase easily forms in the joint and exhibits very high hardness and enhanced brittleness [8,9]. Another problem in the fusion welding of Ti_3Al is that the thermo-mechanically affected zone (TMAZ) is overheated, and thus, the grains are prone to undergo irreversible coarsening and exhibit significantly increasing brittleness. In addition, hot cracks occur in the joints when a thicker slab is welded. However, the potential advantage of solid-phase welding is to avoid casting structures, gas pores, cracks, deformations, and residual stresses caused by material melting during fusion welding.

As one of the solid-phase welding technologies, weld joints are formed on the contact surface through plastic deformation, surface activation, diffusion, recrystallization, and interaction between rubbing bodies. Linear friction welding (LFW) has a series of advantages, including that the joint is a forged structure, the grain-refined structure is dense, the joint quality is high, there is no dust and weld spatter during the welding process, no filler material and gas protection is needed, and there is less material loss and fewer weld defects [10]. Due to the aforementioned advantages, LFW has been widely used in the welding of carbon steel and stainless steel [11–13], aluminum alloys [14,15], titanium alloys [16–19], and nickel-based alloys [20–22], and has become the key technology in the manufacture and repair of Ti-alloy integral discs of aero-engines [23,24]. The LFW process is a thermo-mechanical coupling process that occurs under the action of axial and friction forces, heat generation, phase transformation, and deformation at the welding interface. The phase composition and microstructural characteristics of different materials lead to significant differences in the deformation mechanism and phase transition characteristics in the interface and their vicinity under the thermo-mechanical coupling during LFW. In the LFW process, the main structures of the lamellar-equiaxed dual-state $\alpha + \beta$ type TC4 Ti alloy in the TMAZ are composed of deformed α phase and acicular martensite, while the weld nugget is composed of martensite [25,26]. In the basket weave structure of TC17 LFW joint, the microstructure in the TMAZ is composed of smaller deformed β grains that formed under the effects of force and temperature duing LFW, and in which residual acicular α phase and metastable β phase are scattered, while the WZ is composed of metastable β phase with fine grains [27,28]. The microstructure of the Ti-Al series of Ti_2AlNb alloy includes O phase and α phase. In the LFW joint, the microstructure in the TMAZ is metastable β phase, residual O phase, and α phase, where the O phase structure disappears before the α phase. The main phase in the weld nugget zone is metastable β phase with fine grains [29].

Compared to the above materials, the Ti_3Al alloy has a distinguishable microstructure, since it is mainly composed of β, α_2, and O phases. The structural and physical properties of β, α_2, and O phase are significantly different [30]. In the LFW process, the thermo-mechanical action may have great impact on the deformation and phase transition of those three phases. However, the deformation differences of the three phases in the joint and the effect of microstructure variation caused by heat treatment on the performance of the joint have been rarely reported in the literature [31]. In this study, Standard heat treatment of the Ti_3Al alloy is 980 °C (solution treatment) + 800 °C (aging treatment). LFW was performed on Ti_3Al alloy, and then the joints were subjected to different heat treatments. The microstructural characteristics of the Ti_3Al alloy LFW joint and the effect of heat treatment on the microstructure and mechanical properties of the joints were specifically analyzed.

2. Materials and Methods

2.1. Experimental Materials

The material used in experiment was Ti_3Al-based alloy (Ti-23Al-17Nb (at%)). As it can be seen from Figure 1, the original structure of the alloy is a mixture of α_2, O, and β phases, where the matrix phase is α_2 with an approximate size of 15 μm. Numerous acicular O with a length of 10 μm and alternate β phases are distributed across the equiaxed α$_2$ phase matrix and its chemical composition can be seen in Table 1.

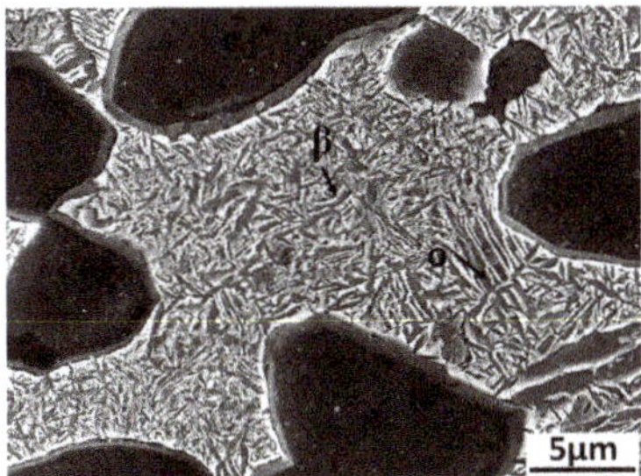

Figure 1. Microstructure of Ti$_3$Al-based alloy.

Table 1. Chemical composition of Ti$_3$Al-based alloy titanium alloy (wt%).

Al	Nb	Ti
12.44	31.71	Bal.

2.2. Experimental Methods

The LFW test was carried out on an LFW-20T equipment, self-developed by AVIC Manufacturing Technology Institute (Beijing, China). The equipment is shown in Figure 2a and the welding principle of the LFW process is shown in Figure 2b. One specimen of a pair to be welded was held by a clamp with a reciprocating motion and the other was held by a clamp having a horizontal motion. During welding, one specimen started to reciprocate with high frequency and small amplitude under the action of the exciting force, while the other gradually moved toward the reciprocating specimen under the action of thrust. Once the specimens come into contact, the convex portions of the interface rub each other. As the frictional pressure increased, the actual contact area increased, and then the frictional force increased rapidly. Accordingly, the interface temperature increased and the friction interface was gradually covered by a layer of high-temperature visco-plastic metal. Under the action of upset pressure, the metal in the welding zone underwent plastic flow and the extruded metal formed a flash. The metal components on the two sides were firmly welded together through mutual diffusion and recrystallization of the metal in the welding zone, and the entire welding process was completed.

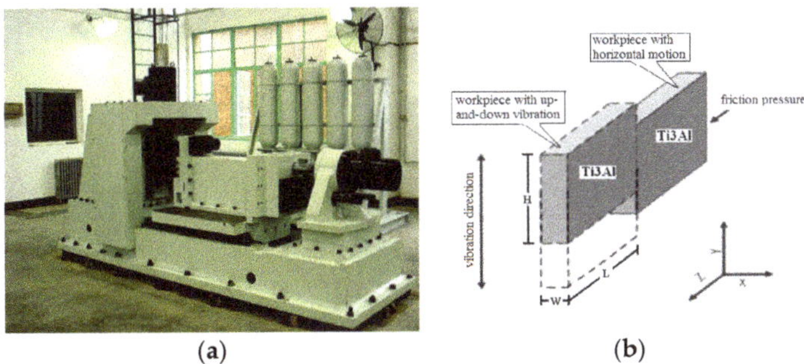

(a)   (b)

Figure 2. (**a**) The LFW-20T equipment; (**b**) Basic mechanism of the LFW process.

The parameters of the Ti$_3$Al LFW process are listed in Table 2.

Table 2. Welding process parameters.

	Amplitude (mm)	Frequency (Hz)	Friction Pressure (T)
Optimized welding process parameters	3	40	4

The cooling rate is very fast after LFW, and the joint is very narrow with severe plastic deformation. So the range from 700 °C–900 °C was used to research the evolution of microstructure and Properties of the Ti$_3$Al joint. After the LFW process, the joints were heated in vacuum at 700 °C, 750 °C, 800 °C, 850 °C, and 900 °C for four hours and were cooled in a furnace. The specimen for microstructure characterization was taken along the direction perpendicular to the welded surface, and then the sample was ground and polished. Aqueous solutions of hydrofluoric acid and nitric acid were adopted successively to etch the sample. Subsequently, the sample was observed using a Zeiss field-emission scanning electron microscope (Jena, Germany) at the test center of the China Aviation Manufacturing Technology Research Institute. The specimen for microhardness measurement was subjected to several processes, including inlaying, polishing, and etching. The microhardness characterization was performed on an Tukon2500 microhardness tester (Wolpert Wilson, IL, USA), where the microhardness of an as-welded joint and a welded joint after heat treatment was measured at a load of 0.2 kg for 15 s. Based on the distance from the center of the joint, at least five points were selected to measure the microhardness of each zone, among the base metal (BM), the TMAZ, and the weld zone (WZ) of the Ti$_3$Al joint. The point-to-point spacing was greater than 0.2 mm. The position for the microstructure and microhardness measurements is shown in Figure 3a. A tensile test piece was prepared according to the HB5143-96 standard, in order to measure the tensile properties of the joints at room-temperature, as shown in Figure 3b. The center line of the joint remained at the center of the tensile test sample and three test pieces were taken from the specimens under different heat treatment conditions.

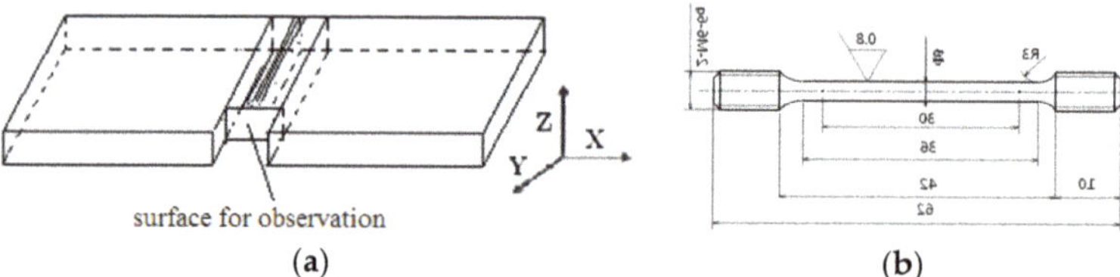

(a)   (b)

Figure 3. (a) Schematic diagram of the intercepting process for the metallographic specimen; (b) Tensile specimen dimensions.

3. Results and Discussions

3.1. Microstructural Characteristics of the As-Welded Joint

The morphology of Ti$_3$Al after LFW can be observed in Figure 4. At the welding interface, under the action of pressure and friction forces, the interfacial temperature increased, then the Ti$_3$Al at the interface reached its molten state, and thus Ti$_3$Al was extruded and formed a flash. After being cooled in air, the flash was grayish white with the same direction. During the LFW process, different materials have flashes with different morphologies, which is highly related to the material physical properties. More specifically, it is easier to form flashes that are extruded as whole bodies from materials with high plasticity and fluidity. As for materials with poor plasticity, during extrusion, the flash separates and curls toward both sides. The morphology of the flash formed by LFW on the Ti$_3$Al alloy is similar to that of ordinary Ti-based alloys, such as TC4, TC11, and TC17, and exhibits integral extrusion (Figure 4). In the linear friction welding process of Ti$_3$Al, the welding temperature can reach over 1200 °C, as illustrated in Figure 5.

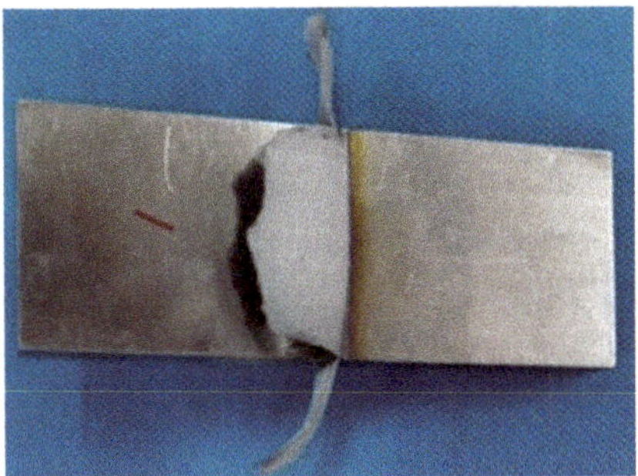

Figure 4. Morphology of the Ti₃Al joint after LFW.

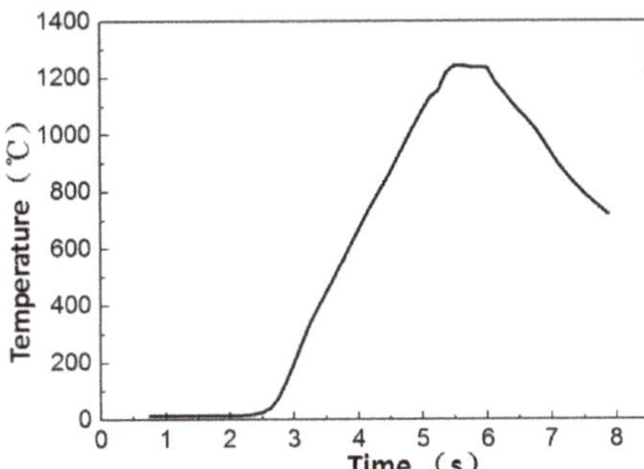

Figure 5. Temperature curve measured during Ti₃Al LFW.

In Figure 6, it can be seen that there were no defects such as voids, cracks, and unwelded regions at the welded interface of the Ti₃Al LFW joint. The Ti₃Al LFW joint can be divided into three zones: the BM zone, the TMAZ, and the WZ. The width of the WZ was about 80 μm and the entire width of the joint was 700 μm. During the LFW process of Ti₃Al, heat was generated at the contact surface under the action of pressure and vibration friction, and then was conducted towards the base metal, which formed a high-to-low temperature gradient from the contact surface to the base material. Meanwhile, under the action of friction pressure, the metal at the interface flowed and extruded from the interface, resulting in a greater deformation of the microstructure. Therefore, the Ti₃Al LFW joint was characterized by a microstructure with gradient variation formed at different temperatures and deformation degrees. The microstructure in the different zones of the joint was quite different. In the BM zone, the morphology, distribution, and quantity of the α_2, O, and β phases demonstrated had not obvious change, as shown in Figure 6a. In the TMAZ near the side of the BM zone, the lamellar O and β phases deformed and after cooling formed a metastable structure, while the morphology of the α_2 phase was not significantly deformed, as shown in Figure 6e. At regions closer to the joint center, the α_2 phase began to deform and the metastable portion increased as the metal was elongated along the direction of the flow, as shown in Figure 6d. In the TMAZ, close to the side of WZ the α_2 phase increased severely and the length-width ratio increased, almost parallel to the joint center, as shown in Figure 6c. At regions closer to the central part of joint's interface, the α_2 phase further reduced, as shown in Figure 6b. In the WZ, the α_2, O, and β phases all transformed to the metastable β phase, except a small amount of deformed a₂ phase, as shown in Figure 6a.

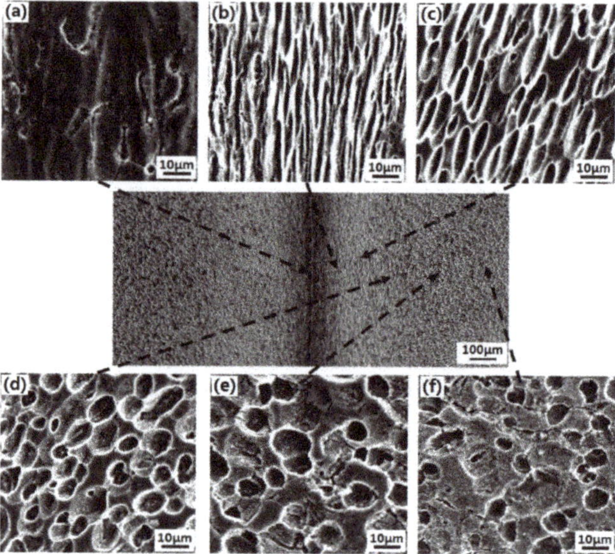

Figure 6. Microstructures in different zones of the as-welded Ti$_3$Al LFW joint: (**a**) in WZ, (**b**) in TMAZ close to WZ, (**c**) in TMAZ, (**d**) in TMAZ close to BM, (**e**) in BM close to TMAZ, and (**f**) in the BM.

3.2. Effects of Heat Treatment on Joint Microstructure

The Ti$_3$Al LFW joints were heat treated at 700 °C, 750 °C, 800 °C, 850 °C, and 900 °C, respectively, and the corresponding microstructures after heat treatment are shown in Figure 7. After being heat treated at 700 °C, the central zone of the joint was composed of small equiaxed grains with a size of around 10 μm. There was a wider grain boundary and dot-like structures less than 1 μm precipitated inside the grains, as shown in Figure 7a. Mount of the dot-like structures (white color point) appeared in the grain and grain boundary reduced in the width, as shown in Figure 7b,c. Under higher temperature, the dot-like structures inside the equiaxed grains began to grow and transform to short acicular structures at 850 °C, as shown in Figure 7a. When the temperature reached 900 °C, the size of the short acicular structures inside the equiaxed grains was close to 3 μm, and exhibited a trend of becoming lamellar, as shown in Figure 7e.

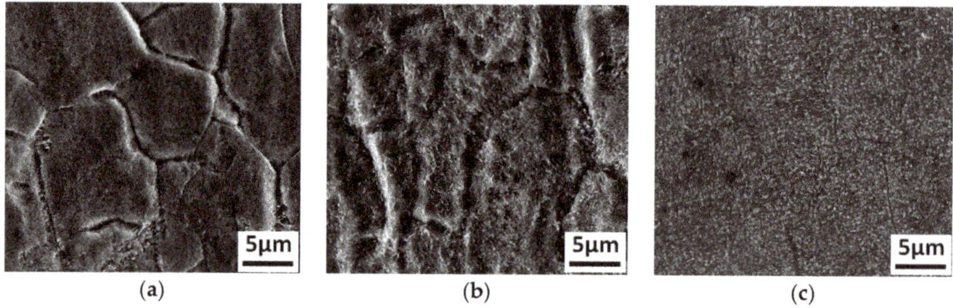

(**a**) (**b**) (**c**)

Figure 7. *Cont.*

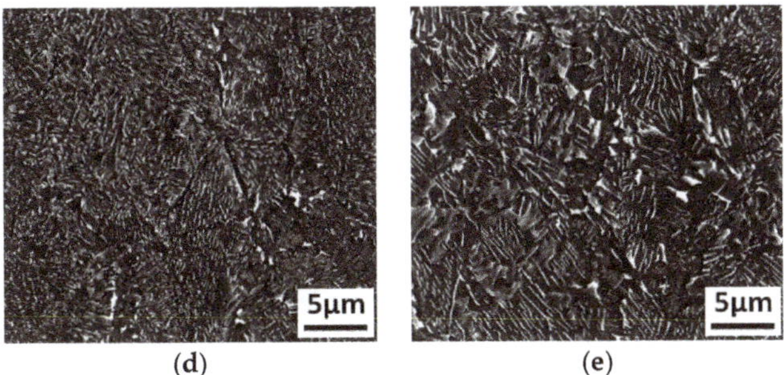

Figure 7. The Ti$_3$Al LFW joint microstructures in the WZ after heat treatments at different temperatures: (a) after heat treatment at 700 °C, (b) after heat treatment at 750 °C, (c) after heat treatment at 800 °C, (d) after heat treatment at 850 °C, and (e) after heat treatment at 900 °C.

The TMAZ of the as-welded joint was mainly composed of deformed α_2 phase and metastable structures. After heat treatment at 700 °C, the morphology of the residual deformed α_2 phase did not change significantly compared to that in the as-welded joint. The dot-like structures precipitated between α_2 phases, while short acicular structures appeared locally, as shown in Figure 8a. As the temperature increased, the size of the dot-like structures increased, while the size of the α_2 phase decreased, which was consistent with the variation in the dot-like structures inside the equiaxed grains at the center of the joint, as shown in Figure 8b. More specifically, at 800 °C, the morphology of the dot-like structure began to transform into short acicular structure, as shown in Figure 8c. It began to exhibit lamellar structure and the size of the α_2 phase decreased further at 850 °C, as shown in Figure 8d. The average length of lamellar structures is up to 5 μm, as shown in Figure 8e.

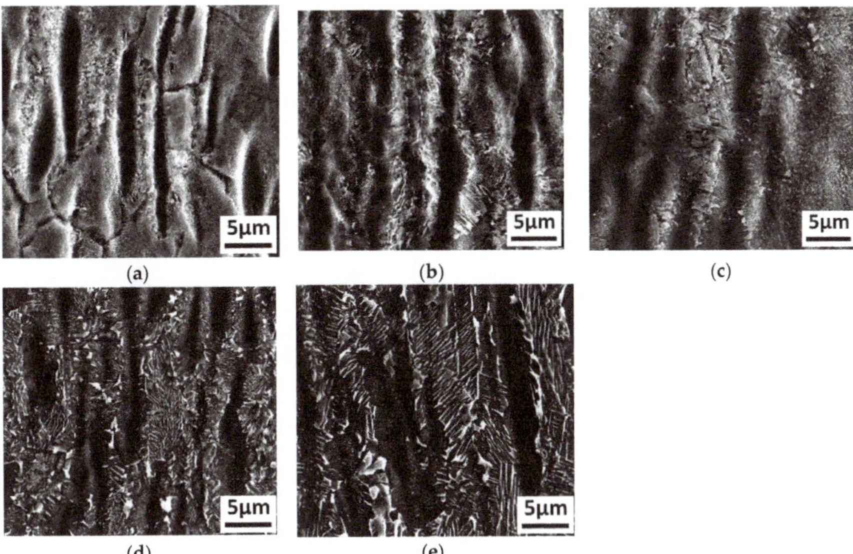

Figure 8. The Ti$_3$Al LFW joint microstructures in the TMAZ after heat treatments at different temperatures: (a) after heat treatment at 700 °C, (b) after heat treatment at 750 °C, (c) after heat treatment at 800 °C, (d) after heat treatment at 850 °C, and (e) after heat treatment at 900 °C.

The microstructure variation in the TMAZ close to the side of the base metal was different than that in the other two zones. The morphology of the α_2 phase did not change significantly with temperature, while the heat treatment led to larger impact on the microstructures between the α_2 phases. At 700 °C, the microstructures between the α_2 phases exhibited dot distribution, as shown in Figure 9a. After heat treatment at 750 °C, the dot-like structures transformed to acicular structures, and then grew rapidly with the increase in temperature, as shown in Figure 9b. At 800 °C, the length of the acicular structures α_2 reached 3 μm but still exhibited a narrow shape, as shown in Figure 9c. At 850 °C, the acicular structures α_2 is larger than that in the 800 °C sample, as shown in Figure 9d. After heat treatment at 900 °C, the length reached approximately 5 μm, as shown in Figure 9e

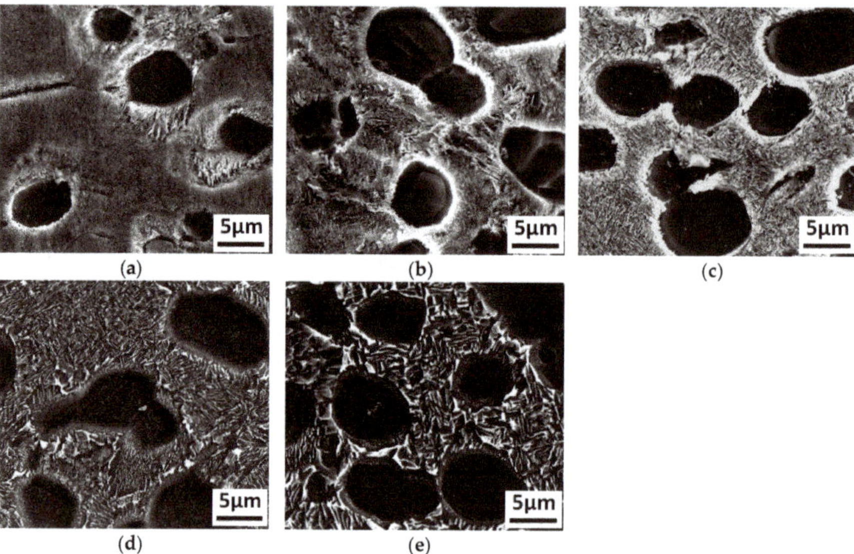

Figure 9. The Ti$_3$Al LFW joint microstructures in the BM zone after heat treatments at different temperatures: (**a**) after heat treatment at 700 °C, (**b**) after heat treatment at 750 °C, (**c**) after heat treatment at 800 °C, (**d**) after heat treatment at 850 °C, and (**e**) after heat treatment at 900 °C.

In the LFW process, under the combined action of axial and friction forces, the contact surface of the Ti$_3$Al alloy undergoes friction, deformation, heat generation, and extrusion of the interface metal. Therefore, the microstructure of the Ti$_3$Al LFW joint is affected by welding process parameters, such as friction pressure, welding amplitude, welding frequency, and cooling rate. The welding temperature of Ti-based alloys during LFW can reach temperatures above 1200 °C. Since the cooling rate after welding is relatively fast, martensite structures can form in TC4 titanium alloys, while metastable β phase can form in Ti17 titanium alloys and near-b titanium alloy [19,32]. From the phase diagram of Ti$_3$Al-based alloys [33], it can be seen that the Ti$_3$Al material has three phase regions: the β + O phase region, the α_2 + β + O phase region, and the β phase region. The microstructures of the as-welded Ti$_3$Al alloy LFW joint indicated that the joint forming between the base material and the interface contains different microstructures, including an α_2 + O + β three-phase region, an α_2 + metastable β dual-phase region, and a metastable β phase region. In the welding process and under the action of friction pressure, deformation occurs in the aforementioned four phase regions ranging from the base metal to the weld interface.

The degree of deformation is related to the amplitude and extent of the softening and flow of the Ti$_3$Al alloy. The closer to the weld interface, the more severe the deformation. In the welding process, two main transitions occur: O phase → β phase and α_2 phase → β phase. In the weld zone, the temperature, which is over 1200 °C, leads to O phase → β and α_2 phase → β phase transitions, but the LFW process is short, so the α_2 phase → β phase transition is incomplete. Thus, there is a small amount of deformed α_2 phase and the extrusion of more softened alloy, which makes the weld spacing narrower after the upsetting force formation process of LFW. The zone is mainly composed of β phase and a small amount of deformed α_2 phase. Due to the higher cooling rate of LFW, the disordered β phase transits into an ordered β phase, while the α_2 phase remains. As the distance from the weld zone increases, the maximum welding temperature decreases, so the corresponding temperature range is located in the $\alpha_2 + \beta$ dual-phase region. Within this temperature region, the O phase transforms to β phase, while the partial α_2 phase transforms to β phase. During the cooling process, the residual deformed α_2 phase can remain until room temperature, therefore the α_2 portion in the WZ is higher than that in the TMAZ close to near the WZ. For the TMAZ close to the BM, the welding temperature is relatively low so the high temperature region is located at the $\alpha_2 + \beta + O$ phase region and the $\beta + O$ phase region. Within this temperature region, the α_2 phase is relatively stable and basically does not transform into β phase. Due to the fact that the O phase has poor stability, the transition of O phase $\rightarrow \beta$ phase occurs easily during the welding process. The higher the temperature, the more complete the transition and the lesser the O phase. Therefore, in the TMAZ near the base metal of the as-welded joint under thermo-mechanical action, the O, β, and α_2 phases in the deformed microstructure do not transform to β phase. The difference in the relationship between temperature and stress/strain in the different regions of the material during the LFW process has a great and large-area impact on the microstructure (morphology, size, and volume fraction of the β, α_2, and O phases) at the different joint regions. After heat treatment, the metastable β phase decomposes into α_2, O, and β phases. The significant differences in the microstructure of each zone of the as-welded joint are responsible for the large differences in the microstructure of each zone of the joint after heat treatment. The more obvious impact is that of the different heat treatment temperatures on the size and morphology of the α_2 phase formed by the decomposition of the metastable β phase.

3.3. Effects of Heat Treatment on Joint Microhardness

The microhardness tests were carried out on as-welded Ti$_3$Al LFW joints and joints after different heat treatments. The variation of microhardness with temperature is shown in Figure 10. In the as-weld condition, the microhardness of the Ti$_3$Al weld zone and the TMAZ was similar. It reached about 375 Hv, much higher than the microhardness of the base metal (300 Hv). After heat treatment at 700 °C, the microhardness in the weld zone increased rapidly to 450 Hv, while that in the TMAZ also increased to 425 Hv, and in the base metal did not change significantly. As the heat treatment temperature increased, the joint microhardness in the WZ and the TMAZ gradually decreased. After heat treatment at 800 °C, the joint microhardness in the TMAZ was close to that of the as-welded joint. When the heat treatment temperature reached 850 °C, the microhardness in the WZ and the TMAZ was much lower than that of the corresponding zones of the as-welded joint, which was reduced to approximately 340 Hv. When the heat treatment temperature was raised to 900 °C, the joint microhardness in the WZ and the TMAZ was close to that of the base metal, which was about 300 Hv. The LFW process is a non-uniform deformation process in which the temperature varies from high to low and the stress/strain also vary from high to low when the location changes from the base metal region to the weld zone. In the secondary process, accompanied with processing-hardening, phase transition, and dynamic recrystallization, there are variations in the quantity, size, and morphology of the α_2, O, and β phases. For instance, the predominant action on the zone close to the base metal is processing-hardening, while the phase transition is predominant in the weld zone. As a result, the microhardness of the TMAZ and the WZ of the Ti$_3$Al joint increases. The heat treatment on the Ti$_3$Al LFW joint is equivalent to the aging treatment on the Ti$_3$Al after rapid cooling and with the deformation in the $O + \beta$, $\alpha_2 + \beta + O$, $\alpha_2 + \beta$, and β phase regions. In the treatment, the α_2 phase precipitates from the metastable β matrix, which produces a strengthening effect and increases the microhardness. However, when the heat treatment temperature is too high, the α phase grows, decreasing the microhardness and reducing the strengthening effect.

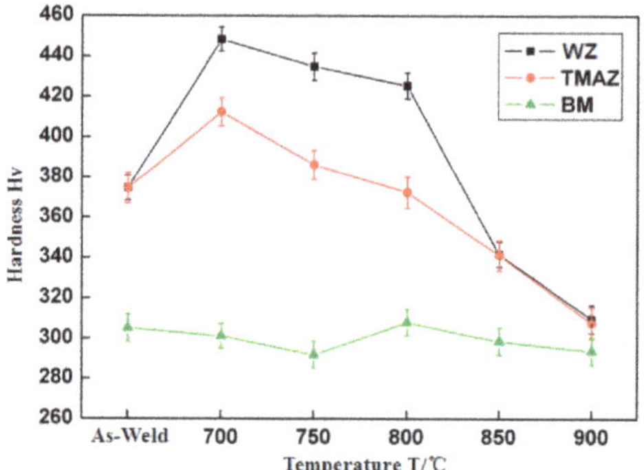

Figure 10. Microhardness of as-welded and heat-treated joints.

3.4. Effects of Heat Treatment on Joint Tensile Properties

After heat treatment at 700 °C, 750 °C, 800 °C, 850 °C, and 900 °C, the tensile strength (σ_b) of the Ti$_3$Al LFW joints at room temperature was 960 ± 6 MPa, 939 ± 7 MPa, 923 ± 10 MPa, 918 ± 9 MPa, and 901 ± 9 MPa, respectively, and tensile strength (σ_b) of joints decreased with the temperature of heat treatment from 700 °C to 900 °C, as show in Figure 11. All fractures occurred in the BM zone of Ti$_3$Al and there was a certain bottleneck shrinking of about 10%. As it can be seen in Figure 12, the macro-fracture was a typical cup-shaped fracture and was composed of a fibrous zone and a shear lip zone. The shear lip zone was more apparent, while the fibrous area exhibited a shallow dimple shape. After being subjected to heat treatment, the microhardness of the WZ and the TMAZ of the Ti$_3$Al LFW joint was higher than that of the base metal, which was one of the reasons for the tensile fracture at the base metal region at room temperature.

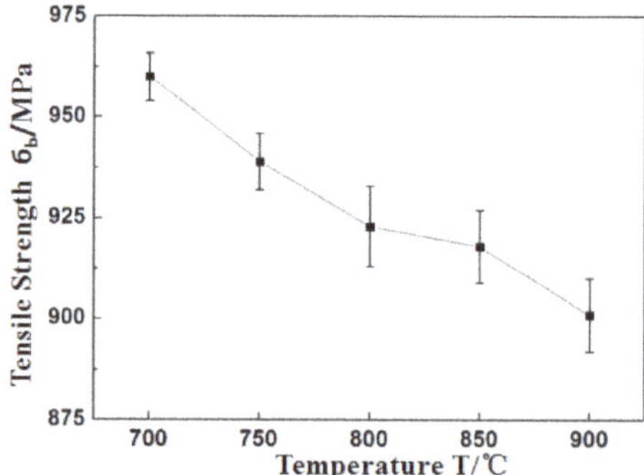

Figure 11. Tensile strength results with heat treatment from 700 °C to 900 °C.

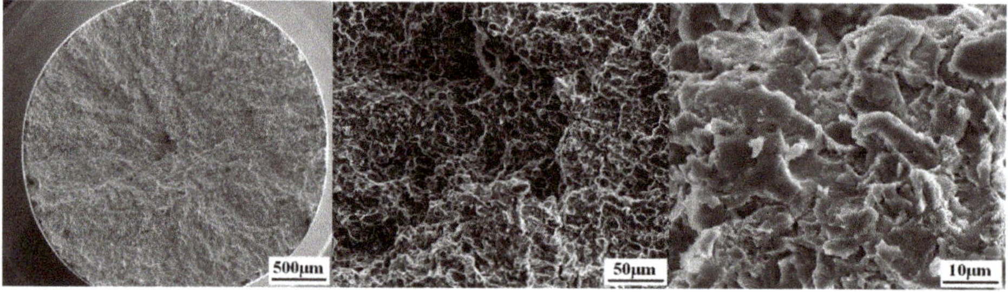

Figure 12. Tensile fracture morphology of the Ti$_3$Al LFW joint at room temperature.

4. Conclusions

(1) The microstructure in the central WZ of Ti$_3$Al LFW joint is metastable β phase. The microstructure in the TMAZ is deformed α$_2$ phase and metastable β phase. As the position becomes closer to the TMAZ near the side of the WZ, the α$_2$ phase gradually decreases, the degree of deformation increases, and the metastable β phase increases.

(2) After being subjected to heat treatment at different temperatures, the microstructure in the WZ of the joint is quite different. At 700 °C, dot-like structures precipitate and their size is less than 1 μm. However, when the heat treatment temperature is above 850 °C, the formed structure is acicular α$_2$ phase with a size of approximately 1 μm, which grows as the heat treatment temperature increases.

(3) The microhardness of the Ti$_3$Al LFW joint after welding is higher than that of the base metal. Heat treatment at 700 °C can significantly enhance the joint microhardness, which then decreases with the increasing heat treatment temperature. After heat treatment at 850 °C, the joint microhardness is lower than that of the as-welded joint. The microhardness of the WZ and the TMAZ is higher than that of the BM zone in the selected heat treatment range.

Author Contributions: Methodology, X.L.; validation, X.L., J.L. and J.H.; investigation, X.L. and J.L.; data curation, X.L. and Y.Z.; writing—original draft preparation, X.L.; writing—review and editing, Y.Z., T.Z. and J.H.; supervision, Y.Z.; project administration, J.T.

Funding: This research was funded by National Science and Technology Major Project, NO.2018ZX04010001 and The APC was funded by National Science and Technology Major Project.

References

1. Zhang, J.; Li, S.; Lang, X.; Cheng, Y. Research and applicatopn of Ti$_3$Al and Ti$_2$AlNb based alloys. *Chin. J. Nonferrous Met.* **2010**, *20*, 336–341.

2. Froes, F.H.; Suryanarayana, C.; Eliezer, D. Review: Synthesis, properties and applications of titanium aluminides. *J. Mater. Sci.* **1992**, *27*, 5113–5140. [CrossRef]

3. Djanarthany, S.; Viala, J.C.; Bouix, J. An overview of monolithic titanium aluminide based on Ti$_3$Al and TiAl. *Mater. Chem. Phys.* **2001**, *3*, 301–319. [CrossRef]

4. Peng, C.; Huang, B.; He, Y. Development of Intermetallic Compound in Ni-Al, Fe-Al and Ti$_3$Al Base System. *Spec. Cast. Nonferrous Alloy.* **2001**, *6*, 27–29.

5. Wu, H.-Q.; Feng, J.-C.; He, J.-S.; Zhou, L. Structure and Crack Forming Susceptibility of TiAl Based Alloy Joints by Electron Beam Welding. *J. Mater. Eng.* **2005**, *4*, 7–10.

6. Wang, G.-Q.; Zhao, Y.; Wu, A.-P.; Zou, G.-S.; Ren, J.-L. Microstructure and high temperature tensile properties of Ti$_3$Al alloys laser welding joint. *Chin. J. Nonferrous Met.* **2007**, *11*, 180301807.

7. David, S.A.; Horton, J.A.; Goodwin, G.M.; Phillips, D.H.; Reed, R.W. Weldability and microstructure of a titanium aluminide. *Weld. J.* **1990**, *69*, 133–140.

8. Baeslack, W.A.; Broderick, T. Effect of cooling rate on the structure and hardness of a Ti-26at% Al-10at% Nb-3at% V-1at% Mo titanium aluminide. *Scr. Metall.* **1990**, *24*, 319–324. [CrossRef]

9. Baeslack, W.A.; Mascorella, T.J.; Kelly, T.J. Weldability of a titanium aluminide. *Weld. J.* **1989**, *68*, 483s–498s.

10. Bhamji, I.; Preuss, M.; Threadgill, P.L.; Addison, A.C. Solid state joining of metals by linear friction welding: A literature review. *Mater. Sci. Technol.* **2011**, *27*, 2–12. [CrossRef]

11. Ma, T.J.; Li, W.-Y.; Xu, Q.Z.; Zhang, Y.; Li, J.L.; Yang, S.Q.; Liao, H.L. Microstructure evolution and mechanical properties of linear friction welded 45 steel joint. *Adv. Eng. Mater.* **2007**, *9*, 703–707. [CrossRef]

12. Grujicic, M.; Yavari, R.; Snipes, J.S.; Ramaswami, S.; Yen, C.-F.; Cheeseman, B.A. Linear friction welding process model for carpenter custom 465 precipitation-hardened martensitic stainless steel. *J. Mater. Eng. Perform.* **2014**, *23*, 2182–2198. [CrossRef]

13. Bhandari, V. Linear Friction Welding of Titanium to Stainless Steel. Master's Thesis, Cranfield University, Cranfield, UK, 2010.

14. Rotundo, F.; Marconi, A.; Morri, A.; Ceschini, A. Dissimilar linear friction welding between a SiC particle reinforced aluminum composite and a monolithicaluminum alloy: Microstructural, tensile and fatigue properties. *Mater. Sci. Eng. A* **2013**, *559*, 852–860. [CrossRef]

15. Song, X.; Xie, M.; Hofmann, F.; Jun, T.S.; Connolley, T.; Reinhard, C.; Atwood, R.C.; Connor, L.; Drakopoulos, M.; Harding, S.; et al. Residual stresses in linear friction welding of aluminium alloys. *Mater. Des.* **2013**, *50*, 360–369. [CrossRef]

16. Wanjara, P.; Jahazi, M. Linear friction welding of Ti-6Al-4V: Processing, microstructure, and mechanical-property inter-relationships. *Met. Mater. Trans. A* **2005**, *36*, 2149–2164. [CrossRef]

17. Vairis, A.; Frost, M. High frequency linear friction welding of a titanium alloy. *Wear* **1998**, *217*, 117–131. [CrossRef]

18. Zhang, T.; Li, J.; Ji, Y.; Sun, C. Structure and mechanical properties of TC4 linear friction welding joint. *Trans. China Weld. Inst.* **2010**, *21*, 53–56.

19. Dalgaard, E.; Wanjara, P.; Gholipour, J.; Cao, X.; Jonas, J.J. Linear friction welding of a near-b titanium alloy. *Acta Mater.* **2012**, *60*, 770–780. [CrossRef]

20. Chamanfar, A.; Jahazi, M.; Gholipour, J.; Wanjara, P.; Yue, S. Maximizing the integrity of linear friction welded Waspaloy. *Mater. Sci. Eng. A* **2012**, *555*, 117–130. [CrossRef]

21. Chamanfar, A.; Jahazi, M.; Cormier, J. A review on inertia and linear friction welding of Ni-based superalloys. *Met. Mater. Trans. A* **2015**, *46*, 1639–1669. [CrossRef]

22. Mary, C.; Jahazi, M. Multi-scale analysis of IN-718 microstructure evolution during linear friction welding. *Adv. Eng. Mater.* **2008**, *10*, 573–578. [CrossRef]

23. García, A.M.M. BLISK fabrication by linear friction welding. In *Advances in Gas Turbine Technology*; Benini, E., Ed.; InTech: Winchester, UK, 2011; pp. 411–434.

24. Kallee, S.W.; Nicholas, E.D.; Russell, M.J. Friction welding of aero engine components. In Proceedings of the 10th World Conference Titanium, (TI-2003), Hamburg, Germany, 13–18 July 2003; Lutjering, G., Albrecht, J., Eds.; Wiley-VCH: Hamburg, Germany, 2003; pp. 2867–2874.

25. Karadge, M.; Preuss, M.; Lovell, C.; Withers, P.J.; Bray, S. Texture development in Ti-6Al-4V linear friction welds. *Mater. Sci. Eng. A* **2007**, *459*, 182–191. [CrossRef]

26. Vairis, A.; Frost, M. On the extrusion stage of linear friction welding of Ti6Al4V. *Mater. Sci. Eng. A* **1999**, *271*, 477–484. [CrossRef]

27. Ma, T.J.; Li, W.Y.; Zhong, B.; Zhang, Y.; Li, J.L. Effect of post-weld heat treatment on microstructure and property of linear friction welded Ti17 titanium alloyjoint. *Sci. Technol. Weld. Join.* **2012**, *17*, 180–185. [CrossRef]

28. Ji, Y.; Liu, Y.; Zhang, T.; Zhang, C. Structure and mechanical properties of TC4/TC17 linear friction welding joint. *Trans. China Weld. Inst.* **2012**, *33*, 109–112.

29. Chen, X.; Xie, F.Q.; Ma, T.J.; Li, W.Y.; Wu, X.Q. Microstructure evolution and mechanicalproperties of linear friction welded Ti2AlNb alloy. *J. Alloys Compd.* **2015**, *646*, 490–496. [CrossRef]

30. Kestner-Weykamp, H.T.; Ward, C.H.; Broderick, T.; Kaufman, M.J. Microstructures andphase relationships in the Ti$_3$Al + Nb system. *Scr. Met.* **1989**, *23*, 1697–1702. [CrossRef]

31. Threadgill, P.L. The prospects for joining titanium aluminides. *Mater. Sci. Eng. A* **1995**, *192–193*, 640–646. [CrossRef]

32. Ballat-Durand, D.; Bouvier, S.; Risbet, M.; Pantleon, W. Multi-scale and multi-technic microstructure analysis of a linear friction weld of the metastable-beta titanium alloy Ti-5Al-2Sn-2Zr-4Mo-4Cr (Ti17) towards a new Post-Weld Heat Treatment. *Mater. Charact.* **2018**, *144*, 661–670. [CrossRef]

33. Raghavan, V. Al-Nb-Ti (Aluminum-Niobium-Titanium). *J. Phase Equilibria Diffus.* **2005**, *26*, 360–368. [CrossRef]

 *materials*

Article

Simulation of Temperature Distribution and Microstructure Evolution in the Molten Pool of GTAW Ti-6Al-4V Alloy

Min Zhang *, Yulan Zhou, Chao Huang, Qiaoling Chu, Wenhui Zhang and Jihong Li

School of Materials and Engineering, Xi'an University of Technology, Xi'an 710048, China;
zhouyulan@163.com (Y.Z.); 18829028936@163.com (C.H.); 13352413558@163.com (W.Z.);
chuqiaoling@xaut.edu.cn (Q.C.); lijihong@xaut.edu.cn (J.L.)
* Correspondence: zhmmn@xaut.edu.cn; Tel.: +86-029-8231-2205

Received: 20 October 2018; Accepted: 11 November 2018; Published: 15 November 2018

Abstract: In this paper, a three-dimensional (3D) finite element model was established by ABAQUS software to simulate the welding temperature field of a Ti-6Al-4V alloy under different welding currents based on a Gaussian heat source model. The model uses thermo-mechanical coupling analysis and takes into account the effects of convection and radiation on all weld surfaces. The microstructure evolution of the molten pool was calculated using the macro-micro coupling cellular automaton-finite different (CA-FD) method. It was found that with the increase of the welding current, the temperature in the central region of the moving heat source was improved and the weld bead became wider. Then, the dendritic morphology and solute concentration of the columnar to equiaxed transition (CET) in the weld molten pool was investigated. It is shown that fine equiaxed crystals formed around the columnar crystals tips during solidification. The coarse columnar crystals are produced with priority in the molten pool and their growth direction is in line with the direction of the negative temperature gradient. The effectiveness of the model was verified by gas tungsten arc welding experiments.

Keywords: temperature field; dendritic morphology; finite element; solute concentration; cellular automaton

---

1. Introduction

Nowadays, titanium alloys, especially Ti-6Al-4V (TC4), have acquired extensive applications in aerospace, aircraft, automotive, biomedical, and chemical industries [1–3]. This is primarily due to their superior performance characteristics, such as low density, good corrosion resistance, high heat resistance, and high specific stiffness and specific strength. In many fields of manufacturing and processing industry, gas tungsten arc welding (GTAW) [4,5] plays a critical role, which is viewed as one of the traditional and significant material processing techniques. The mechanical and physical properties of titanium alloys are greatly affected by the microstructure evolution during molten solidification [6]. In particular, Kou [7] proposed that the dramatic transformation of temperature and solute concentration directly gave rise to an unfavorable influence on the dendritic growth. Salimi [8] developed a 3D transient analytical solution to the heat conduction problem in different plates with a circular moving heat source. The analytical results were validated by the finite element (FE) method and experiments. During metal solidification, to simulate temperature field change and microstructure transition for a laser-based additive manufacturing processing, a successful cellular automaton-finite element (CA-FE) model was proposed by Zhang [9]. Specially, it should be noted that the direct observation and measurement of the dynamic solidification process of the welding molten pool is very difficult to achieve in experiments.

With the continuous advancement of computational technology, several approaches have been developed for modeling microstructure evolution in the weld molten pool. These approaches mainly include Monte Carlo (MC), phase field (PF), and cellular automata (CA). Anderson [10] uncovered the features of crystal growth and studied the relationship between the dendritic growth and undercooling by the MC method. Then, the temperature of the weld process was integrated with the simulation of grain growth in a computationally efficient manner [11]. However, due to lacking a physical basis, this MC method failed to perform quantitative analysis, which also greatly limited its application. Another employed a phase field (PF) method to quantitatively simulate the dendritic growth process. Qin et al. [12] simulated the solidification of a multicomponent and developed the multiphase systems based on the PF model. To simulate microstructure morphology and solute distributions of the Al-4 wt% Cu alloy in a welding molten pool under transient conditions, Wang et al. [13] developed a quantitative phase field model. However, this model needs a huge amount of computational time, resulting from requiring an extremely fine mesh. To this end, Rappaz et al. [14] proposed a CA method which characterizes the discrete temporal and spatial microstructure evolution using a network of regular cells. In the gas tungsten arc welding (GTAW) process, Zhan et al. [15,16] simulated the dendrite morphology based on the CA method. In addition, the CA method has two advantages: low computational cost and ease of coupling with the macroscopic thermal model considering the transfer of heat and mass in complex geometries. During the solidification of the TC4 alloy, Wang et al. [17] investigated the grain microstructure by means of a method where CA was coupled with FE in the molten pool. Chen et al. [18] proposed a 3D CA-FE model to permit the simulation of grain structure solidification during multiple passes of gas tungsten arc welding (GTAW) and gas metal arc welding (GMAW). Here, the CA-FE method was better used for simulating the temperature distribution and microstructure formulation in the molten pool. To date, there has been little investigation on predicting the temperature distribution and dendritic morphology of the columnar to equiaxed transition (CET) and obtaining experimental validation during Ti-6Al-4V alloy gas tungsten arc welding.

In this study, the temperature field and dendritic morphology were simulated by a CA-FD method. The FE software ABAQUS (version 6.13, Dassault SIMULIA, Providence, RI, USA) was used to compute the thermal field evolution under different welding current conditions. Moreover, the CA model was built to simulate the microstructure evolution of the CET process. The simulated results were verified by corresponding experiments.

2. Thermal Modeling of Welding

A visual thermal modeling of the welding process was performed to predict the thermal fields in the solidification area, by using the FE software ABAQUS and its user subroutines. The subroutines were implemented by the FORTRAN programmable language and linked to the ABAQUS software. The entire procedure of the coupling analysis is shown in Figure 1.

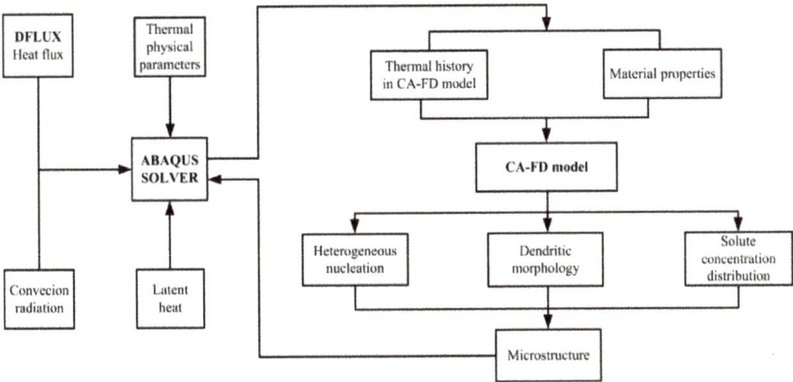

Figure 1. The flow chart of the coupled CA-FD model.

A TC4 alloy plate was used; its chemical compositions are listed in Table 1 and the material properties [19] used for thermal numerical simulation are shown in Figure 2.

Table 1. Chemical composition of the Ti-6Al-4V (TC4) titanium alloy plate.

Element	Ti	Al	V	Fe	O
wt%	Balance	5.5–6.76	3.5–4.5	0.25	0.2

Figure 2. The material properties used to simulate temperature distribution: (**a**) thermal conductivity, (**b**) specific heat capacity, (**c**) elasticity modulus and Poisson ratio, and (**d**) thermal expansion coefficient.

2.1. Meshing and Analysis Settings

The TC4 plate (120 mm × 80 mm × 4 mm) was melted in the centerline by using a gas tungsten arc welding (GTAW) process without filler metal. The dimensions of the simulated welded thin plate were exactly identical to those used in the welding experiments. A total of 14,400 C3D8T elements were used in the simulation. In C3D8T, C represents a solid element, 3D indicates three-dimensional, and 8T denotes eight nodes. The FE model used is shown in Figure 3.

Figure 3. The FE model used in the numerical analysis.

It can be easily seen that the mesh in the central area of the weld bead is relatively fine and the other areas are sparse to obtain higher computation accuracy. In addition, the analysis type uses thermo-mechanical coupling analysis with two analysis steps: a heating analysis step and a cooling analysis step.

2.2. Heat Transfer Equation

The welded sample is viewed as a solid subjected to conduction heat transfer, with boundary conditions to model heat transfer between the sample and the surrounding environment. Then, the three-dimensional transient thermal equation is given as follows [20]:

$$\rho C_\mathrm{P} \frac{\partial T}{\partial t} = \frac{\partial}{\partial x}\left(k\frac{\partial T}{\partial x}\right) + \frac{\partial}{\partial y}\left(k\frac{\partial T}{\partial y}\right) + \frac{\partial}{\partial z}\left(k\frac{\partial T}{\partial z}\right) + Q_\mathrm{v} \tag{1}$$

where T is temperature, ρ is material density, C_P is specific heat, k is thermal conductivity, and Q_v represents volumetric heat flux in $\mathrm{W/m^3}$. To solve the heat equation, the thermal conductivity, density, and specific heat are required.

2.3. Heat Source

The heat source was exerted by a heat flow density imposed on the sample surface interacting with the arc. A Gaussian heat source model was chosen (shown in Figure 4), centered on the arc axis, moving in translation along the x-axis at the welding speed v_s. It should be noted that o represents the coordinate center of the xy plane.

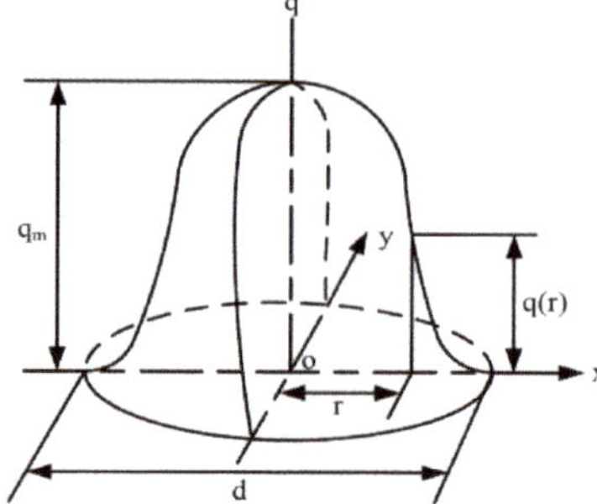

Figure 4. The Gaussian heat source model.

The heat flow density distribution in the surface is then given by [20]:

$$q(r) = \frac{2q_m}{\pi d} \exp \frac{-3r^2}{d^2} \tag{2}$$

where q_m is Gaussian amplitude, d is width, and r represents the distance to the center of the heat source, defined by:

$$r = (X - X_0 - v_s t)^2 + (Y - Y_0)^2 \tag{3}$$

Here, X_0 and Y_0 are the coordinates of the initial position of the heat source and t denotes the time.

The total power transmitted to the sample by this distribution, obtained by integrating Equation (2) for r from 0 to infinity, must match the effective welding power. The relationship between the distribution parameters and process parameters can be given as

$$q_m = \eta \cdot UI \tag{4}$$

where U is the welding voltage, I is the welding current, and η indicates the welding efficiency.

2.4. Boundary and Initial Conditions

At the beginning, the initial and ambient temperatures of the FE model for all simulations were set to 25 °C. The thermal boundary conditions mainly include convection in air, radiation from the surface of the workpiece toward air in light of the Stefan-Boltzmann relationship, and conduction from the workpiece toward the metal support. The heat loss from surface convection and radiation can be given as [20]:

$$q_{conv} = h(T_c - T_0) \tag{5}$$

$$q_{rad} = \varepsilon\sigma \left[(T_c - T_{abs})^4 - (T_0 - T_{abs})^4 \right] \tag{6}$$

Here, T_c is the current temperature, T_0 indicates the ambient temperature, and T_{abs} is the absolute zero temperature. In addition, ε represents the emissivity and σ is the Stefan-Boltzmann constant, which has the value of $5.68 \times 10^{-8} \, J/K^4 m^2 s$. In this paper, the convection coefficient is taken as 8 and the emissivity is 0.85, and the heat transfer coefficient has the value of $h = 10 \, W/m^2 \cdot K$.

2.5. Macro–Micro Coupling of the Temperature Field

The calculation of the welding heat transfer process is the basis of the microstructure simulation. However, the calculation of the heat transfer process of the weld pool is performed on a macro scale, while the calculation of the dendrite growth based on the CA method is carried out on a micro scale. Therefore, it is necessary to establish a macro-micro coupling model for temperature field calculation. The macroscopic temperature field was solved by the finite difference (FD) method using the ABAQUS finite element software. The CA model was built to simulate the microstructure evolution of the columnar to equiaxed transition (CET) process. The temperature of the CA element is affected by the temperature of the macro elements around it. It is related to the distance from the central node of the element to the surrounding macro elements as shown in Figure 5.

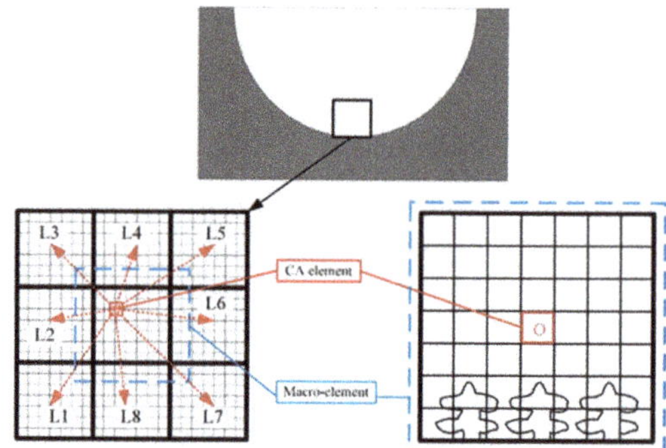

Figure 5. The macroscopic and microscopic coupling analysis.

The temperature value of the CA element can be expressed by the following formula [16]:

$$To = \sum_{i=1}^{N} Li^{-1} Ti / \sum_{i=1}^{N} Li^{-1} \tag{7}$$

where To is the temperature of the micro-element O; Ti is the macro-element temperature around the point O; Li is the distance from the point O to the surrounding macro-element; and N is the number of macro-elements around the micro-element, the value of which is 8.

3. Modeling of the Dendritic Growth

The growth of columnar grains and equiaxed grains was simulated in the modeling. The CET can be calculated when the volume fraction for equiaxed grains reached a given limit at the solidification front. The thermal properties of the material used in the present simulation are shown in Table 2 [18].

Table 2. Material properties' parameters used in the microstructure simulation.

Property	Variable	Value
Liquidus temperature	T_{L}	1703 °C
Solidus temperature	T_{S}	1678 °C
Partition coefficient	k_0	0.95
Diffusion coefficient in liquid	D_{L}	5×10^{-9} m^2/s
Diffusion coefficient in solid	D_{S}	5×10^{-13} m^2/s
Liquidus slope	m_{L}	-1.4
Maximum nucleation density	$n_{\max}$	4×10^9/m^3
Standard deviation of undercooling	ΔT_σ	0.5 °C
Maximum undercooling	$\Delta T_{\max}$	2 °C
Gibbs–Thomson coefficient	Γ	3.66×10^{-7} m · K
Initial concentration	C_0	6 at.%

3.1. Dendritic Nucleus Model

During the solidification process of the molten pool, the interface between the molten pool of liquid and the solid substrate is the nucleation surface affiliated with the grains' growth in the molten pool. A heterogeneous nucleation method was used to simulate the solidification evolution process of the grains in the molten pool. A quasi-continuous nucleation model was established based on the Gaussian distribution function. The term $dn/d(\Delta T)$ was used to describe the variation of the grains' nucleation density, and the total density of nuclei $n(\Delta T)$ formed at a certain undercooling ΔT was given as [21]:

$$n(\Delta T) = \int_0^{\Delta T} \frac{dn}{d(\Delta T)} d(\Delta T) \tag{8}$$

The change rate of nucleation density varies with the supercooling degree satisfied with Gaussian distribution. It can be calculated by the following expression:

$$\frac{dn}{d(\Delta T)} = \frac{n_{max}}{\sqrt{2\pi}\Delta T} \exp\left[-\frac{1}{2}\left(\frac{\Delta T - \Delta T_N}{\Delta T_\sigma}\right)^2\right] \tag{9}$$

where n_{max} is the maximum nucleation density and ΔT_σ is the standard deviation of undercooling. In the calculation process, the nucleation point location was randomly chosen.

3.2. Solute Diffusion Model

For binary alloys or multicomponent alloys, the solute diffusion also plays an important role on dendrite growth in the weld molten pool and the solute concentration gradient is the driving force for solute diffusion in the solid and liquid phases. Then, the solute concentration in the solid and liquid phases is determined by solving the governing equation for each phase separately, as follows:

$$\frac{\partial C_i}{\partial t} = D_i \nabla^2 C_i + C_i \cdot (1 - k)\frac{\partial f_s}{\partial t} \tag{10}$$

where C is the solute concentration, D represents the solute diffusivity, the subscript i denotes a solid or liquid, and k is the partition coefficient.

At the solid–liquid interface, the solute partition between the liquid and solid is given by:

$$C_s{}^* = k \cdot C_l{}^* \tag{11}$$

where $C_s{}^*$ and C_l^* denote the interface solute concentrations in the solid and liquid phases, respectively.

Based on the counting method advanced by Nastac [22], the interface curvature of a cell with a solid fraction can be derived by calculating the nearest and second-nearest neighboring cells:

$$K = \frac{1}{l_{CA}}\left(1 - 2\frac{f_s + \sum_{j=1}^{N} f_s(j)}{N + 1}\right) \tag{12}$$

where l_{CA} represents the length of the CA cell side, N is the number of the nearest and the second-nearest neighboring cells, and $f_s(j)$ is the solid fraction of neighboring cells.

The liquid concentration in the interface cell is given as [23]:

$$C_l = C_l{}^* - \frac{1 - f_s}{2} l_{CA} \cdot G_c \tag{13}$$

where G_c is the concentration gradient in front of the solid-liquid interface, and the interface equilibrium composition $C_l{}^*$ is obtained by:

$$C_l{}^* = C_0 + \frac{1}{m_l}[T^* - T^{eq}{}_l + \Gamma K] \tag{14}$$

where C_0 indicates the initial solute concentration, T^* is the interface equilibrium temperature calculated by Equation (1), T_l^{eq} is the equilibrium liquidus temperature at C_0, m_l is the liquidus slope, Γ is the Gibbs-Thomson coefficient, and K is the average curvature of the liquid-solid interface.

3.3. Undercooling

Kurz et al. [24] developed the KGT (Kurz-Giovanola-Trivedi) model. It was introduced to simulate and calculate the growth process of the dendrite tip. De-Chang et al. [25] proposed that the undercooling of the solid–liquid interface mainly includes temperature, concentration, and curvature in the CA model. The anisotropy of the interface energy has a great effect on the curvature undercooling, so the model must take the interface anisotropy into consideration. At the same time, the degree of undercooling in the solid–liquid interface prerequisite is calculated as [26]:

$$\Delta T(t_n) = T' - T_{i,j} + m_l(C_0 - C_l^*) + \Gamma K(t_n)\{1 - 15G_x \cos[4(\theta - \theta_0)]\} \tag{15}$$

where T' is the temperature at the interface, $T_{i,j}$ is the temperature of the undercooled melt, C_0 is the initial solute concentration, C_l^* is the solute concentration of the liquid at the interface, and m_l is the slope of the liquidus. Besides, Γ is the Gibbs–Thompson coefficient and $K(t_n)$ is the interface curvature calculated from Equation (12). G_x is the anisotropy intensity of the liquid–solid interface.

$$\theta_0 = \cos^{-1}\left(\frac{\partial f_s/\partial x}{\left((\partial f_s/\partial x)^2 + (\partial f_s/\partial y)^2\right)^{1/2}}\right) \tag{16}$$

$$\theta = \arctan\left(\frac{\partial f_s/\partial y}{\partial f_s/\partial x}\right) \tag{17}$$

Then, θ_0 represents the angle between the growth direction of the dendrites and the positive direction of the coordinate axis, and it can be calculated from Equation (16). Likewise, θ is also the angle between the normal of the solid–liquid interface and the positive direction of the coordinate axis, which is derived from Equation (17). As the heat undercooling is relatively small, the effect of dynamics on high dendritic growth is not taken into account and the solute diffusion in the solid phase can be neglected.

3.4. Selection of Time Step

In order to ensure stability in the calculation of the composition field and the advance rate of the solid–liquid interface, the time step is determined by the following Equation [27]:

$$\Delta t \leq \frac{1}{5}\min\left(\frac{\Delta x^2}{D_l}, \frac{\Delta x}{V_{max}}\right) \tag{18}$$

where Δx is the grid size, D_l is the solute diffusion coefficient in the liquid, and V_{max} represents the welding speed.

3.5. Experimental Details

For the purpose of validating the feasibility and accuracy of the presented model, the welding experiment was automatically carried out by an YC-400TX GTAW machine (Panasonic Welding Systems (Tangshan) Co., Ltd, Tangshan, China). The welding parameters used are shown in Table 3. It should be pointed that the welding efficiency in Table 3 was set as 0.8 [28]. Then, the specimens were mechanically ground with 120-grit SiC paper, were etched in a solution composed of 400 mL H_2O + 40 g KOH + 40 mL H_2O_2, and purged with an ultrasonic cleaner in order to remove surface contaminants.

Table 3. Welding procedure parameters of Ti–6Al–4V alloy plate gas tungsten arc welding (GTAW).

Parameters	Welding Speed	Welding Voltage	Welding Current	Welding Efficiency
Value	5 mm/s	13.8 V	75 A	0.8

An accurate simulation for the moving heat source of the weld process is a prerequisite to correctly predict the temperature distribution and dendritic growth. The parameters' calibration for the heat source, including the peak temperature, shape, and dimensions of the welding molten pool, is performed by optimizing parameters such as the welding voltage and the welding current. The parameter optimization is based on satisfying the macrograph experimentally observed for a welding cross section.

The calculated temperature field was validated by means of the K-type thermocouple measurement of the transient temperature captured at corresponding test points during the actual welding process, as shown in Figure 6a. In order to evaluate the accuracy for the simulated temperature distribution results, the test platform of the welding thermal cycle was built as shown in Figure 6b. It primarily consisted of a stored energy welding machine (self-developed), welding sample, K-type thermocouple, temperature measurement module (TR-W500, KEYENCE Corporation, Osaka, Japan), and computer. Since it is very hard to put the thermocouple very close to the welding sample surface without surpassing the temperature limitation for the thermocouple, the temperature distribution was measured at a smaller distance from the heat affected zone (HAZ). Hence, the K-type sheathed thermocouples were directly embedded in the welding sample surface by using the stored energy welding machine. The embedded thermocouples were located at four points on the workpiece surface; as shown in Figure 6c, these four points were evenly spaced along the welding direction.

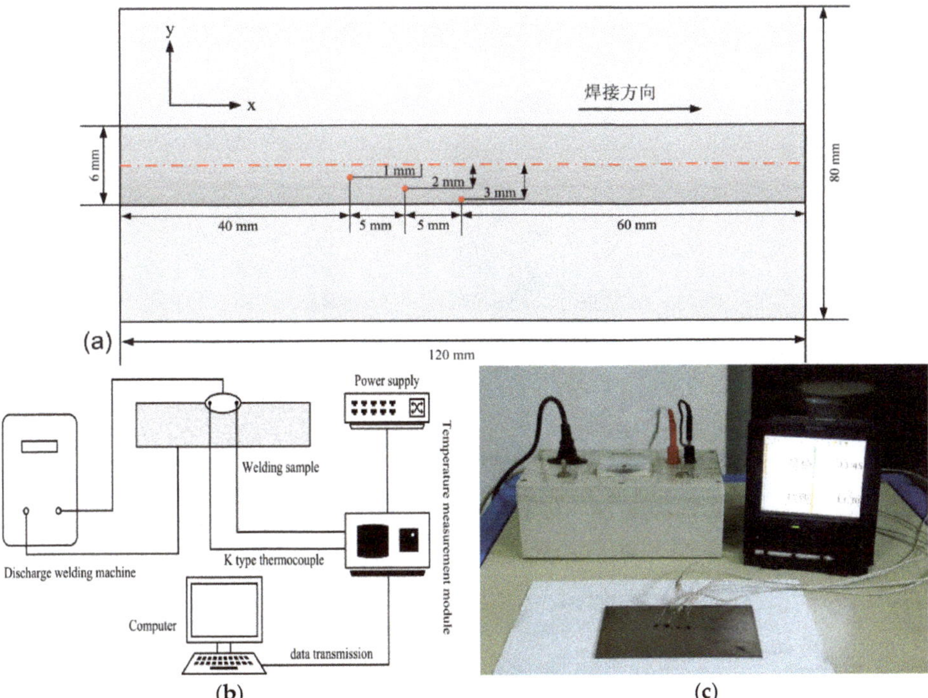

Figure 6. Testing of the welding thermal cycle: (**a**) the welding sample geometry, (**b**) the diagram of the test platform, and (**c**) the picture of the test platform.

4. Results and Discussions

4.1. Temperature Distribution and Weld Bead Geometry

The temperature distributions in the welding process were calculated under different welding currents, including 60 A, 75 A, and 90 A. The specific temperature distribution results are shown in Figure 7. It should be noted that the welding voltage is 13.8 V, the welding speed is 5 mm/s, and the welding efficiency is 0.8. According to Equation (4), the welding input power can be derived.

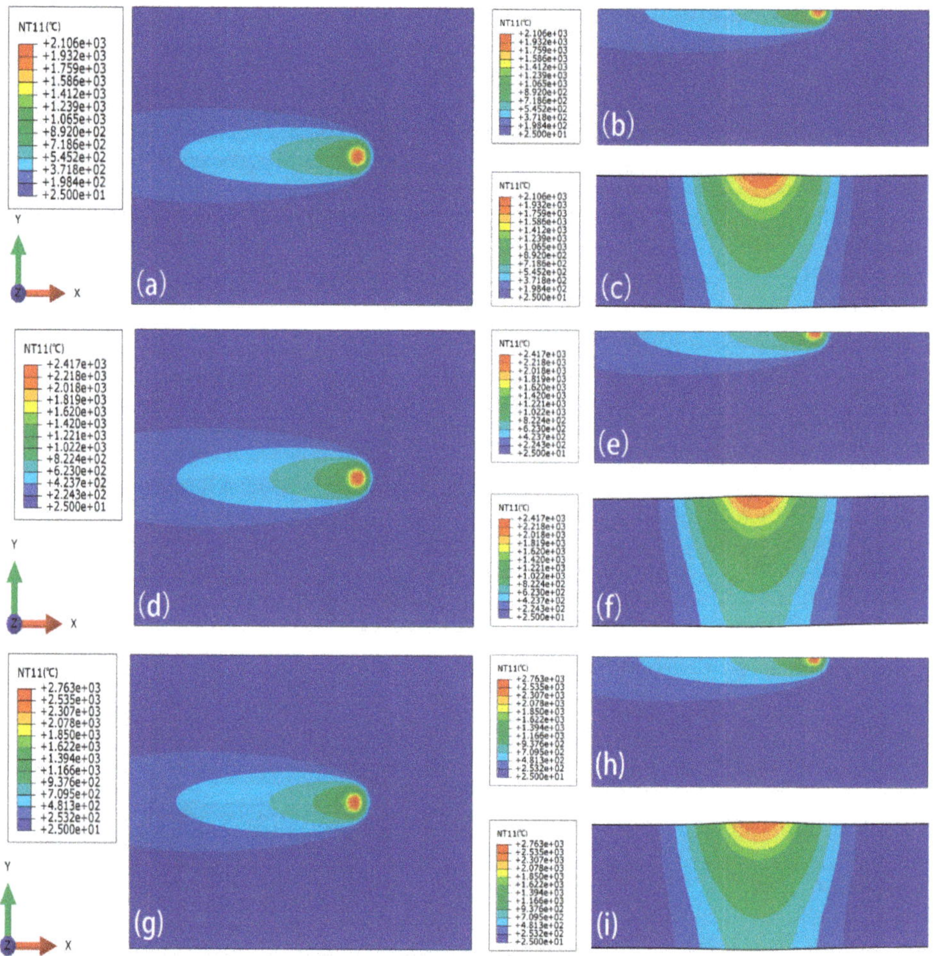

Figure 7. The temperature fields under different welding currents at t = 16 s, with (**a**) the xy plane at I = 60 A, (**b**) the half xy plane at I = 60 A, (**c**) the yz plane at I = 60 A; (**d**) the xy plane at I = 75 A, (**e**) the half xy plane at I = 75, (**f**) the yz plane at I = 75 A, (**g**) the xy plane at I = 90 A, (**h**) the half xy plane at I = 90 A, and (**i**) the yz plane at I = 90 A.

From Figure 7, it can be seen that as the welding heat source moves, the molten pool advances stably and the shape of the molten pool remains substantially unchanged. When the welding current is gradually increased, the temperature in the central region of the moving heat source is increased. To further investigate the relationship between the welding current and the welding bead geometry, the cross sections of the welding bead under different welding currents were observed, as shown in Figure 8. Then the bottom width of weld bead is represented by W_D.

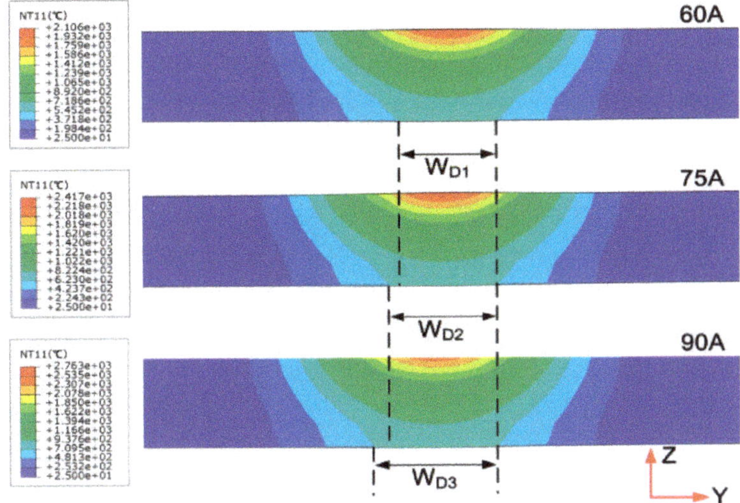

Figure 8. The cross sections of the welding bead under different welding currents.

From Figure 8, it can be found that as the welding current increases, the temperature of the central region of the moving heat source increases and the weld transverse cross section becomes wider, i.e., $W_{D_1} > W_{D_2} > W_{D_3}$. In addition, the temperature in the central region of the moving heat source is larger than the melting point of the titanium alloy (1650 °C).

In order to further analyze the effect of the welding current on the temperature distribution during the welding, the thermal cycle curves were obtained, as shown in Figure 9. From Figure 9, it can be found that due to the rapid heating rate of the welding, the curve rises extremely fast and the temperature rapidly reaches a peak; the closer the weld bead is, the faster the temperature rises and the higher the peak temperature. During the cooling phase, the temperature drops relatively slowly. In addition, the simulated temperature values under different distances and currents are listed in Table 4.

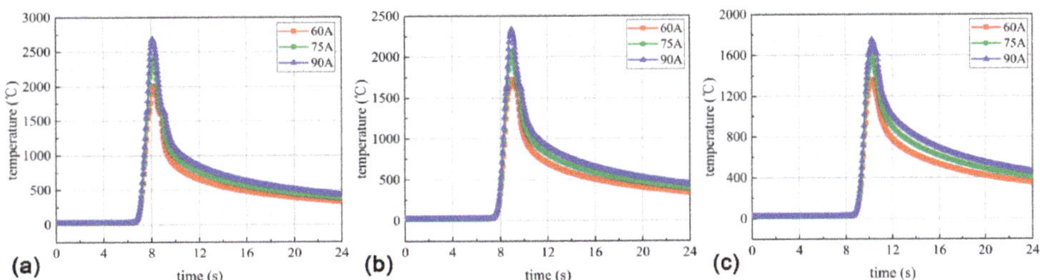

Figure 9. The simulated thermal cycle curves at different distances from the weld center: (**a**) 1 mm, (**b**) 2 mm, and (**c**) 3 mm.

Table 4. The simulated temperature maximum values under different distances and currents.

Parameters	Different Currents		
Different Distances	**I = 60 A**	**I = 75 A**	**I = 90 A**
L = 1 mm	2013.27 °C	2394.68 °C	2681.89 °C
L = 2 mm	1723.94 °C	2063.25 °C	2337.25 °C
L = 3 mm	1420.60 °C	1653.74 °C	1780.63 °C

From Table 4, it can be found that the temperature value is lower than the melting point of the titanium alloy (1650 °C) under L = 3 mm and I = 60 A. However, the temperature is larger than the melting point of the titanium alloy under L = 3 mm and I = 90 A. Then, the welding heat input P can be calculated thus:

$$P = \frac{\eta U I}{V} \tag{19}$$

Here, U is the welding voltage, I is the welding current, V represents the welding speed, and η is the welding efficiency. According to Equation (19), the welding current is positively correlated with the welding heat input when U and V are determined. It should be noted that the welding efficiency is 0.8 and the welding voltage is 13.8 V in Equation (19). Then, the welding heat inputs under I = 60 A, I = 75 A, and I = 90 A are 132.48 J·mm^{-1}, 165.60 J·mm^{-1}, and 198.72 J·mm^{-1}, respectively. When the welding heat input is too small, welding defects are easily caused; but when the welding heat input is too large, coarse columnar crystals are generated in the weld bead. This leads to the increase of the brittleness of the welded joints. Therefore, within the range of reasonable welding heat input, the weld grain size is more uniform. Based on the analysis mentioned above, the welding thermal cycle test is performed under the welding current I = 75 A.

The calibration between the simulated results and the experimental observations is presented as shown in Figure 10. The comparative analysis of both the simulated results and experimental observations was performed. Figure 10a shows that the experimental and simulated time–temperature curves match well, including the heating rate, peak temperature, and cooling rate. Here, the heating and cooling rates are calculated by dividing the difference between the initial and current temperature values by the fixed time step. The comparison between the simulated weld bead geometry and the measured macrograph of a polished and chemically etched weld bead transverse cross section is shown in Figure 10b.

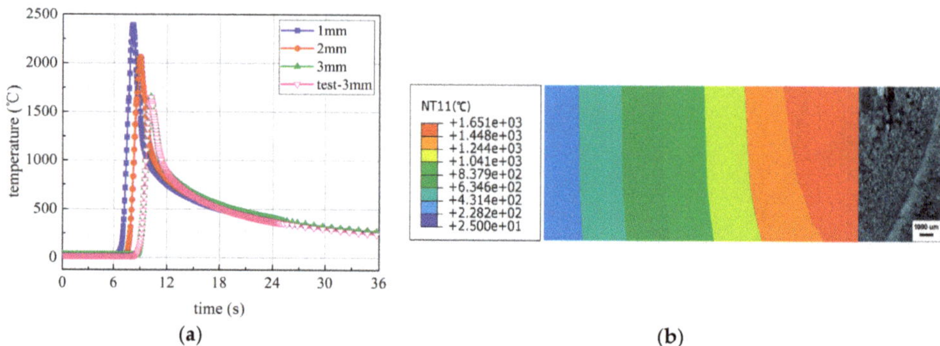

Figure 10. The comparison between the simulated results under I = 75 A and the experimental observations: for (**a**) thermal cycle curves and (**b**) weld bead geometry.

According to the temperature contour above the melting point of 1650 °C, the fusion zone is determined in the FE model. From Figure 10b, it can be found that the simulated fusion boundary isotherm is in good accordance with the experimental fusion boundary and penetration depth. It is preliminarily concluded that the weld bead transverse cross section is better when the welding current is 75 A.

4.2. Calculations and Measurements of the CET

For further evaluating the validity and rationality of the obtained welding process parameters from the welding temperature field analysis, these parameters act as the heat input parameters of the CA-FD coupling model. Then, the microstructure evolution is calculated by the CA-FD coupling model. The welding voltage is 13.8 V and the welding current is 75 A in this calculation, and the calculation areas are divided into rectangular macroscopic grids. In addition, these macroscopic grids consist of uniform microscopic grids, the grid size is 5 μm, and the time step is 0.5 ms. Of importance, the iteration number is set as 130, 372, and 454, respectively. The grains' growth tends to be dramatically changed with the influence of heat dissipation direction. The growth morphology of the dendrites in the weld pool varies with time during the solidification, as shown in Figure 11.

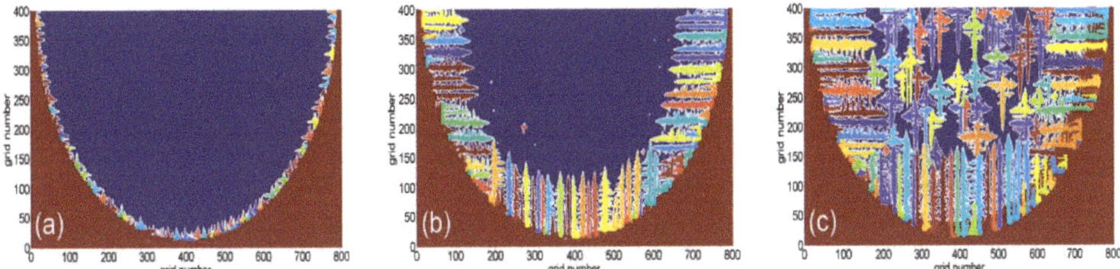

Figure 11. The morphology of grains in the weld pool at different times: (**a**) t = 0.065 s, (**b**) t = 0.186 s, and (**c**) t = 0.227 s.

From Figure 11a, it can be seen that the columnar crystals are formed at the edge of the weld pool at the beginning of solidification. With the progress of solidification, these columnar crystals continuously grow and some fine equiaxed crystals form at their tips, as shown in Figure 11b. The solidification layer continues to move inward and the solid-phase cooling capability is gradually weakened. It should be pointed that when the constitutional supercooling is high enough, the nonuniform nucleation is activated and the equiaxed crystals' nucleation starts in the center of the molten pool. From Figure 11c, it is shown that the growth of the columnar crystals is greatly affected by the growth of the central equiaxed grains. This leads to the arrest of the longitudinal growth for the partial columnar crystals, whereas their radial growth and secondary dendrite arms are intensified. The transformation from the columnar crystals to the equiaxed crystals is eventually achieved. When the grain growth direction is perpendicular to the heat dissipation direction of the molten pool wall, columnar crystals are formed. If the heat dissipation is along all directions around the melt, equiaxed crystals are formed. It can be concluded that the morphology of the dendrites exhibits randomness and asymmetry under the effect of the altering temperature field in the molten pool.

The solute concentration is the key factor of the liquid–solid state during the dendritic nucleation and growth. When the solute field is changed, the direction and morphology of the grain growth will be changed. Hence, the corresponding liquid solute concentration and solid solute concentration, as shown in Figure 12, are discussed. From Figure 12a,c,e, it can be seen that the release of solutes in the solid phase makes the liquid phase solute concentration located at the front of solid–liquid interface increase. Simultaneously, the solute diffusion rate in the liquid phase of the dendritic tip is much smaller than the dendrites' growth speed at the solid–liquid interface. In addition, when the competitive growth of columnar and equiaxed crystals occurs, the solute fields formed by the growth of equiaxed grains and columnar crystals will affect each other.

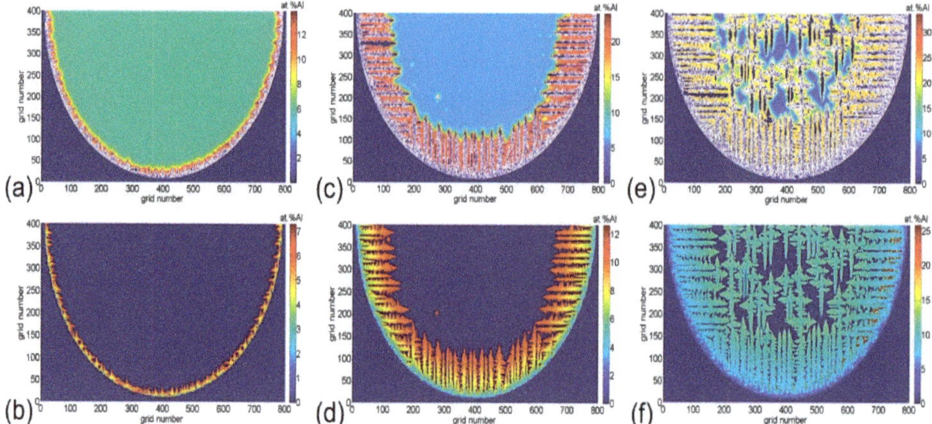

Figure 12. The solute distribution for solid and liquid phases in the weld pool at different times: (**a**) liquid phase at t = 0.065 s, (**b**) solid phase at t = 0.065 s, (**c**) liquid phase at t = 0.186 s, (**d**) solid phase at t = 0.186 s, (**e**) liquid phase at t = 0.227 s, and (**f**) solid phase at t = 0.227 s.

In light of Figure 12b,d,f, it can be seen that the growth of dendrites is always followed by the segregation phenomenon when observing the distribution of solute concentration in the solid phase during the CET process. At the beginning of the CET, the region of highest solute concentration is the columnar crystal tip. When the CET transformation is conducted at the final stage, the equiaxed crystals grow well and the solid solute concentration reaches 15%. At this point, it is prone to forming regional segregation after solidification. Therefore, the evolution of columnar–equiaxed crystals is not only influenced by the growth of columnar crystals being hindered by the equiaxed grains, but also the interaction between the solid and liquid solute concentrations.

The cross section of the TC4 alloy welded joint is selected for the purpose of testing the validity of the simulated microstructure evolution results. It was vertical to the direction of the weld moving heat source and was ground and polished. The welding samples were etched by 10 mL HF + 20 mL HNO$_3$ + 50 mL H$_2$O. Microstructure observation was conducted using an inverted GX671 metallographic microscope produced by OLYMPUS (Tokyo, Japan). Owing to the whole welding pool having good symmetry, the microscopic morphology of only half of the weld pool is shown in Figure 13a.

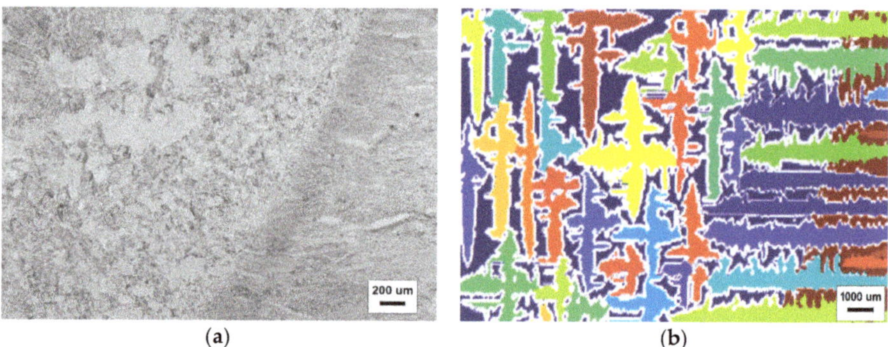

Figure 13. The micrograph of the TC4 alloy in the weld pool: (**a**) the test result and (**b**) the simulated result.

From Figure 13a, it can be seen that the equiaxed crystals mainly concentrate in the central region of the weld pool and the CET appears in the region near the fusion line. It can be concluded that the simulated results are in good agreement with the metallographic tests.

5. Conclusions

In this study, based on the CA-FD model coupled with the FE model, the temperature distribution and dendritic morphology of the molten pool for TC4 alloy plates welded by GTAW were analyzed theoretically and experimentally. In addition, the effect of the welding current on weld bead geometry was analyzed in detail. In summary, the main conclusions inferred from the above analysis are as follows:

(1) In order to ensure the accuracy of the simulated temperature distribution, the developed FE model took nonlinear thermal analysis, the temperature dependency of the thermal materials' properties, and a moving heat source into consideration. Furthermore, the convection and radiation conditions were also considered in this model.

(2) During the GTAW process, the temperature distribution in a macro region around the molten pool was calculated by the developed FE model under different welding currents. It was found that the transverse cross section of the weld bead was better when the welding current was 75 A. The obtained current parameter acted as the input parameter of the CA-FD coupling model.

(3) Then, the effect of several process conditions on the solidification microstructure was investigated by the CA-FD model, especially solidification time and temperature. It is shown that the coarse columnar crystals are produced with priority in the molten pool and their growth direction is in line with the direction of the negative temperature gradient. With the increase of temperature and solute concentration at the front of the solid-liquid interface, the columnar crystal grains show the trend of transforming to equiaxed crystals.

(4) This work contributes to the understanding of microstructure evolution and temperature characteristics in the molten pool. It can provide a fundamental basis for the selection of welding process parameters for GTAW processing of the TC4 alloy. The present model will be further enhanced to include the effect of fluid flow on dendrite growth in the molten pool.

Author Contributions: Conceptualization, M.Z. and Y.Z.; Data curation, C.H.; Formal analysis, W.Z.; Investigation, Y.Z.; Methodology, J.L.; Resources, J.L.; Software, W.Z.; Supervision, M.Z.; Validation, C.H.; Writing—original draft, Y.Z.; Writing—review & editing, Q.C.

Funding: This research was funded by the Natural Science Foundation of China, grant number 51274162, and the Education Department of Shaanxi Provincial Government scientific research projects of key laboratory, grant number 15JS082, and the Education Department of Shaanxi Provincial Government service local special projects, grant number 16JF021.

Conflicts of Interest: The authors declare no conflict of interest.

References

1. Lutjering, G.; Williams, J.C. *Titanium*, 2nd ed.; Springer: Berlin, Germany, 2007; pp. 2–4, ISBN 978-3-540-71397-5.
2. Wang, D.C. Development and application of high-strength titanium alloys. *Chin. J. Nonferr. Met.* **2010**, *20*, 958–963.
3. Zhao, Y.Q.; Ge, P. Current situation and development of new titanium alloys invented in China. *J. Aeronaut. Mater.* **2014**, *34*, 51–61.
4. Babu, N.K.; Raman, S.G.S.; Mythili, R.; Saroja, S. Correlation of microstructure with mechanical properties of TIG weldments of Ti–6Al–4V made with and without current pulsing. *Mater Charact.* **2007**, *58*, 581–587. [CrossRef]
5. Wang, L.M.; Lin, H.C. The characterization of corrosion resistance in the Ti-6Al-4V alloy fusion zone following a gas tungsten arc welding process. *J. Mater. Res.* **2009**, *24*, 3680–3688. [CrossRef]
6. David, S.A.; Vitek, J.M. Correlation between solidification parameters and weld microstructures. *Int. Mater. Rev.* **1989**, *34*, 213–245. [CrossRef]
7. Kou, S.; Sun, D.K. Fluid flow and weld penetration in stationary arc welds. *Metall. Trans. A* **1985**, *16*, 203–213. [CrossRef]
8. Salimi, S.; Bahemmat, P.; Haghpanahi, M. A 3D transient analytical solution to the temperature field during dissimilar welding processes. *Int. J. Mech. Sci.* **2014**, *79*, 66–74. [CrossRef]

9. Zhang, J.; Liou, F.; Seufzer, W.; Taminger, K. A coupled finite element cellular automaton model to predict thermal history and grain morphology of Ti-6Al-4V during direct metal deposition (DMD). *Addit. Manuf.* **2016**, *11*, 32–39. [CrossRef]

10. Anderson, M.P.; Srolovitz, D.J.; Grest, G.S.; Sahni, P.S. Computer simulation of grain growth—I. kinetics. *Acta Metall.* **1984**, *32*, 783–791. [CrossRef]

11. Wei, H.L.; Elmer, J.W.; DebRoy, T. Crystal growth during keyhole mode laser welding. *Acta Mater.* **2017**, *133*, 10–20. [CrossRef]

12. Qin, R.S.; Wallach, E.R.; Thomson, R.C. A phase-field model for the solidification of multicomponent and multiphase alloys. *J. Cryst. Growth* **2005**, *279*, 163–169. [CrossRef]

13. Wang, L.; Wei, Y.; Zhan, X.; Yu, F. A phase field investigation of dendrite morphology and solute distributions under transient conditions in an Al-Cu welding molten pool. *Sci. Technol. Weld. Join.* **2016**, *21*, 446–451. [CrossRef]

14. Rappaz, M.; Gandin, C.A. Probabilistic modeling of microstructure formation in solidification processes. *Acta Metall. Mater.* **1993**, *41*, 345–360. [CrossRef]

15. Zhan, X.; Wei, Y.; Dong, Z. Cellular automaton simulation of grain growth with different orientation angles during solidification process. *J. Mater. Process. Tecnol.* **2008**, *208*, 1–8. [CrossRef]

16. Zhan, X.; Dong, Z.; Wei, Y.; Ma, R. Simulation of grain morphologies and competitive growth in weld pool of Ni–Cr alloy. *J. Cryst. Growth* **2009**, *311*, 4778–4783. [CrossRef]

17. Wang, Z.J.; Luo, S.; Li, W.Y.; Song, H.W.; Deng, W.D. Simulation of microstructure during laser rapid forming solidification based on cellular automaton. *Math. Probl. Eng.* **2014**, *2014*. [CrossRef]

18. Chen, S.; Guillemot, G.; Gandin, C.A. Three-dimensional cellular automaton-finite element modeling of solidification grain structures for arc-welding processes. *Acta Mater.* **2016**, *115*, 448–467. [CrossRef]

19. Liu, Q.; Li, X.; Jiang, Y. Numerical simulation of EBCHM for the large-scale TC4 alloy slab ingot during the solidification process. *Vacuum* **2017**, *141*, 1–9. [CrossRef]

20. Ahn, J.; He, E.; Chen, L.; Wimpory, R.C.; Dear, J.P.; Davies, C.W. Prediction and measurement of residual stresses and distortions in fibre laser welded Ti-6Al-4V considering phase transformation. *Mater. Des.* **2017**, *115*, 441–457. [CrossRef]

21. Gandin, C.A.; Rappaz, M. A coupled finite element-cellular automaton model for the prediction of dendritic grain structures in solidification processes. *Acta Metall. Mater.* **1994**, *42*, 2233–2246. [CrossRef]

22. Nastac, L. Numerical modeling of solidification morphologies and segregation patters in cast dendritic alloys. *Acta Mater.* **1999**, *47*, 4253–4262. [CrossRef]

23. Wang, W.; Lee, P.D.; McLean, M. A model of solidification microstructures in nickel-based superalloys: predicting primary dendrite spacing selection. *Acta Mater.* **2003**, *51*, 2971–2987. [CrossRef]

24. Kurz, W.; Giovanola, B.; Trivedi, R. Theory of microstructural development during rapid solidification. *Acta Metall.* **2010**, *34*, 823–830. [CrossRef]

25. Tsai, D.C.; Hwang, W.S. Numerical simulation of solidification morphologies of Cu-0.6Cr casting alloy using modified cellular automaton model. *Trans. Nonferr. Met. Soc. China* **2010**, *20*, 1072–1077. [CrossRef]

26. Zhu, M.F.; Stefanescu, D.M. Virtual front tracking model for the quantitative modeling of dendritic growth in solidification of alloys. *Acta Mater.* **2007**, *55*, 1741–1755. [CrossRef]

27. Beltran-Sanchez, L.; Stefanescu, D.M. A quantitative dendrite growth model and analysis of stability concepts. *Metll. Mater. Trans. A* **2004**, *35*, 2471–2485. [CrossRef]

28. Dal, M.; Masson, P.L.; Carin, M. Estimation of fusion front in 2D axisymmetric welding using inverse method. *Int. J. Therm. Sci.* **2012**, *55*, 60–68. [CrossRef]

Article

Transient Liquid Phase Bonding of Ti-6Al-4V and Mg-AZ31 Alloys Using Zn Coatings

Abdulaziz AlHazaa [1,2,*]**, Ibrahim Alhoweml** [1]**, Muhammad Ali Shar** [2]**, Mahmoud Hezam** [2]**, Hany Sayed Abdo** [3,4] **and Hamad AlBrithen** [1,2,5]

[1] Research Chair for Tribology, Surface, and Interface Sciences, Department of Physics and Astronomy, College of Science, King Saud University, P.O. Box 2455, Riyadh 11451, Saudi Arabia; ialhoweml@ksu.edu.sa (I.A.); Brithen@ksu.edu.sa (H.A.)
[2] King Abdullah Institute for nanotechnology, King Saud University, P.O. Box 2455, Riyadh 11451, Saudi Arabia; mashar@ksu.edu.sa (M.A.S.); mhezam@ksu.edu.sa (M.H.)
[3] Center of Excellence for Research in Engineering Materials (CEREM), King Saud University, P.O. Box-800, Riyadh 11421, Saudi Arabia; habdo@ksu.edu.sa
[4] Mechanical Design and Materials Department, Faculty of Energy Engineering, Aswan University, Aswan 81521, Egypt
[5] National Center for Nanotechnology, King Abdulaziz City for Science and Technology, P.O. Box 6086, Riyadh 11442, Saudi Arabia
* Correspondence: aalhazaa@ksu.edu.sa

Received: 6 February 2019; Accepted: 1 March 2019; Published: 6 March 2019

Abstract: Ti-6Al-4V and Mg-AZ31 were bonded together using the Transient Liquid Phase Bonding Process (TLP) after coating both surfaces with zinc. The zinc coatings were applied using the screen printing process of zinc paste. Successful bonds were obtained in a vacuum furnace at 500 °C and under a uniaxial pressure of 1 MPa using high frequency induction heat sintering furnace (HFIHS). Various bonding times were selected and all gave solid joints. The bonds were successfully achieved at 5, 10, 15, 20, 25, and 30 min. The energy dispersive spectroscopy (EDS) line scan confirmed the diffusion of Zn in both sides but with more diffusion in the Mg side. Diffusion of Mg into the joint region was detected with significant amounts at bonds made for 20 min and above, which indicate that the isothermal solidification was achieved. In addition, Ti and Al from the base alloys were diffused into the joint region. Based on microstructural analysis, the joint mechanism was attributed to the formation of solidified mixture of Mg and Zn at the joint region with a presence of diffused Ti and Al. This conclusion was also supported by structural analysis of the fractured surfaces as well as the analysis across the joint region. The fractured surfaces were analyzed and it was concluded that the fractures occurred within the joint region where ductile fractures were observed. The strength of the joint was evaluated by shear test and found that the maximum shear strength achieved was 30.5 MPa for the bond made at 20 min.

Keywords: TLP Bonding; Mg alloy; Ti alloy; coatings; Zinc; shear strength

---

1. Introduction

The growing concerns regarding fuel consumption within the aerospace and transportation industries led to the development of fuel-efficient systems to overcome significant engineering challenges. Mg-AZ31 and Ti-6Al-4V alloys are separately used in the automotive industries due to their excellent physical and mechanical properties such as high specific strength, low mass density, and good machinability and workability [1–3]. Ti-6Al-4V alloy covers more than 50% of industrial titanium in the market due to its balance between having high specific strength and good corrosion resistance. On the other hand, the Mg-AZ31 alloy is one of the most popular magnesium alloys. These two alloys are increasingly used in similar sectors. For example, in the automotive industry, titanium has been mainly used in high temperatures zones, and high strength requirement areas, such as exhaust systems, suspension springs, valve springs, valves, and connecting rods. Mg alloys are used in steering hanger beam, steering wheels, transmission outer parts, and seat frames. Therefore, fabricating a joint assembly that combines both alloys is of high interest. However, the vast difference between their melting points makes welding them using commercial methods like fusion welding unsuccessful. Moreover, the binary phase diagram of Mg-Ti system expects very limited mutual solubility. The phase diagram also indicates that no intermetallic compound (IMC) could form between the two metals. It was reported that the solubility of Mg in titanium is 1.6 at% at 765 °C where the solubility of Ti in magnesium is 0.12 at% the same temperature [4]. This means magnesium cannot be welded directly to titanium by solid state diffusion bonding under conventional conditions. Bonding various types of alloys was successfully achieved using transient liquid phase bonding (TLP). In this TLP method, the differences in the physical and mechanical properties of the two alloys can be overcome by inserting an interlayer or applying coatings such as between the two mating surfaces prior to the bonding process [5–7]. The challenge of the TLP bonding is to choose a suitable interlayer material that can interact with both materials and form a liquid phase at the selected bonding temperature so a higher diffusion rate can be achieved [8,9]. An important effect of forming a liquid phase at the joint region is the disruption of the native oxide layer that usually forms at metallic surfaces especially for light metals like Ti, Mg, and Al. Studies have shown that the formation of a metallic liquid phase during the bonding process disrupts the formation and growth of stable oxide films. Such oxide formation could prevent successful bonding [10,11]. Many successful examples of dissimilar joints produced by TLP bonding were reported in the scientific literature. Those joints were not possible to achieve using a commercial fusion bonding technique including high precision and well localized welding such as laser or electron beam welding. For instance, the Al7075 alloy was bonded to the Ti-6Al-4V alloy using Cu interlayer and Cu coatings with the Sn interlayer [12,13]. Mg-AZ31 was bonded to the 316 austenitic stainless-steel using the Cu interlayer [14]. Al was bonded with Mg using the Ni interlayer [15].

TLP bonding of Mg AZ31 to Ti-6Al-4V were reported in two studies [16,17]. The first study used Ni coatings. Bonding Mg AZ31 to Ti-6Al-4V using Ni coatings resulted in eutectic formation between the Mg AZ31 and Ni at the Mg side, but, at the Ti side, there was no Ni/Ti eutectic formation occurrence. The bond at the Ti side was interpreted as a result of solid-state diffusion, which is a slow process and needs a long bonding time. Moreover, there is a need of much higher temperature to facilitate the inter-diffusion between Ti and Ni, which is higher than the melting point of the Mg alloy [16]. Another study used a combination of Cu coatings and Cu coatings with a Sn interlayer and reported the formation of Sn_5Ti_6 and Mg_2Cu IMC's at the joint region where the fracture occurs [17]. Research used Spark Plasma Sintering technology to bond magnesium to titanium with various amount of Al impurities in magnesium and found that the joints occurred as a result of Al diffusion into the joint region where Ti_3Al formed [18].

Generally, for TLP bonding, it is desirable to form solid solutions at the joint region rather IMC's in order to gain high strength and avoid the formation of cracks at the interfaces between the formed IMC's and base alloys. Therefore, it will be interesting to fabricate interlayer/coatings that can react with both dissimilar surfaces and form solid solution across the joint region. Zinc seems to be a potential interlayer to bond Mg with Ti alloys since Zn has good solubility in both Mg and Ti. Zn forms eutectic reaction with Ti that could result in forming Zn and $Zn_{15}Ti$ at 418 °C where peritectic points were also reported between the two metals at 486 °C [19]. On the other hand, the ternary phase diagram of Mg and Zn shows eutectic point at 340 °C at the Mg rich region and a eutectic point 364 °C at the Zn rich region where the melting point of Zn is 419.6 °C. Zn is considered to be an alloying element for Mg that improves the castability and corrosion behavior of the magnesium alloy (AZ31) [20]. It was reported that, in the range of 375 °C to 575 °C, the activation energy and pre-exponential factor of the impurity diffusion of Zn in Mg is 109.8 kJ/mol and 10^{-5} m^2/s [21]. More studies showed that Zn diffuses in the Mg matrix faster than many other alloying elements like Al and even faster than the self-diffusion of Mg [22].

Zn interlayers were already used to bond Mg to Al and showed significant improvement in bond quality compared to direct bonding of Mg to Al. The shear strength of the bonded aluminum to magnesium was reported to have a maximum value of 83 MPa, which is twice the maximum value of the shear strength produced by direct bonding of aluminum to magnesium [23]. The zinc interlayer was not used before to bond the Ti alloy with dissimilar materials like Mg alloys. Therefore, the aim of this research study is to apply zinc coatings on the mating surfaces in order to facilitate the bonding between Ti-6Al-4V and Mg-AZ31 alloys. Bonding times were chosen as a variable in order to investigate the effect of bonding time on the bond formation and strength of the joints. A bonding temperature of 500 °C was selected because it is above the eutectic temperatures between Mg and Zn on one hand. On the other hand, the phase diagram study of Ti-Zn suggests that solid solution with IMC's such as TiZn3 and Ti_2Zn_3 can be formed at 500 °C [24].

2. Experimental Procedure

Commercial magnesium (AZ31) and titanium (Ti-6Al-4V) alloys were purchased from Goodfellow (Cambridgeshire, UK) in a form of extruded cylindrical rods. The diameter of the rods was selected to be 10 mm to fit the bonding machine specifications. Each rod was sliced by the Diamond cutter to produce many identical samples of a thickness of about 5 mm. Every sample was grinded starting from 80 grit to 1000 grit using SiC sand paper. Magnesium alloy samples were grinded with acetone instead of water to reduce the corrosion of the surfaces during the grinding process. Afterward, the samples were ultrasonically cleaned in ethanol medium for 20 min and then kept in a desiccator.

2.1. Zn Coatings Using the Screen Printing Technique

Additionally, 30 g of zinc micro-powder (Sigma Aldrich, St. Louis, MO, USA) was dispersed in an organic vehicle composed of 6 g of α-terpineol (Acros Organics, Morris Plains, NJ, USA) and 4 g of 10% weight ethyl cellulose (Sigma Aldrich) in ethanol. The mixture was mixed and grinded by a mortar-pestle for 30 min, which was followed by vigorous magnetic stirring for another 30 min. The mixture was homogenized using a rod homogenizer for 30 to 40 min. The resulting paste was deposited on each metallic disc using a screen printing technique [25]. The screen-printing process was performed using a polyester screen mesh of 32T mesh count (80 mesh/inch), 100 μm thread diameter, and a 210 μm mesh opening. The coating process was optimized to produce a thickness of 10 μm on each surface.

2.2. Transient Liquid Phase Bonding

Bonding experiments between the Mg-AZ31 alloy and the Ti-6Al-4V alloy were performed in a high frequency induction heating sintering system (HFIHS). Figure 1a shows the HFIHS made by ELTek Co. (Gyeonggi-do, Seol, Korea) where Figure 1b shows the insider image where the bond occurs. The uniaxial pressure was fixed for all experiments to be 1 MPa. Once the vacuum reaches 10^{-2} mbar, a heating rate of 300 °C per minute was applied until reaching a temperature of 500 °C. The holding (bonding) time at 500 °C is considered to be the variable. For each bonding experiment once the bonding time was reached, the bonded sample was cooled down under vacuum to an ambient temperature and then taken out from the chamber. There were two sets of bonded samples. The first set was used for metallographic observations and micro-hardness measurements, and the second set was used for a shear strength test and a fractured surfaces analysis. Therefore, the bonded samples that belong to the first set were cut perpendicular to the joint plane and then mounted in Bakelite powder (Buehler, Lake Bluff, IL, USA) for better handling. The bonded samples were ground to 1200 grit and then they were polished in Al_2O_3 suspension down to 1 μm. A JSM-7600F scanning electron microscope (SEM, Joel, Tokyo, Japan) was used to characterize the microstructure and perform line scan and EDS mapping of the joints. X-ray Photoelectron Spectroscopy (PHI VersaProbe XPS

Microprobe, Chigasaki, Japan) was used to analyze the interface at the joint line close to the Ti-6Al-4V side. The second set of the samples were fractured using a universal tensile testing machine (Instron 3360, Norwood, MA, USA). A special design for the cylindrical bonded samples was made to fit with the tensile machine in order to measure the shear strength, as shown in Figure 2. SEM and EDS analysis were used to study the fractured surfaces. The X-ray Diffraction (XRD) measurements were carried out using CuKα radiation source (wavelength = 0.154 nm) operated at 45 kV and 40 mA, and the signal was collected using a PIXcel detector (PAnalytical, Almelo, The Netherlands) for 2θ range of 5 to 80° with 0.02° step size to identify the phases present at the fractured surfaces.

Figure 1. Photos of (**a**) high-frequency induction heating sintering (HFIHS) machine used for bonding experiments. (**b**) Inside look at the vacuum chamber. (**c**) Sample photo of the two pieces of a bond after cutting.

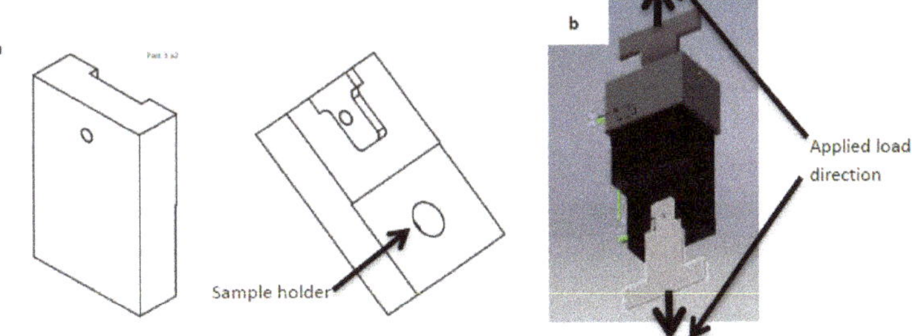

Figure 2. Fixture of the shear testing for bonded samples. (**a**) Back side and front side. (**b**) Full view of the fixture.

3. Results and Discussions

3.1. Evolution of the Interfacial Layer

Figure 3 shows SEM micrographs of the joints produced at different bonding times. The width of the joint region remains almost constant in Figure 3a–c, which indicates that the maximum width of the liquid zone was already reached. Therefore, the bonds made at 5, 10, and 15 min are expected to be in the liquid zone homogenization, according to the TLP bonding process [6]. On the other hand, for bonds made at 20, 25, and 30 min shown in Figure 3d–f, the width of the joint region was reduced due to loss of solute by diffusion and the isothermal solidification. By observing the relation between the width of the joint region and the bonding time, it can be concluded that the joint was produced mainly by forming a solidified melt. This means no intermetallic compound (IMC) layer was formed at the joint region. Otherwise, a proportional relation between the joint region width and the bonding time will be observed [12,26,27]. The diffusion rate of zinc in magnesium can be calculated from the frequency factor and activation energy available in the literature [21] to be 1.03×10^{-5} m^2/s. This value is higher than the diffusion rate of Al in Mg at the same temperature [22]. On the other hand, the diffusion coefficient data for zinc in titanium is not available in the literature even though it is expected to be much lower [24]. Since the bonding temperature used is 500 °C and the melting point of Zn is 419.5 °C, the mechanism of bonding is expected to start with complete melting of the Zn coatings, which is followed by diffusion of Zn in Mg and dissolution of Mg in the molten Zn where a eutectic reaction between Mg and Zn occurs at the Mg side of the joint. The dissolution of Ti in the molten Zn and diffusion of Zn in Ti are expected to proceed as well, but with much slower rates. A line scan of Ti, Mg, Zn, and Al was taken across the joint region at various bonding times in order to study the diffusion mechanism for the various elements. Figure 4 shows the EDS line scan across the joint region following the vertical lines that appeared in Figure 3a,c,d,f. From Figure 4, it can be seen that there is a noticeable diffusion of Al from the base alloys into the joint region. A peak of Al is observed for all bonds, which suggests that Al contributes to the joining mechanism either as IMC's or as dissolved solute in Zn-Mg solid solution. This observation agrees with a recent study that used Spark Plasma Sintering technology to bond Mg with Ti without interlayers and with various Al contents in the Mg [18]. The study showed that Al diffused from the Mg base alloy to the interface forming Ti$_3$Al IMC. From Figure 4, it can be seen that the dominant diffusion is the diffusion of Mg into the joint region where Zn was also diffusing away with more into the Mg side. Figure 4c,d show that the Mg line occupied the joint region intersecting with the Ti line. This means that the joint region at bonds made for 20 min and above was rich in magnesium. Isothermal solidification may be reached. On the other hand, the figure shows that Ti diffusion into the joint region is limited.

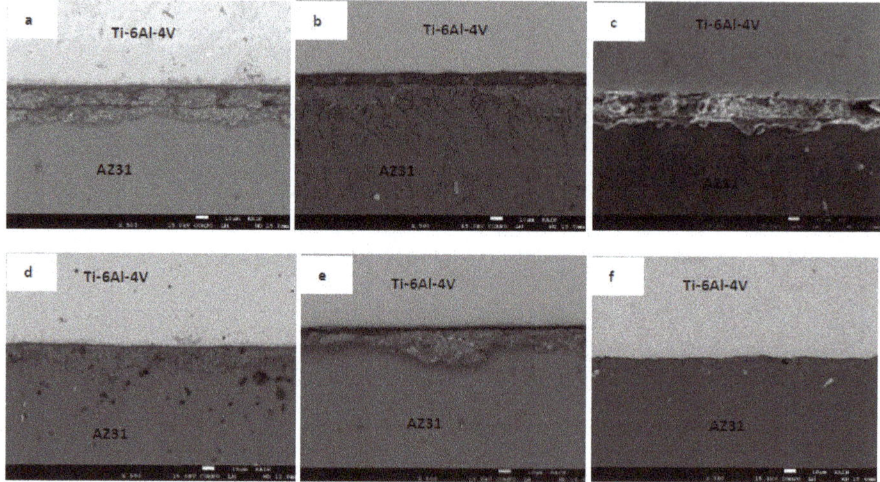

Figure 3. Backscattered SEM images of the joint regions for the bonds made at (**a**) 5, (**b**) 10, (**c**) 15, (**d**) 20, (**e**) 25, and (**f**) 30 min. Scale bar: 10 μm.

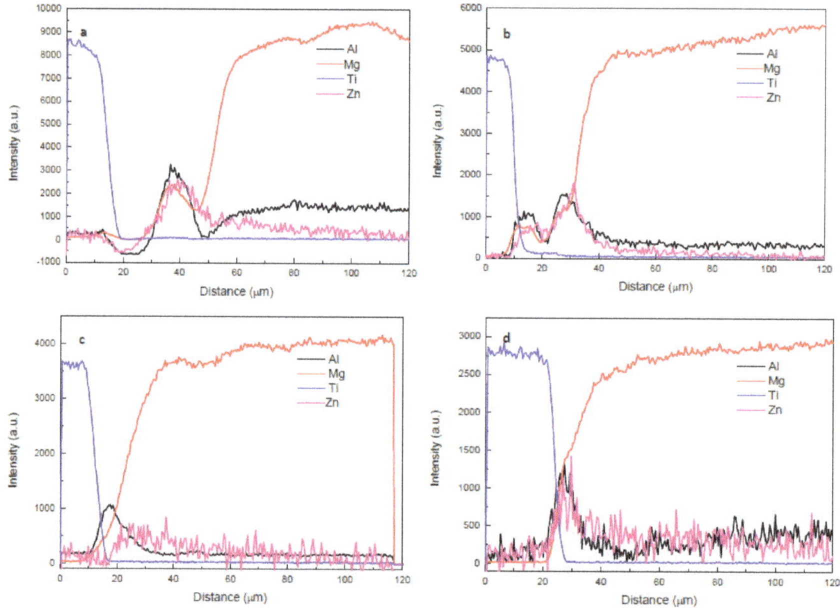

Figure 4. EDS line scans of Mg, Ti, Zn, and Al across the joint region for bonds made at (**a**) 5, (**b**) 15, (**c**) 20, and (**d**) 30 min.

3.2. Morphology and Composition of the Joint Region

In order to investigate the morphology and composition of the joint region and their change with respect to bonding time, we have chosen three bonds for detailed EDS spot analysis. Figures 5 and 6 show EDS spot analysis of various locations in the joint region for bonds made at 5 min and 30 min, respectively. For spectrum 3 that was taken far away from the joint region in the Mg side for the bond made at 5 min, the weight percentages of Mg, Ti, Zn, and Al were measured to be 96.7 wt%, 0.13 wt%,

3.17 wt%, and 2.2 wt%, respectively. The presence of Zn indicates that the diffusion of Zn through the Mg base alloy was noticeable even during the shortest bonding time. The detection of traces of Ti indicates that Ti was able to diffuse into the Mg base alloy regardless of the limited solubility between Ti and Mg. For spectrum 4 in Figure 5, the weight percentages of Mg, Ti, Zn, and Al were measured as 63.1 wt%, 0.53 wt%, 36.4 wt%, and 9.8 wt%, respectively. These values are expected when compared to the measured values in spectrum 3 where more Zn is expected to diffuse away to the Mg base alloy. However, it seems that, for a bond made at 5 min, the remaining Zn at the joint region is excessive. The weight percentage of Mg, Ti, Zn, and Al, which were taken from spectrum 5 were measured to be 28.2 wt%, 61.9 wt%, 9.9 wt%, and 10.5 wt%, respectively. This spot analysis was taken from a joint region that is near the Ti/joint interface with about 10 μm from the Ti side. The measured elemental composition indicates diffusion/dissolution of Ti in the molten Zn, which was originally occupied at the joint region. Furthermore, a noticeable presence of Mg in this region due to the diffusion of Mg in the molten Zn is confirmed. Al was detected in various locations at the joint region and its weight percent was measured to be around 10%. This measurement confirms the diffusion of Al from the base alloy to the joint region. A similar kind of analysis was done with selected spots for the bond made at 30 min, as shown in Figure 6. The EDS analysis for spectrum 3 (near the Ti/joint interface) gave weight percentages of Mg, Ti, Zn, and Al as 18.4 wt%, 72.9 wt%, 1.7 wt%, and 1.1 wt%. With the high percentages of Ti and Mg and low percentage of Zn, the isothermal solidification is expected to be complete. This trend is understood for the TLP process where more Zn diffused away from the joint region. For spectrum 4 (at the joint region), 90.6 wt%, 0.9 wt%, 6.5 wt%, and 2.7 wt% are the percentages of Mg, Ti, Zn, and Al. Lastly, for spectrum 6, which was taken close to the Mg/joint interface, the weight percentages of Mg, Ti, Zn, and Al were measured to be 93.1 wt%, 0.37 wt%, 6.4 wt%, and 4.4 wt%. In comparison with the bond made at 5 min, the joint region for the bond made at 30 min was seen to be occupied with Mg and Ti where less Zn was present. The reduction of the weight percent of Al to be less than 5% compared to the amount detected in the joint region taken from the bond made for 5 min could be attributed to the increase of Ti and Mg percentages. This indicates that the diffusion of Mg and Ti to the joint region is time-dependent where the quantities increase with the increase of bonding time. On the other hand, the diffused quantity of Al in the joint region does not increase much with increasing bonding time. This could be due to the fact that Al is only an alloying element with only 3 wt% in the Mg alloy and 6 wt% in the Ti side. Therefore, the source of Al from the base alloys is limited.

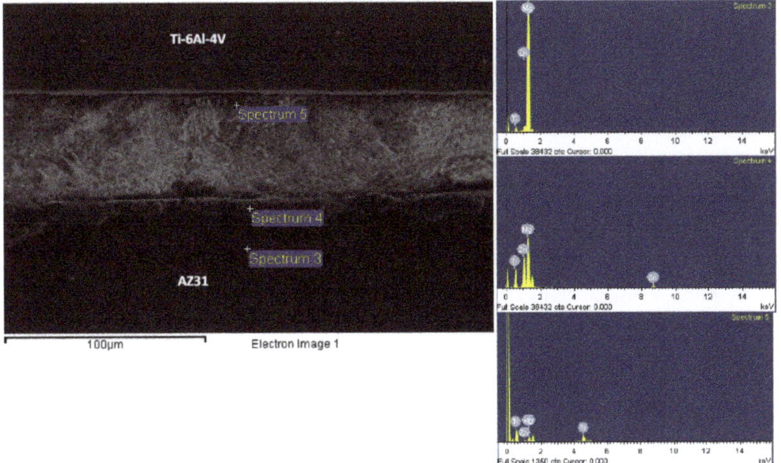

Figure 5. SEM micrograph of the bond made at 5 min and corresponding EDS spot analysis.

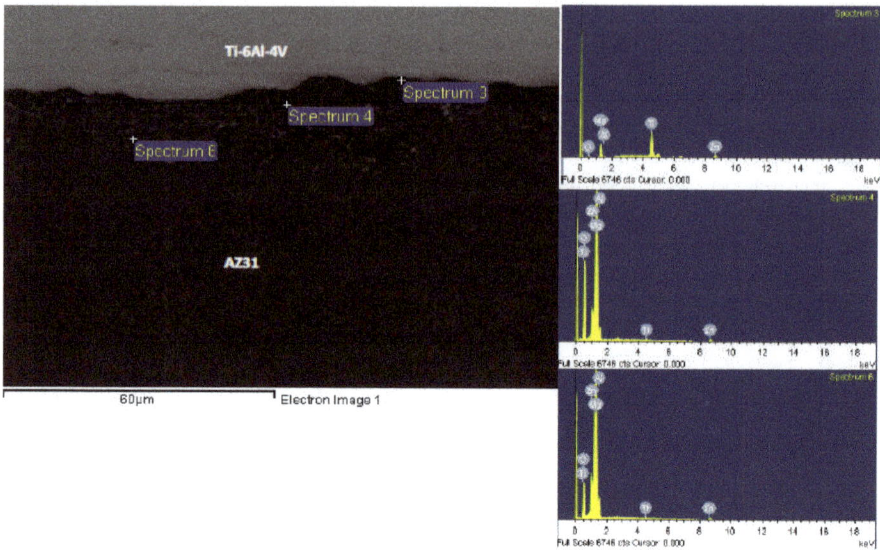

Figure 6. SEM micrograph of the bond made at 30 min and corresponding EDS spot analysis.

4. Analysis of the Fractured Surfaces

Bonds made at 500 °C with different bonding times were fractured in order to study the morphology and composition of fractured surfaces and, therefore, determine the mechanism and type of fracture. The bonds were fractured using a shear test, which will be discussed in Section 5. Figure 7a,b show the fractured surfaces for bond made at 5 min where Figure 7a represents the Mg fractured surface and Figure 7b represents the Ti fractured surface. Table 1 shows the corresponding elemental composition obtained by EDS spot analysis. In Figure 7a, there are lamellar structure (eutectic like regions) as well as scattered white regions and dark regions in the Mg side. These specific regions were analyzed by EDS. To follow the labeling in the figure, A1 (eutectic) reveals the presence of 46.9 wt% and 47.0 wt% that corresponded to 67.1 at% and 25.0 at% atomic percent for Mg and Zn, respectively. It should be noted that in the Mg-Zn binary phase diagram, the atomic ratio of Zn should be 28.1% in order for the eutectic reaction to occur. Mg and MgZn IMC's are predicted to form as a result of the eutectic reaction. Therefore, the atomic composition that is corresponding to the lamellar structure A1 (eutectic) can be attributed to eutectic MgZn and Mg phases. The presence of 7.9 at% of Al in this Mg-Zn eutectic region can be explained by the Al-Mg-Zn ternary phase diagram where Al can be dissolved in this eutectic region. The A2 (white) region is richer in Zn compared to A1 (eutectic) region. This region could contain other IMC's based on Zn and Mg. The percentage of Al in this region is not significantly different from the A1 (eutectic) region. Therefore, no Al based IMC's is expected to form in the A2 region. The dark region presented by A3 (dark) consists of high quantity of Mg (~94%) with less than 5% of both Zn and Al. This region consists of the Mg-based alloy where most of Zn was diffused away and isothermal solidification in this region took place. In Figure 7b, which is the fractured surface of the Ti for the bond made at 5 min, two distinctive regions were identified by the backscattered SEM and analyzed by EDS spot analysis. The B1 (white) region gave a composition of about 97% of Ti. Therefore, this region represents the Ti surface with less than 4% of Mg, Zn, and Al. The B2 (dark) region consists mainly of Mg with 72.4%. Much less of Zn and Ti were detected in the B2 (dark) region as 7.3% and 4.6%, respectively. The composition of this region is close to the composition of A3 (dark) in the Mg fractured surface. In addition, there is a significant amount of Al (~15.3%) that was detected in this region. The presence of a higher percentage of Al is understood since the solubility of Al in Mg rich phase is high. It is concluded that A3 and B2 are solid solution phases based on rich

Mg where B1 is a solid solution based on Ti. Since it is known that the mutual solubility of Ti and Mg is very limited, it is expected that Mg and Ti were both diffused through Zn.

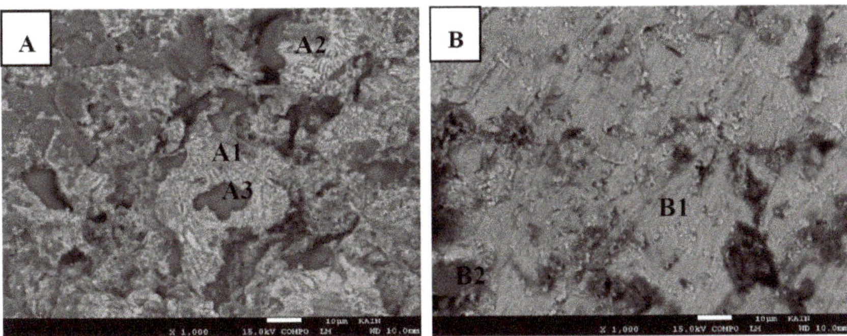

Figure 7. The fractured surfaces of bond made at 5 min. (**A**) Mg side. (**B**) Ti side.

Table 1. Corresponding EDS weight/atomic composition for Figure 7.

Region	Mg (wt%/at%)	Zn (wt%/at%)	Ti (wt%/at%)	Al (wt%/at%)
A1 (eutectic)	46.9/67.1	47.0/25.0	0	6.1/7.9
A2 (white)	39.9/60.8	53.7/30.4	0	6.4/8.8
A3 (dark)	87.1/93.5	10.5/4.2	0	2.4/2.3
B1 (white)	0.9/1.8	1.5/1.1	96.6/96.1	1.1/1.8
B2 (dark)	61.4/72.7	16.6/7.3	7.7/4.6	14.4/15.3

Figure 8a,b show the fractured surfaces for bond made at 20 min where Figure 8a represents the Mg side and Figure 8b represents the Ti side. Table 2 shows the corresponding elemental composition obtained by EDS spot analysis. Less lamellar structure regions are present in Figure 8a compared to Figure 7a. This can be due to more diffusion of Mg into the joint region. Therefore, the solid solution based on Mg became the major structure, as predicted from the Mg-Zn phase diagram. EDS spot analysis used to determine the composition of the white region and dark region appeared in the backscattered SEM micrograph. The white regions A2 (white) observed in Figure 7 seem to disappear for a bond made at 20 min. This region is rich in Zn. Therefore, with more bonding time, Zn diffused away from the joint region. The C2 (dark) region that is rich in Mg looks like the A3 (dark) in terms of composition except that more Al is present for a bond made at 20 min. In the Ti side, the D1 (white) region is rich in Ti but has significantly more Zn compared to the B1 (white) region. This indicates that, at a longer time, more mutual diffusion between Ti and Zn occurs. The D2 (dark) is a rich Mg region present at the Ti fractured surface and it showed more Mg and less Zn compared to B2 (dark). Figure 9 and the corresponding elemental composition shown in Table 3 reveals similar information for a bond made at 30 min. When comparing the elemental compositions for the various regions in Figure 9 with Figure 8, it can be seen that there is no significant difference among them. The bond made at 20 min could reach the isothermal solidification stage. Therefore, no major mechanisms or changes in compositions were revealed at longer bonding times.

Table 2. Corresponding EDS weight/atomic composition for Figure 8.

Region	Mg (wt%/at%)	Zn (wt%/at%)	Ti (wt%/at%)	Al (wt%/at%)
C1 (white)	65.5/81.3	30.1/13.9	0	4.3/4.9
C2 (dark)	88.0/92.4	6.8/2.6	0	5.2/4.9
D1 (white)	2.1/4.2	13.6/10.1	84.1/85.3	0.2/0.4
D2 (dark)	72.6/81.7	4.2/3.1	13.7/8.3	3.9/4.5

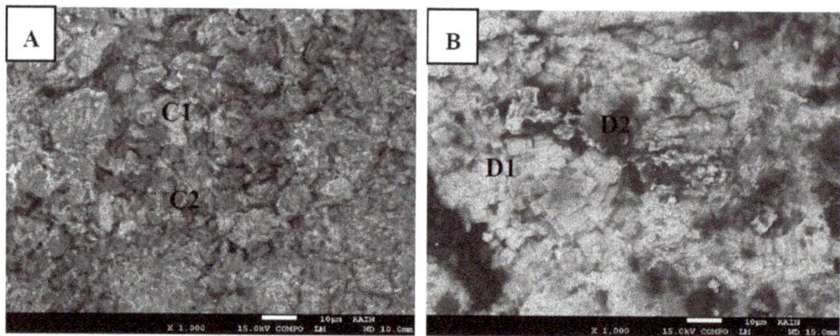

Figure 8. The fractured surfaces of bond made at 20 min. (**A**) Mg side. (**B**) Ti side.

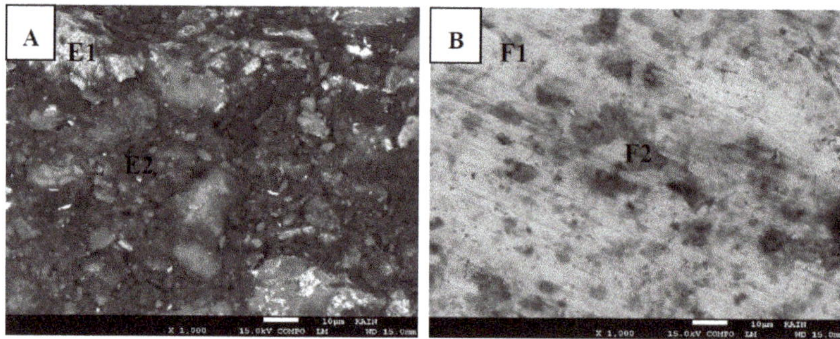

Figure 9. The fractured surfaces of bond made at 30 min. (**A**) Mg side. (**B**) Ti side.

Table 3. Corresponding EDS weight/atomic composition for Figure 9.

Region	Mg (wt%/at%)	Zn (wt%/at%)	Ti (wt%/at%)	Al (wt%/at%)
E1 (white)	62.8/76.9	27.4/12.5	0.3/0.2	9.5/10.5
E2 (dark)	85.3/89.0	5.2/2.0	0	9.5/8.9
F1 (white)	2.8/5.4	5.7/4.1	91.4/90.2	0.2/0.3
F2 (dark)	78.8/87.7	3.9/1.6	15.3/8.6	2.1/2.1

XRD analysis was used to identify the phases formed at the fractured surfaces for bonds made at 5 min and bonds made at 30 min, as shown in Figure 10. The fractured surfaces for the magnesium side for the bond made at 5 min (Figure 10a) and the bond made at 30 min (Figure 10c) show similar patterns. High intensity peaks of Mg were observed. The patterns also showed the Zn and $MgZn_2$ phases. No Ti-related peaks were observed on the Mg fractured surface. On the other hand, the titanium side of the fractured surfaces shows a strong peak of Mg, as seen in Figure 10b,d. Furthermore, the patterns from the Ti fractured surfaces show the presence of Ti, Zn, and $MgZn_2$. From the XRD patterns, only $MgZn_2$ IMC was detected at both fractured surfaces. No indication of IMC's based on Ti-Zn or Ti-Al, which indicates that, at the selected bonding conditions, the only stable phase that can be formed is the $MgZn_2$. The XRD patterns agreed with the SEM/EDS observations from the fractured surface, which reveal a considerable amount of Mg at both fractured surfaces. There is no significant difference and no new compounds detected by XRD through fractures at the bond made at 5 min and fracture for the bond made at 30 min. The mechanism of the bonds was not changed with changing bonding time except the change in concentrations of the elements composing the joints where the fracture occurs. Although Al was detected by SEM/EDS in a considerably noticeable amount, Al did not form IMC's at the joint and, therefore, it could be present as a solute in the Mg-Zn eutectic and at the Mg-rich phase, which aligned with previous studies [28]. XRD analysis did not detect Al or Al-related IMC's.

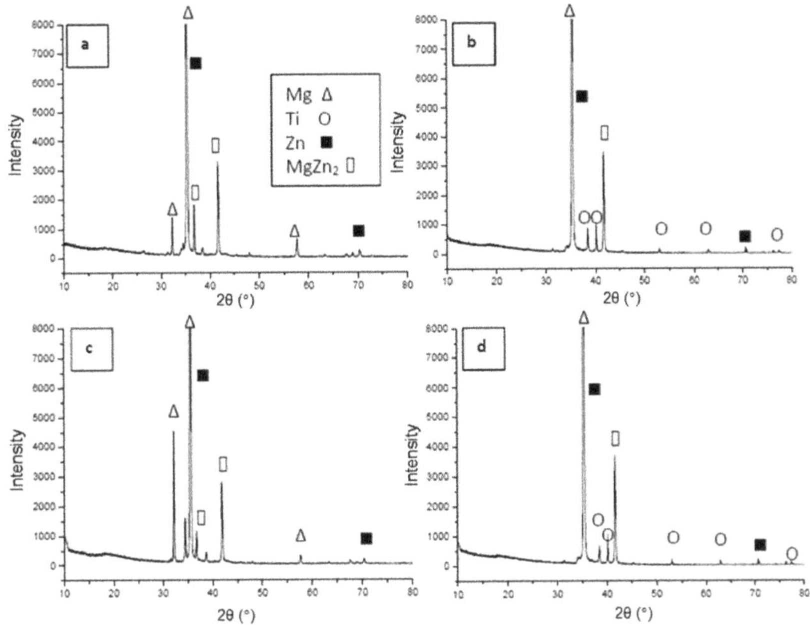

Figure 10. XRD patterns of the fractured surfaces for the bond made at 5 min. (**a**) Mg side and (**b**) Ti side and XRD patterns of the fractured surfaces for the bond made at 30 min. (**c**) Mg side and (**d**) Ti side.

5. Shear Strength and Micro-Hardness Measurements

The shear test was conducted for the bonds made at various bonding times. Table 4 shows the maximum load and maximum shear strength applied against each bond at the fracture point. There is an increase of the shear strength with the increase of bonding time from the bond made at 5 min to the bond made at 20 min where the maximum strength achieved was 30.5 MPa. On the other hand, when bonding time increases for more than 20 min, a slight decrease of shear strength was noticed. The optimum measured shear strength among all bonds was seen to be related to the bond made at 20 min. The load vs. extension was plotted for three bonds as seen in Figure 11 to reveal the nature of the fracture. The shear tests graph shown in the figure indicate that elongation (extension) occurs before the fracture, which means that the fracture is a ductile fracture. The extension before the fracture was measured to be 1.4 mm, 1.9 mm, and 2.25 mm for 5, 20, and 30 min bonds, respectively. The ductile fracture observed in the shear test measurements could be more evidence for the nature of the material at the joint where the fracture propagates. IMC's are usually brittle in nature while solid solution is ductile in nature. Since microstructural analysis along with XRD analysis showed that, the joint regions mainly consist of solid solutions of Mg, Ti, Zn and Al with no major formation of IMC's, the fracture is expected to propagate along the grains of the solid solution. The fracture occurs within the joint region mainly occupied by Mg and Zn where some Ti was detected in the joint region due to the diffusion of Ti into Zn and into MgZn. Therefore, the mechanisms of joining starts with Mg-Zn eutectic formation followed by diffusion of Zn into the Mg side and little diffusion of Zn into the Ti side. This process coincides with the diffusion/dissolution of both Mg and Ti into the joint region in order to form a solid solution.

The eutectic reaction between Mg and Zn occurs at 340 °C at the Mg-rich region (~68 at% Mg), and also occurs at 419.5 °C at the Zn-rich region (~92 at% Zn). A two eutectic points that are well below our selected bonding temperature (500 °C) would highly speed the process of TLP bonding. TLP bonding usually starts with inter-diffusion of the interlayer and base materials followed by eutectic reaction and then isothermal solidification. It ends with homogenization of the joint region. Therefore, in our case, the isothermal solidification was believed to be complete for the bonds made at 20 min where less than 7 wt% of Zn was present in the Mg side of the fracture surface seen in Table 2. When comparing the Zn content of Table 2 to the Zn content in Table 3, it can be noted that little reduction of Zn was measured for the bond made at 30 min. This indicates that the bonds made at more than 20 min were in the homogenization stage of the TLP bonding, which is a stage after isothermal solidification [6]. Al was also seen to diffuse into this joint region in all bonds made, which is proven by EDS analysis in Figure 4. This observation agrees with the work done to join Mg to Ti using the spark plasma sintering process (SPS) [18]. However, Al in our research project did not contribute to the joining process by forming any noticeable IMC's. The only detected IMC that was formed in the joint region is the $MgZn_2$, which indicate that, at the given bonding conditions, this compound is the most stable due to the fast eutectic reaction between Zn and Mg. Ti was detected at the joint region, which implies that Ti diffuses into the molten Zn and then solidifies in the matrix. This is expected by studying the Ti-Zn reaction interfaces. The time and temperature for Ti diffusion in Zn were reported to be 30 min and 500 °C, respectively [24]. This is also because both Mg and Ti have very limited mutual solubility, which makes it difficult for Ti and Mg to diffuse through each other even at a bonding temperature of 500 °C.

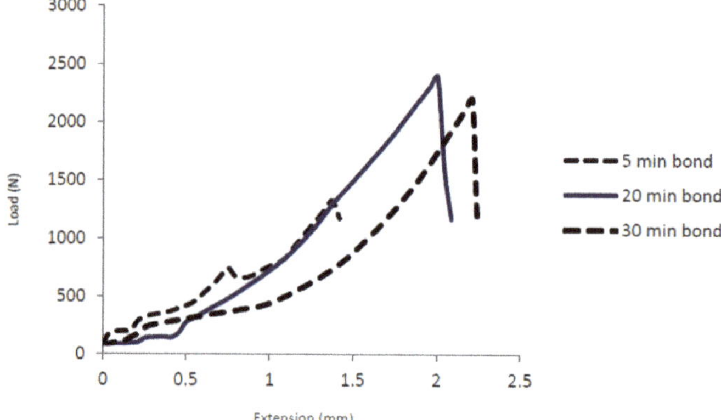

Figure 11. Load vs extension for the shear test of the 5-, 20-, and 30-min bonds.

Table 4. Maximum load and shear strength of the bonds made at different bonding times.

Sample	5 min	10 min	15 min	20 min	25 min	30 min
Force (N)	1324	1513	1954	2396	2253	2197.6
Strength (MPa)	16.8	19.2	24.8	30.5	28.7	28.0

6. Conclusions

This research has shown that, despite significant differences in their physical and mechanical properties, Mg-AZ31 and Ti-6Al-4V alloys can be joined by using the TLP bonding method. The screening printing process was applied for the first time in the TLP process. This process can replace other complex processes of coatings like electroplating, thermal coatings, and physical vapour deposition (PVD). At a bonding temperature of 500 °C and various bonding times of 5, 10, 15, 20, 25, and 30 min, the joints were successfully achieved. Microstructural analysis showed the presence of Mg and Ti at the joint region as well as Zn diffused away from the joint region. There was no reaction layer formed at the joint region that changed in size with the change of bonding time, which indicates that the joining process does not rely on a growth of intermetallic compound layers. The detection of Mg in a large amount compared to Zn at the joint region indicates that, after mutual melting (Mg-Zn eutectic formation), zinc diffused away where Mg diffused to the joint region and where isothermal solidification occurred. Isothermal solidification can occur because of inter-diffusion between the coating material and the base alloy at a temperature above the eutectic temperature. On the other hand, the mechanism of bonding at the Ti side relies on the diffusion of molten Zn into Ti and the dissolution/diffusion of Ti into the joint region where Zn is present. Ti will preferably diffuse into the zinc-rich region at the joint and then will solidify in the melt. Al also diffuses into the joint region and contributes to the solidified melt since the solubility of Al in both Zn and Mg is high. XRD confirms the detection of $MgZn_2$ IMC. However, this IMC is not the major phase at the joint region. The shear strength analysis confirms the ductile nature of the joint and gives a maximum shear strength of 30.5 MPa for the bond made at 20 min where isothermal solidification is completed. The slight reduction of the joint strength for the 25-min and 30-min bonds could be due to the softening of the Mg alloy at the homogenization stage.

Author Contributions: The proposal of this research project was made by A.A. and H.A. The experimental works were designed by A.A. Zinc coatings were designed and performed by M.H. and I.A. The bonding experiments were carried out by H.S.A. and I.A. The analysis and characterizations were carried out by M.A.S. and I.A. The results and data were analyzed by A.A. and assisted by M.H. The manuscript was written by A.A. and the final draft was revised by H.A.

Funding: This research received fund from Vice Deanship of Scientific Research Chairs, King Saud University, Riyadh, Saudi Arabia.

Acknowledgments: The authors are grateful to the Deanship of Scientific Research, king Saud University for funding through Vice Deanship of Scientific Research Chairs.

Conflicts of Interest: The authors declare no conflict of interest.

References

1. Joost, W.J.; Krajewski, P.E. Towards magnesium alloys for high-volume automotive applications. *Scr. Mater.* **2017**, *128*, 107–112. [CrossRef]
2. Kulekci, M.K. Magnesium and its alloys applications in automotive industry. *Int. J. Adv. Manuf. Technol.* **2008**, *39*, 851–865. [CrossRef]
3. Sachdev, A.K.; Kulkarni, K.; Fang, Z.Z.; Yang, R.; Girshov, V. Titanium for Automotive Applications: Challenges and Opportunities in Materials and Processing. *JOM* **2012**, *64*, 553–565. [CrossRef]
4. Murray, J.L. The Mg-Ti (Magnesium-Titanium) System. *Bull. Alloy Phase Diagr.* **1986**, *7*, 245–248. [CrossRef]
5. Cook, G.O., III; Sorensen, C.D. Overview of transient liquid phase and partial transient liquid phase bonding. *J. Mater. Sci.* **2011**, *46*, 5305–5323. [CrossRef]
6. Tuah-Poku, I.; Dollar, M.; Massalski, T.B. A Study of the Transient Liquid Phase Bonding Process Applied to a Ag/Cu/Ag Sandwich Joint. *Metall. Trans. A* **1988**, *19A*, 675. [CrossRef]
7. Illingworth, T.C.; Golosnoy, I.O.; Clyne, T.W. Modelling of transient liquid phase bonding in binary systems—A new parametric study. *Mater. Sci. Eng. A* **2007**, *445–446*, 493–500. [CrossRef]
8. Azqadan, E.; Ekrami, A. Transient liquid phase bonding of dual phase steels using Fe-based, Ni-based, and pure Cu interlayers. *J. Manuf. Process.* **2017**, *30*, 106–115. [CrossRef]
9. Assadi, H.; Shrzadi, A.A.; Wallach, E.R. Transient Liquid Phase Diffusion Bonding under a temperature gradient: Modeling of the interface morphology. *Acta Mater.* **2011**, *49*, 31–39. [CrossRef]
10. Cooke, K. Diffusion Bonding and Characterization of a Dispersion Strengthen Aluminum Alloy. Ph.D. Thesis, University of Calgary, Calgary, AB, Canada, 2011.
11. Jin, Y.J.; Khan, T.I. Effect of bonding time on microstructure and mechanicalproperties of transient liquid phase bonded magnesium AZ31 alloy. *Mater. Des.* **2012**, *38*, 32–37. [CrossRef]
12. AlHazaa, A.; Khan, T.I.; Haq, I. Transient liquid phase (TLP) bonding of Al7075 to Ti–6Al–4V alloy. *Mater. Charact.* **2010**, *61*, 312–317. [CrossRef]
13. Alhazaa, A.N.; Khan, T.I. Diffusion bonding of Al7075 to Ti–6Al–4V using Cu coatings and Sn–3.6Ag–1Cu interlayers. *J. Alloys Compd.* **2010**, *494*, 351–358. [CrossRef]
14. Elthalabawy, W.; Khan, T.I. Eutectic bonding of austenitic stainless steel 316L to magnesium alloy AZ31 using copper interlayer. *Int. J. Adv. Manuf. Technol.* **2011**, *55*, 235–241. [CrossRef]

15. Zhang, J.; Luo, G.; Wang, Y.; Shen, Q.; Zhang, L. An investigation on diffusion bonding of aluminum and magnesium using a Ni interlayer. *Mater. Lett.* **2012**, *83*, 189–191. [CrossRef]

16. Atieh, A.M.; Khan, T.I. TLP bonding of Mg AZ31 to Ti-6Al-4V using pure Ni electro-deposited coats. *J. Mater. Process. Technol.* **2014**, *214*, 3158–3168. [CrossRef]

17. AlHazaa, A.N. Effect of bonding temperature on the microstructure and strength of the joint between magnesium AZ31 and Ti-6Al-4V alloys using Cu coatings and Sn interlayers. *Key Eng. Mater.* **2017**, *735*, 34–41. [CrossRef]

18. Pripanapong, P.; Umeda, J.; Imai, H.; Takahashi, M.; Kondoh, K. Tensile Strength of Ti/Mg Alloys Dissimilar Bonding Material Fabricated by Spark Plasma Sintering. *Int. J. Eng. Innov. Res.* **2016**, *5*, 2277–5668.

19. Murray, J.L. The Titanium-Zinc system. *Bull. Alloy Phase Diagr.* **1984**, *5*, 52–56. [CrossRef]

20. Cao, X.; Jahazi, M.; Immarigeon, J.P.; Wallace, W. A review of laser welding techniques for magnesium, alloy. *J. Mater. Process. Technol.* **2006**, *171*, 188–204. [CrossRef]

21. Bermudez, B.; Sohn, K. Diffusion Couple Investigation of the Mg-Zn System. In *Magnesium Technology*; Springer: Berlin/Heidelberg, Germany, 2012; pp. 323–327.

22. Kammerer, C.; Kulkami, N.; Warmack, R.; Belova, K.P.I.; Murch, G.; Sohn, Y. Impurity Diffusion Coefficients of Al and Zn in Mg determined from solid-to-solid diffusion couples. In *Magnesium Technology*; Springer: Berlin/Heidelberg, Germany, 2014; pp. 505–509.

23. Zhao, L.M.; Zhang, Z.D. Effect of Zn alloy interlayer on interface microstructure and strength of diffusion-bonded Mg–Al joints. *Scr. Mater.* **2008**, *58*, 283–286. [CrossRef]

24. Vassilev, G.P.; Liu, X.J.; Ishida, K. Reaction kinetics and phase diagram studies in the Ti–Zn system. *J. Alloy. Compd.* **2004**, *375*, 162–170. [CrossRef]

25. Li, M.; Li, Y.T.; Li, D.W.; Long, Y.T. Recent developments and applications of screen-printed electrodes in environmental assays—A review. *Anal. Chim. Acta* **2012**, *734*, 31–44. [CrossRef] [PubMed]

26. Miriyev, A.; Levy, A.; Kalabukhov, S.; Frage, N. Interface evolution and shear strength of Al/Ti bi-metals processed by a spark plasma sintering (SPS) apparatus. *J. Alloy. Compd.* **2016**, *678*, 329–336. [CrossRef]

27. Xu, L.; Cui, Y.Y.; Hao, Y.L.; Yang, R. Growth of intermetallic layer in multilaminated Ti/Al diffusion couples. *Mater. Sci. Eng. A* **2016**, *435*, 638–647.

28. Materials Science International Team MSIT. Al-Mg-Zn (Aluminium—Magnesium—Zinc). In *Light Metal Systems. Part 3. Landolt-Börnstein—Group IV Physical Chemistry (Numerical Data and Functional Relationships in Science and Technology)*; Effenberg, G., Ilyenko, S., Eds.; Springer: Berlin/Heidelberg, Germany, 2018; Volume 11A3.

Article

Stress Concentration Induced by the Crystal Orientation in the Transient-Liquid-Phase Bonded Joint of Single-Crystalline Ni_3Al

Hongbo Qin [1,2,*], Tianfeng Kuang [1,2], Qi Li [1], Xiong Yue [1], Haitao Gao [1], Fengmei Liu [1] and Yaoyong Yi [1,*]

[1] Guangdong Provincial Key Laboratory of Advanced Welding Technology, Guangdong Welding Institute (China-Ukraine E.O. Paton Institute of Welding), Guangzhou 541630, China

[2] Key Laboratory of Guangxi Manufacturing System and Advanced Manufacturing Technology, Guilin University of Electronic Technology, Guilin 541004, China

* Correspondence: qinhb@guet.edu.cn (H.Q.); yiyy@gwi.gd.cn (Y.Y.)

Received: 8 August 2019; Accepted: 26 August 2019; Published: 28 August 2019

Abstract: Single-crystalline Ni_3Al-based superalloys have been widely used in aviation, aerospace, and military fields because of their excellent mechanical properties, especially at extremely high temperatures. Usually, single-crystalline Ni_3Al-based superalloys are welded together by a Ni_3Al-based polycrystalline alloy via transient liquid phase (TLP) bonding. In this study, the elastic constants of single-crystalline Ni_3Al were calculated via density functional theory (DFT) and the elastic modulus, shear modulus, and Poisson's ratio of the polycrystalline Ni_3Al were evaluated by the Voigt–Reuss approximation method. The results are in good agreement with previously reported experimental values. Based on the calculated mechanical properties of single-crystalline and polycrystalline Ni_3Al, three-dimensional finite element analysis (FEA) was used to characterize the mechanical behavior of the TLP bonded joint of single-crystalline Ni_3Al. The simulation results reveal obvious stress concentration in the joint because of the different states of crystal orientation between single crystals and polycrystals, which may induce failure in the polycrystalline Ni_3Al and weaken the mechanical strength of the TLP bonded joint. Furthermore, results also show that the decrease in the elastic modulus of the intermediate layer (i.e., polycrystalline Ni_3Al) can relieve the stress concentration and improve the mechanical strength in the TLP bonded joint.

Keywords: Ni_3Al; transient liquid phase bonding; stress concentration; crystal orientation; finite element analysis

---

1. Introduction

Owing to their extremely high tensile strength, toughness, endurance strength, fatigue strength, corrosion resistance, and oxidation resistance at high temperatures (such as $\geq$980 °C), nickel(Ni)-based superalloys have been widely applied in aviation, aerospace, and military fields [1–3]. Because of the poor casting performance of Ni-based superalloys and the complexity of components, Ni-based superalloys usually need to be welded for their applications. The shortcomings of fusion welding and brazing in the connection of Ni-based superalloys (such as shrinkage stresses induced by rapid precipitation, formation of brittle phases, and the tendency to crack due to local fragility, etc.) limit the industrial application of the process for Ni-based superalloys [4,5]. Paulonis et al. proposed a transient liquid phase (TLP) bonding method [4], where in a thin layer of intermediate alloy with a lower melting temperature is employed as the connecting material. During TLP bonding, a low-melting liquid phase is formed between the parent material and the intermediate layer by heating under vacuum conditions; the liquid phase is then homogeneously diffused and therefore isothermally solidified, finally forming a joint with a uniform microstructure. Owing to its reasonable welding temperature, low applied pressure requirement, and ability to form a welded joint with excellent mechanical properties, TLP bonding is widely considered as one of the most ideal welding methods for Ni-based superalloys [5–7].

Grain boundaries, as surface imperfections, adversely affect the high-temperature mechanical properties of metals, especially their high-temperature endurance strength. Thus, the high-temperature mechanical properties of single-crystalline alloys are substantially better than those of polycrystalline alloys. Accordingly, single-crystalline alloys are extensively used as the latest generation of Ni-based superalloys. In the TLP bonding of homogenous Ni-based single-crystalline alloys, because the intermediate-layer alloy is a foil-belt amorphous or powdery Ni-based alloy, a region of polycrystalline Ni-based alloy is unavoidably formed in the joint after TLP bonding; the single-crystalline Ni-based alloy (i.e., parent alloy) is usually connected by a polycrystalline Ni-based alloy (i.e., intermediate layer) after TLP bonding. Extensive investigations on the compositions, heat treatments, mechanical properties, and simulations of single-crystalline Ni-based alloys have been reported [8–15]. In addition, some experimental studies that focus on the bonding process, intermediate-layer alloys, or joint microstructures in the welding joints of single-crystalline Ni-based alloys have recently been published [7,16–20]. Thus far, because of limitations in experimental characterization methods (e.g., the magnitude and distribution of stress and strain in the TLP bonded joint cannot be experimentally measured), few studies have been conducted to analyze the influence of the different states of crystallographic orientations between the single-crystalline parent material and the polycrystalline intermediate layer on the mechanical behavior of a TLP bonded joint.

Ni_3Al-based alloys belong to Ni-based superalloys, which have excellent mechanical and physical properties [21]. In the present study, the mechanical properties of single-crystalline and polycrystalline Ni_3Al were calculated by first-principles calculations based on density functional theory (DFT). In addition, the mechanical behavior of the TLP bonded joint of single-crystalline Ni_3Al under a simple tension load was studied via the finite element (FE) method, and the stress concentration caused by the different states of crystallographic orientations between the single-crystalline parent alloy and the polycrystalline intermediate layer was evaluated.

2. Simulation Methods and Details

Intermetallic compound Ni_3Al has a face-centered cubic lattice structure and *Pm-3m* space group; its lattice constant is $a = 3.572$ Å [22], as shown in Figure 1a. In this study, the CASTEP program [23] was used to perform first-principles calculations based on DFT. In the lattice-structure optimization and elastic-constant calculation, the local density approximation proposed by Ceperley and Alder was applied to investigate the exchange-correlation potential [24]. In addition, Vanderbilt ultra-soft pseudo potentials [25] and the Broyden–Fletcher–Goldfarb–Shanno algorithm [26] were also used during the lattice-structure optimization. The energy cutoff was taken as 600 eV, the *k*-points were set to $10 \times 10 \times 10$, and the convergence tolerance of energy was set as 5.0×10^{-6} eV/atom. The self-consistent field had a convergence accuracy of 5.0×10^{-7} eV/atom, and the maximum ionic Hellmann–Feynman force was 0.01 eV/Å. The stress deviation during the calculation was less than 0.02 GPa.

The length of the TLP bonding sample, the diameter of the joint, and the thickness of the intermediate layer were 66 mm, 5 mm, and 80 μm, respectively, as shown in Figure 1b,c.

A simple tension load of 50 MPa was applied to the ends of the sample at room temperature. Given the axisymmetry of the sample, a one-fourth (1/4) symmetric FE model was established for three-dimensional (3D) finite element analysis (FEA) (Figure 1d), which can drastically reduce calculations and save time. The FE model comprised the parent alloy (i.e., single-crystalline Ni_3Al) and an intermediate-layer alloy (i.e., polycrystalline Ni_3Al), and there were 50,688 elements and 221,201 nodes. Obviously, the mesh density of the TLP bonded joint in the model, as further shown in the magnification of the indicated zone in Figure 1d, was not sufficiently fine to accurately analyze the localized stress and strain distribution; thus, the submodel method was adopted in the FEA on the basis of Saint-Venant's principle, as displayed in Figure 1e. The location of the submodel is the same as the indicated zone in Figure 1d, and the submodel comprises 186,850 elements and 775,736 nodes. All FE calculations were performed with the ABAQUS 2018 software, and quadratic complete integration (C3D20) and full Newton iteration were employed to accurately solve the stress–strain relationship.

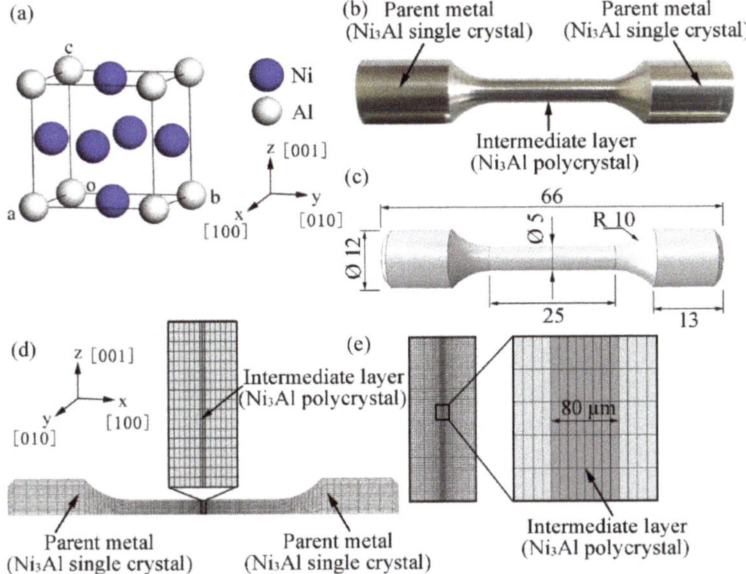

Figure 1. TLP bonded joint sample: (**a**) crystal structure of cubic Ni₃Al, (**b**) a sample prepared in experiments, (**c**) the geometry of the sample for FE modeling (mm), and (**d**) the one-fourth 3D FE model and (**e**) submodel.

3. Results and Discussion

3.1. Mechanical Properties of Ni₃Al

In the elastic stage, the stress and strain relationship can be described by Hooke's law: $\sigma_{ij} = D_{ijkl}\varepsilon_{ij}$, where D_{ijkl} denotes the elastic constants. According to the symmetry of the crystal lattice, the elastic stress–strain matrix $[D_{ijkl}]$ of single-crystalline Ni₃Al can be expressed as

$$\begin{pmatrix} \sigma_{11} \\ \sigma_{22} \\ \sigma_{33} \\ \sigma_{12} \\ \sigma_{13} \\ \sigma_{23} \end{pmatrix} = \begin{pmatrix} D_{1111} & D_{1122} & D_{1133} & 0 & 0 & 0 \\ & D_{2222} & D_{2233} & 0 & 0 & 0 \\ & & D_{3333} & 0 & 0 & 0 \\ & & & D_{1212} & 0 & 0 \\ & Sym. & & & D_{1313} & 0 \\ & & & & & D_{2323} \end{pmatrix} \begin{pmatrix} \varepsilon_{11} \\ \varepsilon_{22} \\ \varepsilon_{33} \\ \gamma_{12} \\ \gamma_{13} \\ \gamma_{23} \end{pmatrix}$$

where σ_{ij}, ε_{ij}, and γ_{ij} are the stresses, normal strains, and shearing strains respectively. The stiffness matrix $[D_{ijkl}]$ was calculated by linearly fitting four small strains (±0.001 and ±0.003) under nine deformation conditions (see Supplementary Figure S1); the calculated elastic constants are listed in Table 1. The flexibility matrix $[S_{ijkl}]$ was calculated as the inverse matrix of the stiffness matrix $[D_{ijkl}]$, i.e., $[S_{ijkl}] = [D_{ijkl}]^{-1}$; the results for elastic constants S_{ijkl} are given in Table 2.

Table 1. Elastic constants D_{ijkl} for single-crystalline Ni$_3$Al.

Elastic Constant	D_{1111}	D_{1122}	D_{1133}	D_{2222}	D_{2233}	D_{3333}	D_{1212}	D_{1313}	D_{2323}
Present work	240.10	160.03	160.03	240.10	160.03	240.10	123.83	123.83	123.83
Experiment [27]	224.3	148.6	148.6	224.3	148.6	224.3	125.8	125.8	125.8

Table 2. Elastic constants S_{ijkl} for single-crystalline Ni$_3$Al.

Elastic Constant	S_{1111}	S_{1122}	S_{1133}	S_{2222}	S_{2233}	S_{3333}	S_{1212}	S_{1313}	S_{2323}
Present work	0.009	−0.004	−0.004	0.009	−0.004	0.009	0.008	0.008	0.008

According to the Voigt–Reuss–Hill approximation method [28], the bulk modulus (B) and shear modulus (G) of polycrystalline Ni$_3$Al can be calculated by Equations (1) and (2), respectively:

$$B = \frac{1}{2}\left[\frac{1}{3S_{1111} + 6S_{1122}} + \frac{1}{3}(D_{1111} + 2D_{1122}) \right], \tag{1}$$

$$G = \frac{1}{2}\left[\frac{15}{4S_{1111} - 4S_{1122} + 3S_{1212}} + \frac{1}{5}(D_{1111} - D_{1122} + 3D_{1212}) \right]. \tag{2}$$

The calculated values of B and G are 186.72 GPa and 78.86 GPa respectively. Moreover, the elastic modulus (Young's modulus, E) and Poisson's ratio (v) can be evaluated by Equations (3) and (4) respectively:

$$E = \frac{9BG}{3B + G}, \tag{3}$$

$$v = \frac{3B - E}{6B}. \tag{4}$$

The calculated values of E and v for polycrystalline Ni$_3$Al are 207.38 GPa and 0.315 respectively; these values are similar to the experimental values $E = 203.1$ GPa and $v = 0.305$ at room temperature [29].

3.2. Stress Concentration Induced by the State of Crystal Orientation

The calculated elastic constants D_{ijkl} and the aforementioned mechanical properties were used in FEA for a tension load of 50 MPa, which is much less than the yield strength of Ni$_3$Al (>350 MPa [30]). Covalent bonding and ionic bonding are well known to lead to brittleness in intermetallic compounds. Herein, the crystal plane (111) was selected to evaluate the chemical bonds between Ni–Ni and Ni–Al atoms, and the electron density in the (111) planes was calculated via DFT (Figure 2). From the overlap of the electron densities of Ni–Ni atoms, it is inferred that covalent bonds exist. Meanwhile, the electron charge density around Ni atoms is 1.0, whereas that close to Al atoms is 0.0, indicating that the electrons of Al atoms move toward the Ni atoms and that Ni and Al atoms are bonded together by ionic bonds. Therefore, we assumed that Ni$_3$Al exhibits some brittleness.

According to the maximum principal stress theory, in a simple tension test, failure occurs if the first principal stress (σ_1) reaches the elastic limit stress. Figure 3a1,a2 display the distribution of σ_1 in the submodel and the intermediate layer, where the σ_1 is maximally localized at the edge of the intermediate layer (refer to the red contour). The maximum σ_1 and average σ_1 are 331.9 and 200.1 MPa respectively, and the ratio of the maximum σ_1 to average σ_1 is 1.66. Clearly, different states of crystal orientation induce stress concentration in the intermediate layer, which may lead to failure in this layer. Moreover, Pugh et al. [31] proposed that the ratio between the bulk modulus and the shear modulus (i.e., B/E) can be used to evaluate the ductility of a material; they also proposed that ductile materials have B/E values greater than 1.75. The calculated B/G ratio for Ni$_3$Al in this study is 2.37; we therefore speculate that Ni$_3$Al also has some ductility, which can also be verified from the experimental results reported elsewhere [32].

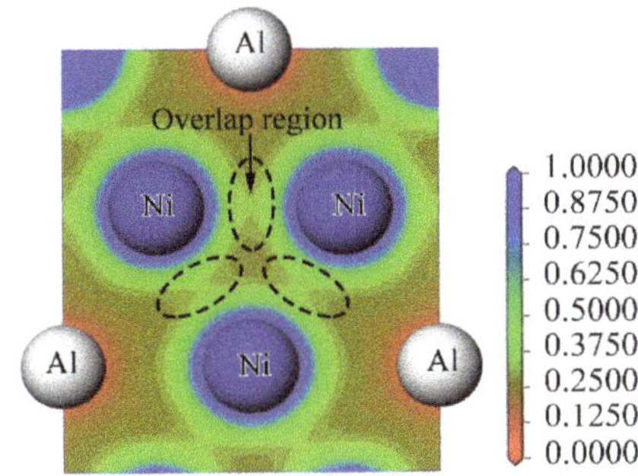

Figure 2. Distribution of electron density in the (111) crystal plane.

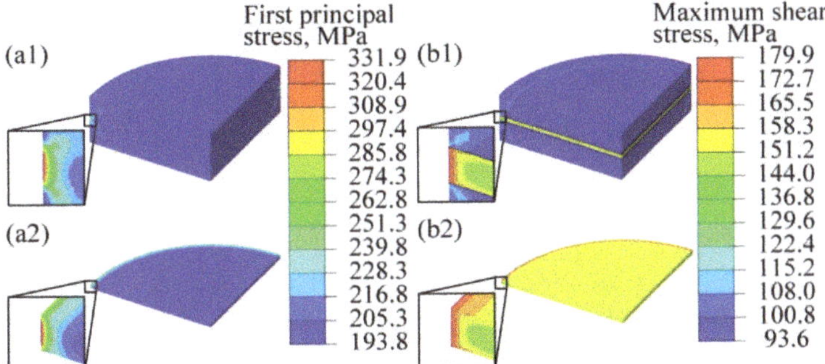

Figure 3. Simulation results of first principal stresses and shear stress: first principal stress in (**a1**) the TLP bonded joint and (**a2**) the intermediate layer; and shear stress in (**b1**) the TLP bonded joint and (**b2**) the intermediate layer.

The maximum shear stress theory assumes that failure occurs when the maximum shear stress reaches the yield point measured in the simple tension test, which is often used during the strength design of ductile materials. Because the maximum principal stress theory mentioned above is not always suitable for ductile materials, the maximum shear stress was used to evaluate the failure behavior of the TLP bonded joint. The maximum shear stress in each element can be expressed as

$$\tau_{\max} = \frac{\sigma_1 - \sigma_3}{2}, \tag{5}$$

where σ_3 is the third principal stress. Figure 3b1,b2 show the distribution of the maximum shear stresses in the submodel of a TLP bonded joint, where the maximum shear stress is located at the edge of the intermediate layer and at the interfaces between the intermediate layer and the parent alloy (refer to the red contour in the figure). In the intermediate layer, the maximum and average $\tau_{\max}$ are 179.9 MPa and 155.9 MPa respectively, and the ratio between the maximum $\tau_{\max}$ and the average $\tau_{\max}$ is 1.15. In addition, our calculation shows that if the intermediate layer is treated as a single alloy with the same orientation as the parent alloy, no stress concentration occurs in the joint and the $\tau_{\max}$ is 100 MPa.

To consider the influence of the second principal stress σ_2 on the stress concentration of a complex stress system, the von Mises stress (equivalent stress, σ_{eq}) was also calculated in this study. The von Mises hypothesis (i.e., the fourth strength theory) [33] holds that a material begins to yield when σ_{eq} reaches the yield strength and that σ_{eq} is given by Equation (6):

$$\sigma_{eq} = \sqrt{\frac{(\sigma_1 - \sigma_2)^2 + (\sigma_2 - \sigma_3)^2 + (\sigma_3 - \sigma_1)^2}{2}}. \tag{6}$$

Meanwhile, to evaluate the stress triaxiality, which affects the initiation and growth of micro-voids and cracks [34] on the failure of the TLP bonded joint, the damage equivalent stress (σ_{eq}^*) in the joint was also investigated. It can be written as follows [34]:

$$\sigma_{eq}^* = \sigma_{eq}\left[\frac{2}{3}(1+v) + 3(1-2v)(T_\sigma)^2\right]^{\frac{1}{2}},\tag{7}$$

where T_σ is stress triaxiality, and

$$T_\sigma = \sigma_m/\sigma_{eq},\tag{8}$$

$$\sigma_m = (\sigma_1 + \sigma_2 + \sigma_3)/3,\tag{9}$$

where σ_m is the hydrostatic stress. Figure 4 shows the simulation results for σ_{eq} and σ_{eq}^* in the TLP bonded joint. Obviously, both σ_{eq} and σ_{eq}^* in the intermediate layer are greater than those in the parent alloy. The σ_{eq} and σ_{eq}^* maximally distribute at the edges of the intermediate layer and at the interfaces between the intermediate layer and parent alloy (refer to the red contour in the figure), the locations of which are similar to those of the maximum τ_{max}. The maximum σ_{eq} and σ_{eq}^* in the intermediate layer are 350.7 GPa and 338.0 GPa respectively. If the intermediate layer is treated as a single alloy with the same orientation as the parent alloy, then no stress concentration occurs and the values of σ_{eq} and σ_{eq}^* are 200 MPa (see Table S1).

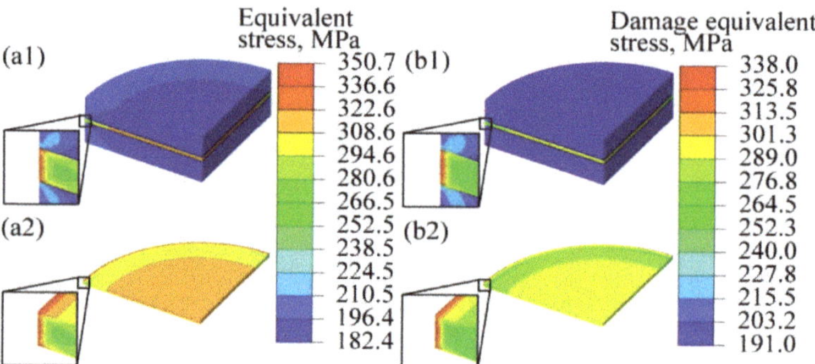

Figure 4. Simulation results of the von Mises equivalent stress and the damage equivalent stress: von Mises equivalent stress in the (**a1**) TLP bonded joint and (**a2**) intermediate layer, and damage equivalent stress in the (**b1**) TLP bonded joint and (**b2**) intermediate layer.

Elastic modulus is the principal mechanical parameter for structural materials, which can be adjusted and controlled by adding alloyed elements or performing heat treatments. Figure 5 presents the influence of the elastic modulus of the intermediate layer (E_{inter}) on the stress concentration, wherein the elastic modulus ranges from 0.85E to 1.15E (E = 207.378 GPa). Clearly, with the decreasing elastic modulus of the intermediate layer, the maximum stresses of first principal stress, maximum shear stress, equivalent stress, and damages equivalent stress decrease linearly. The ratio between the maximum stress and the average stress is also reduced, indicating that reducing E_{inter} can relieve the stress concentration. Also, the average stresses decrease linearly with decreasing elastic modulus of the intermediate layer (see Supplementary Figure S2). Therefore, in addition to increasing the mechanical strength of the intermediate layer, the simulation result in this study suggests that appropriately reducing the elastic modulus of the intermediate layer can also improve the mechanical strength of a TLP bonded joint. To verify this opinion, additional experimental work was also carried out.

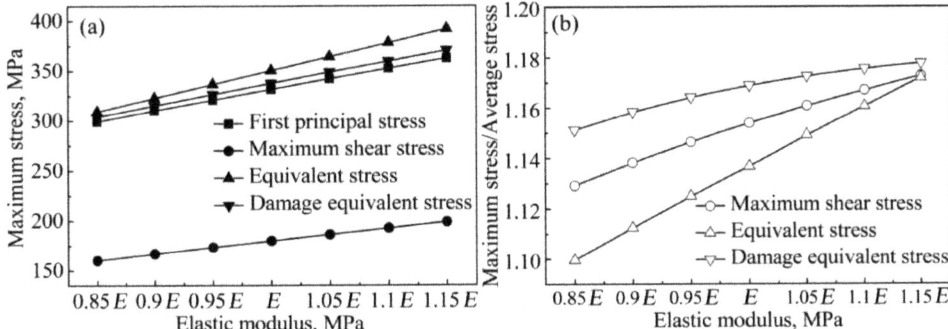

Figure 5. Influence of elastic moduli on (**a**) the maximum stresses and (**b**) the ratio between the maximum stress and the average stress in the TLP bonded joint.

In the experiment, single-crystalline Ni_3Al-based superalloy (named IC10 alloy) samples were prepared by TLP bonding process in a high vacuum diffusion furnace (Centorr Vacuum Industries, Workhorse II, Nashua, NH, USA) and the bonding temperature, time, and pressure were 1250 °C, 6 h, and 5 MPa respectively. The samples of TLP bonded joints and their geometry have been given in Figure 1b,c above, and the grain boundary and crystal orientation in a typical zone in the joint are presented in Supplementary Figure S3. By nanoindentation testing, as illustrated in Supplementary Figure S4, the elastic modulus and hardness of the parent alloy and intermediate-layer alloy were measured. Test results show that the average elastic modulus of parent alloy (E_{parent}) and the average elastic modulus of intermediate-layer alloy E_{inter} are 221.4 GPa and 208.5 GPa respectively (see Supplementary Figure S5), and the ratio of E_{inter} to E_{parent} (i.e., E_{inter}/E_{parent}) is 0.942. Then, to reduce E_{inter} and E_{inter}/E_{parent}, post weld heat treatment (PWHT) was carried out in a heat treatment furnace (SG-QF1400, Shanghai, China). During PWHT, samples of TLP bonded joints were heated to about 1200°C and held for 6 h, and then cooled down in air to room temperature. Because that Boron was added to the intermediate-layer alloy as a melting-point depressant, the atomic percent of Boron in the intermediate-layer alloy was much higher than that in the parent alloy before PWHT. In the PWHT, Boron atoms diffused from the intermediate layer to the parent alloy. As the result of solution strengthening, after PWHT, E_{parent} increased to 223.7 GPa while E_{inter} decreased to 205.4 GPa, and E_{inter}/E_{parent} was reduced to 0.918. According to the result of tensile testing, the average tensile strength after PWHT is 819.0 MPa at room temperature, which is higher than before PWHT (i.e., 798.0 MPa) as shown in Supplementary Figure S6. Clearly, the decrease of E_{inter} or E_{inter}/E_{parent} can improve the mechanical strength of a TLP bonded joint, which supports the simulation result in the manuscript. Notably, the ideal strength of the TLP bonded joint should be the same as or practically be as close as possible to the strength of the parent alloy. Usually, achieving a mechanical strength of a TLP bonded joint greater than 90% of the mechanical strength of a single crystal is difficult in experiments. To further improve the mechanical strength of TLP bonded joints, sufficient attention should be devoted to the stress concentration caused by the different states of crystal orientation between the parent alloy and intermediate layer.

4. Conclusions

Elastic constants of single-crystalline Ni_3Al as well as the elastic modulus, shear modulus, and the Poisson's ratio of polycrystalline Ni_3Al were calculated via DFT, and the calculation results were subsequently verified against previously reported experimental data. Based on the calculated mechanical properties of both single-crystalline and polycrystalline Ni_3Al, 3D FEA was used to characterize the mechanical behavior of the TLP bonded joint of single-crystalline Ni_3Al under a simple tension load. The simulation results revealed obvious stress concentrations in the joint as a result of different states of crystal orientation between the parent metals (single crystals) and intermediate layer (polycrystals), which will lead to failure in the polycrystalline Ni_3Al and thereby weaken the mechanical strength of the TLP bonded joint. The maximum values of the first principal stress, maximum shear stress, equivalent stress, and the damages equivalent stress in the joint decrease linearly with decreasing elastic modulus of the intermediate layer; thus, reducing the elastic modulus of the intermediate layer can relieve the stress concentration and benefit the mechanical reliability of a TLP bonded joint, which can be verified by experiments.

Supplementary Materials: The following are available online at http://www.mdpi.com/1996-1944/12/17/2765/s1, Figure S1: The strain conditions for deriving nine elastic stiffness coefficients; Figure S2: Relationship between the average stress and elastic modulus of intermediate layer; Figure S3: The grain boundary and crystal orientation in the TLP bonded joint: (a) SEM-EBSD image of the joint, (b) grain boundary in the joint; (c) all Euler map; and (d) IPF-Y0 map. The instrument used is FEI Quanta 650F+HKL Channel 5; Figure S4: Nanoindentation test: (a) SEM image of a typical zone in TLP bonded joint; and (b) schematic of a location of nanoindentation array (Berkovich indenter, instrument: Agilent G200); Figure S5: The influence of the post weld heat treatment (PWHT) on the elastic modulus (a) and hardness (b) of parent alloy and intermediate-layer alloy; Figure S6: The tensile strengths of TLP bonded joints before and after PWHT (three samples were measured for each point; instrument: universal testing machine, GP-TS2000M); Table S1: The stress in the joint ignoring the crystal orientation of intermediate layer.

Author Contributions: H.Q. and Y.Y. conceived and designed the research; H.Q., T.K., and Q.L. performed the first principles calculation and analyzed the data; Q.L., X.Y., and H.G. carried out the experimental study; T.K. wrote the manuscript; X.Y. and F.L. reviewed and edited the manuscript. All authors read and approved the final manuscript.

Funding: This research was funded by Implementing Innovation-Driven Development Capacity Building Special Funds project of Guangdong Academy of Sciences (2018GDASCX-1005 & 2018GDASCX-0113), Key Program for International Cooperation of Science and Technology (China-Ukraine, 2015DFR50310).

Conflicts of Interest: The authors declare no conflict of interest.

References

1. Caron, P.; Khan, T. Evolution of Ni-based superalloys for single crystal gas turbine blade applications. *Aerosp. Sci. Technol.* **1999**, *8*, 513–523. [CrossRef]
2. Zhang, P.; Hu, C.; Ding, C.G.; Zhu, Q.; Qin, H.Y. Plastic deformation behavior and processing maps of a Ni-based superalloy. *Mater. Des.* **2015**, *65*, 575–584. [CrossRef]
3. Smith, T.M.; Rao, Y.; Wang, Y.; Ghazisaeidi, M.; Mills, M.J. Diffusion processes during creep at intermediate temperatures in a Ni-based superalloy. *Acta. Mater.* **2017**, *141*, 261–272. [CrossRef]
4. Paulonis, D.F.; Duvall, D.S.; Owczarski, W.A. Diffusion bonding utilizing transient liquid phase. U.S. Patent 3678570, 25 July 1972.
5. Khakian, M.; Nategh, S.; Mirdamadi, S. Effect of bonding time on the microstructure and isothermal solidification completion during transient liquid phase bonding of dissimilar nickel-based superalloys In738lc and nimonic 75. *J. Alloys Compd.* **2015**, *653*, 386–394. [CrossRef]
6. Cook, G.O.; Sorensen, C.D. Overview of transient liquid phase and partial transient liquid phase bonding. *J. Mater. Sci.* **2011**, *46*, 5305–5323. [CrossRef]
7. Yang, Z.W.; Lian, J.; Cai, X.Q.; Wang, Y.; Wang, D.P.; Liu, Y.C. Microstructure and mechanical properties of Ni_3Al-based alloy joint transient liquid phase bonded using Ni/Ti interlayer. *Intermetallics* **2019**, *109*, 179–188. [CrossRef]
8. Ai, C.; Li, S.S.; Zhao, X.B.; Zhou, J.; Guo, Y.J.; Sun, Z.P.; Song, X.D.; Gong, S.K. Influence of solidification history on precipitation behavior of TCP phase in a completely heat-treated Ni_3Al based single crystal superalloy during thermal exposure. *J. Alloys Compd.* **2017**, *722*, 740–745. [CrossRef]
9. Luan, X.H.; Qin, H.B.; Liu, F.M.; Dai, Z.B.; Yi, Y.Y.; Li, Q. The mechanical properties and elastic anisotropies of cubic Ni_3Al from first principles calculations. *Crystals* **2018**, *8*, 307. [CrossRef]
10. Shah, D.M. Orientation dependence of creep behavior of single crystal γ' (Ni_3Al). *Scripta Metal.* **1983**, *17*, 997–1002. [CrossRef]
11. Wang, W.; Jiang, C.B.; Lu, K. Deformation behavior of Ni_3Al single crystals during nanoindentation. *Acta Mater.* **2003**, *51*, 6169–6180. [CrossRef]

12. Wen, Z.X.; Zhang, D.X.; Li, S.W.; Yue, Z.F.; Gao, J.Y. Anisotropic creep damage and fracture mechanism of nickel-base single crystal superalloy under multiaxial stress. *J. Alloys Compd.* **2017**, *692*, 301–312. [CrossRef]
13. Reed, R.C.; Tao, T.; Warnken, N. Alloys-by-design: Application to nickel-based single crystal superalloys. *Acta Mater.* **2009**, *57*, 5898–5913. [CrossRef]
14. Fleischmann, E.; Miller, M.K.; Affeldt, E.; Glatzel, U. Quantitative experimental determination of the solid solution hardening potential of rhenium, tungsten and molybdenum in single-crystal nickel-based superalloys. *Acta Mater.* **2015**, *87*, 350–356. [CrossRef]
15. Huang, M.; Zhu, J. An overview of rhenium effect in single-crystal superalloys. *Rare Metals* **2016**, *35*, 127–139. [CrossRef]
16. Ghoneim, A.; Ojo, O.A. On the influence of boron-addition on tlp bonding time in a Ni$_3$Al-based intermetallic. *Intermetallics* **2010**, *18*, 582–586. [CrossRef]
17. Ojo, O.A.; Ghoneim, O.J.; Hunedy, J. Transient liquid phase bonding of single crystal Ni3Al-based intermetallic alloy. In Proceedings of the International Brazing and Soldering Conference, Las Vegas, NV, USA, 22–25 April 2012.
18. Chai, L.; Huang, J.H.; Hou, J.B.; Lang, B.; Wang, L. Effect of holding time on microstructure and properties of transient liquid-phase-bonded joints of a single crystal alloy. *J. Mater. Eng. Perform.* **2015**, *24*, 2287–2293. [CrossRef]
19. Guo, W.; Wang, H.Y.; Jia, Q.; Peng, P.; Zhu, Y. Transient liquid phase bonding of nickel-base single crystal alloy with a novel Ni-Cr-Co-Mo-W-Ta-Re-B amorphous interlayer. *High Temp. Mat Pr-isr.* **2017**, *36*, 677–682. [CrossRef]
20. Sheng, N.C.; Liu, J.D.; Jin, T.; Sun, X.F.; Hu, Z.Q. Precipitation behaviors in the diffusion affected zone of tlp bonded single crystal superalloy joint. *J. Mater. Sci. Technol.* **2015**, *31*, 129–134. [CrossRef]
21. Polkowski, W.; Pęczek, E.; Zasada, D.; Komorek, Z. Differential speed rolling of Ni$_3$Al based intermetallic alloy—effect of applied processing on structure and mechanical properties anisotropy. *Mater. Sci. Eng. A* **2015**, *647*, 170–183. [CrossRef]
22. Mohan Rao, P.V.; Suryanarayana, S.V.; Satyanarayana Murthy, K.; Nagender Naidu, S.V. The high-temperature thermal expansion of Ni$_3$Al measured by X-ray diffraction and dilation methods. *J. Phys. Condens. Mat.* **1989**, *1*, 5357–5361.
23. Segall, M.D.; Lindan, P.J.D.; Probert, M.J.; Pickard, C.J.; Hasnip, P.J.; Clark, S.J.; Payne, M.C. First-principles simulation: Ideas, illustrations and the CASTEP code. *J. Phys. Condens. Mat.* **2002**, *14*, 2717–2744. [CrossRef]
24. Ceperley, D.M.; Alder, B.J. Ground state of the electron gas by a stochastic method. *Phys. Rev. Lett.* **1980**, *45*, 566–569. [CrossRef]
25. Vanderbilt, D. Soft self-consistent pseudopotentials in a generalized eigenvalue formalism. *Phys. Rev. B* **1990**, *41*, 7892–7895. [CrossRef] [PubMed]
26. Pfrommer, B.G.; Côté, M.; Louie, S.G.; Cohen, M.L. Relaxation of crystals with the quasi-Newton method. *J. Comput. Phys.* **1997**, *131*, 233–240. [CrossRef]
27. Kayser, F.X.; Stassis, C. The elastic constants of Ni$_3$Al at 0 and 23.5 °C. *Phys. Status Solidi A* **1981**, *64*, 335–342. [CrossRef]
28. Hill, R. The elastic behaviour of a crystalline aggregate. *Proc. Phys. Soc. A* **1952**, *65*, 349. [CrossRef]
29. Prikhodko, S.V.; Yang, H.; Ardell, A.J.; Carnes, J.D.; Isaak, D.G. Temperature and composition dependence of the elastic constants of Ni$_3$Al. *Metall. Mater. Trans. A* **1999**, *30*, 2403–2408. [CrossRef]
30. Yoo, M.H.; Liu, C.T. Effect of prestress on tensile yield strength of a Ni$_3$Al alloy. *J. Mater. Res.* **2011**, *3*, 845–847. [CrossRef]
31. Pugh, S.F. XCII. Relations between the elastic moduli and the plastic properties of polycrystalline pure metals. *Dublin Philos. Mag. J. Sci.* **1954**, *45*, 823–843. [CrossRef]
32. Aoki, K.; Izumi, O. On the ductility of the intermetallic compound Ni$_3$Al. *J. Jpn. Inst. Met.* **1977**, *41*, 170–175. [CrossRef]
33. Kolupaev, V.A. *Equivalent Stress Concept For Limit State Analysis*; Springer: Darmstad, Germany, 2018; Volume 86, pp. 16–18.
34. Lemaitre, J. *A Course On Damage Mechanics*, 2nd ed.; Springer: Berlin, Germany, 2012; pp. 39–46.

Article

Microstructural Characterization and Mechanical Behavior of NiTi Shape Memory Alloys Ultrasonic Joints Using Cu Interlayer

Wei Zhang [1], Sansan Ao [1], Joao Pedro Oliveira [2], Zhi Zeng [3,*], Yifei Huang [1] and Zhen Luo [1]

[1] School of Material Science and Engineering, Tianjin University, Tianjin 300072, China; zhang.vv@foxmail.com (W.Z.); ao33@tju.edu.cn (S.A.); yifeihuang@tju.edu.cn (Y.H.); lz_tju@163.com (Z.L.)
[2] UNIDEMI, Departamento de Engenharia Mecânica e Industrial, Faculdade de Ciências e Tecnologia, Universidade Nova de Lisboa, 2829-516 Caparica, Portugal; jp.oliveira@campus.fct.unl.pt
[3] School of Mechanical and Electrical Engineering, University of Electronic Science and Technology of China, Chengdu 221116, China
* Correspondence: zhizeng@uestc.edu.cn; Tel.: +86-159-0814-7867

Received: 4 September 2018; Accepted: 23 September 2018; Published: 26 September 2018

Abstract: NiTi shape memory alloys (SMAs) are a class of functional materials which can be significantly deformed and recover their original shape via a reversible martensitic phase transformation. Developing effective joining techniques can expand the application of SMAs in the medical and engineering fields. In this study, ultrasonic spot welding (USW), a solid-state joining technique, was used to join NiTi sheets using a Cu interlayer in between the two joining sheets. The influence of USW process on the microstructural characteristics and mechanical behavior of the NiTi joints was investigated. Compared with conventional fusion welding techniques, no intermetallic compounds formed in the joints, which is extreme importance for this particular class of alloys. The joining mechanisms involve a combination of shear plastic deformation, mechanical interlocking and formation of micro-welds. A better bonding interface was obtained with higher welding energy levels, which contributed to a higher tensile load. An interfacial fracture mode occurred and the fracture surfaces exhibited both brittle and ductile-like characteristics with the existence of tear ridges and dimples. The fracture initiated at the weak region of the joint border and then propagated through it, leading to tearing of Cu foil at the fracture interface.

Keywords: NiTi shape memory alloys; microstructural characterization; failure behavior; fracture morphology; ultrasonic spot welding

---

1. Introduction

Near equiatomic NiTi is one of the most important shape memory alloys (SMAs) due to its excellent functional properties, namely shape memory effect and superelasticity, combined with high corrosion resistance, as well as, biocompatibility [1,2]. The functional properties originate from a reversible phase transformation between austenite with a B2 cubic structure and a B19′ monoclinic martensite [3,4]. When the material is deformed in the austenitic phase it can exhibit superelastic properties, that is, it can undergo a significant deformation during loading with full recovery to its original shape upon unloading [5]. The excellent mechanical and functional properties exhibited by these alloys make this material widely desired in both medical and engineering fields [6–9]. To achieve successful fabrication of complex parts of NiTi, it is necessary to develop effective and efficient processing technologies due to the poor machinability of these alloys. In recent years, various welding technologies have been used to join NiTi SMAs both to themselves and to other conventional engineering alloys such as stainless steels [10,11], Ti6Al4V alloys [12–14] and Cu-based alloys [15].

Joining techniques such as resistance spot welding [11,16], arc welding [17] and laser welding [10,12] are some examples which are capable to produce defect-free joints. However, a major drawback of the NiTi joints produced by traditional fusion welding techniques is the formation of brittle intermetallic compounds (IMCs) in the weld region, such as Ti_2Ni, which tend to reduce the mechanical strength of joints [17–20]. In addition, fusion welding methods can also contribute to significant changes in the transformation temperatures which can impair the potential applications of the joints [19,21].

Considering the possible formation of IMCs in the weld metal, the addition of an interlayer is suggested as a potential solution to adjust the chemical compositions in the weld region and improve the mechanical properties of NiTi joints [12,22]. Cu is a soft metal with a melting point lower than NiTi, and it shows not only high thermal and electrical conductivity, good corrosion resistance and ductility but also a good metallurgical compatibility with NiTi [23–25]. For this reason, Cu interlayers have been used in dissimilar laser welding of NiTi to titanium alloys or stainless steel [14,26,27] to limit the mixing of the base material (BM) and increase the mechanical properties of the joints. It has been found that proper selection of the thickness of the Cu interlayer can enhance the mechanical properties of joints by reducing the amount of brittle Ni-Ti-based IMCs [14].

The thermal history experienced by NiTi during welding can significantly affect its shape memory and superelastic properties [28]. Thus, it is necessary to reduce the heat input of the welding process to restrict the thermophysical deterioration in the weld zone [26]. Solid state joining techniques are known for their low heat input and possibility to avoid solidification defects. For example, friction welding has been carried out on NiTi SMAs in recent years [29–31]. However, it has been reported for dissimilar joints of NiTi to stainless steel obtained by friction welding that high welding times can promote the formation of brittle phases at the weld interface [30]. Currently, there is a need to develop other solid-state techniques that can successfully join NiTi to itself and to other relevant engineering materials.

Ultrasonic spot welding (USW) is a rapidly developing non-melting joining method which is widely used in plastic forming, electronics and automotive fields [32,33]. As compared to the fusion welding processes, such as resistance and laser welding, USW can produce high strength joints without metal depletion or reduced extension of the heat affected zone. Such translates into almost no detrimental effects produced on the BM. USW is especially suitable for achieving effective joints of miniature components [34,35], such as metallic foils, wires and plates, which make this process particularly interesting to weld materials with lower weldability. Traditionally, studies have been mainly focused on the joining of light materials for weight reduction in industrial products [32–35]. However, knowledge on the use of USW in NiTi SMAs is currently extremely limited. Thus, carrying out USW on NiTi is a very worthwhile investigation since this technique has great potential for the fabrication of variable electromagnetic switches, radiator fins and other components based on NiTi SMAs.

In the current study, the operating procedures, weld morphology, chemical compositions, interface characteristics and the tensile shear properties of the ultrasonic spot welded NiTi joints with Cu interlayer were analyzed and discussed. Additionally, the effects of welding energy on microstructural characteristics and failure behavior were investigated in detail.

2. Materials and Methods

2.1. Materials

Ni-Ti shape memory alloy sheets (50.8 at.%), 150 μm thick, were used as the BM. Copper (99.9% purity) foils with a thickness of 20 μm were chosen as the interlayer material. The as-received NiTi alloy sheets were subjected to cold rolling and subsequently stress relieving annealing by heat treatment at 400 °C for 45 min in Ar atmosphere to stabilize the phase transformation temperatures. The DSC result of the BM showed that the transformation start and finish temperatures of austenite and martensite are −45.43, −12.5, 22.91 and −58.75 °C, respectively, which suggests that the NiTi alloy was fully austenitic at room temperature, presenting superelastic behavior. The NiTi sheets were machined into rectangular specimens of 60 mm length and 15 mm width (8 mm width for tensile test of NiTi BM), and the Cu foil was processed into a square shape of 15 mm × 15 mm, which is equal to the overlapped region of the BM specimens. Prior to welding, the oxide layer on the NiTi surface was removed by using a mixed solution of 7.5% HF, 20% HNO_3 and 72.5% H_2O for 40–50 s.

2.2. Ultrasonic Spot Welding

USW was performed using a SONICS MSC4000-20. The schematic diagram of the USW system is shown in Figure 1a, which was comprised of ultrasonic generator, transducer, amplitude transformer, sonotrode tip and anvil. The size of the sonotrode tip was a square of 8 mm × 8 mm, consisting of 10 × 10 gridding knurls with a spacing and a depth of 0.8 mm and 0.4 mm, respectively. Figure 1b depicts the relative welding positions of both the NiTi and the Cu interlayer. The ultrasonic vibration direction during the USW process was perpendicular to the longitudinal direction of the NiTi sheets. In this study, the energy control mode was used, during which process the welding time and power input were adjusted automatically. The main process parameters selected were a frequency of 20 kHz, a welding amplitude of 55 μm and a constant clamping pressure of 0.38 MPa, with different energies of 500, 700 and 1000 J, respectively. The USW process was performed at the center of the overlapped position, as shown in Figure 1b.

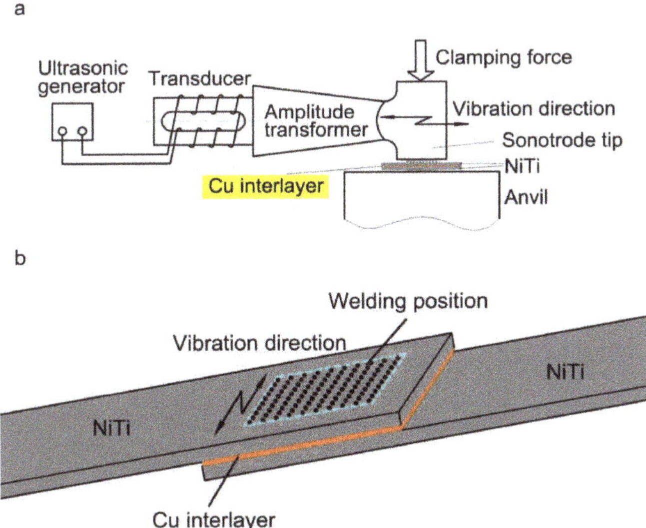

Figure 1. (**a**) Schematic illustration of USW process; (**b**) Schematic diagram showing the welding position of NiTi with Cu interlayer.

2.3. Microstructural Characterization

The surface profiles of the NiTi welds obtained using different welding parameters were examined using a digital microscope (VHX-2000C, KEYENCE, Osaka, Japan). Additionally, the metallographic observations were performed on the transverse section of the weld, which was parallel to the ultrasonic vibration direction. After mounting the cross sections of the welded joints, mechanical polishing using sandpapers from 240 to 2000 grits, followed by polishing solution agent was performed. The etching process was conducted in an acid solution consisting of 3% HF, 14% HNO_3 and 82% H_2O (in volume) with an immersion time of 20–30 s. The microstructures and chemical compositions of the cross sections for different welds were characterized via an optical microscope (OM; GX71, OLYMPUS, Tokyo, Japan) and with a scanning electron microscope (SEM; SU1510, HITACHI, Tokyo, Japan) equipped with an energy dispersive spectroscopy (EDS) analysis system.

2.4. Evaluation of Mechanical Properties

Tensile lap shear tests of different joints were performed using an electro-mechanical universal testing machine (AG-100KNA, SHIMADZU, Kyoto, Japan) and the displacement speed was set at 0.1 mm/min. The tensile performance was expressed as joint failure load, determined as the average of three specimens at each welding energy condition. The micro-morphologies and chemical compositions of the fracture surfaces were analyzed by SEM and EDS. In this present work, the trial experiments had shown that in all the welded joint failure always occurred along the bonding interface. Therefore, it was decided to investigate the phase composition of this region using X-ray diffraction (XRD; D8 Advanced, BRUKER, Karlsruhe, Japan) analysis. XRD was performed at 40 kV and 40 mA with Cu-Kα radiation.

3. Results and Discussions

3.1. Surface Morphology of Welds

Representative surface morphology at the sonotrode tip and anvil sides of different samples after the USW process are exhibited in Figure 2. Figure 3 shows the depths of surface indentations at the edge of samples as depicted with red rectangles.

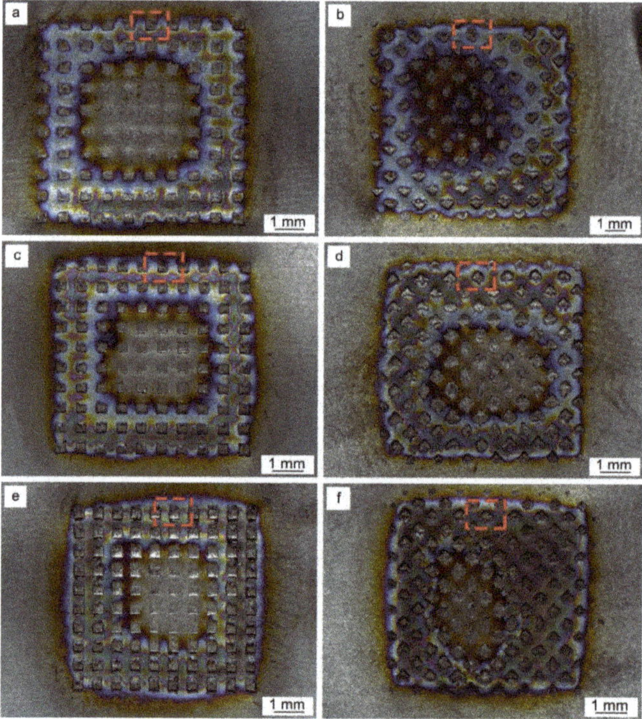

Figure 2. Surface morphology of different samples obtained with different welding energies: (**a**) sonotrode tip and (**b**) anvil sides of 500 J; (**c**) sonotrode tip and (**d**) anvil sides of 700 J; (**e**) sonotrode tip and (**f**) anvil sides of 1000 J. The red squares indicate the regions analyzed to determine the surface indentation depth depicted in Figure 3.

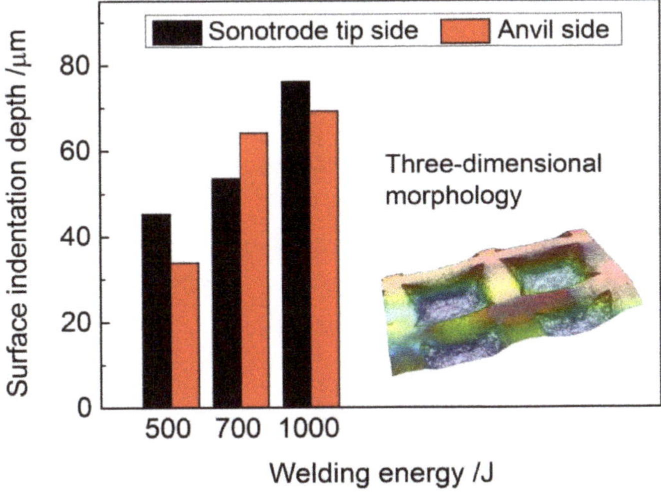

Figure 3. Indentation depths of different samples.

From Figure 2, it can be observed that the indentations of sonotrode tip and anvil were distinct on the specimen surfaces, and distinct levels of oxidation were observed on the weld material surface during welding process, due to NiTi's temperature sensibility and the open structure of the ultrasonic welder. Higher welding energy translates into a higher temperature experienced by the material making it more prone to surface oxidation. With increasing welding energy, the indentation depth in both sides exhibited an increasing trend, as shown in Figure 3, and the insert image shows the three-dimensional morphology of the surface indentation. During the USW process, continuous shear vibrations were applied to the NiTi BM by the action of sonotrode, which quickly generated frictional heat at the weld interface, leading to the rising of temperature and the material was softened, thus shear plastic deformation took place on the material surface [32,36–38], resulting in the formation of indentations. Under the same welding parameters, the depth of indentations depends on the accessibility of the sonotrode knurl penetrating into the NiTi material surface, during which a tighter engagement was provided, leading to more relative slippages and friction at the weld interface [36]. Therefore, with the welding energy increasing from 500 J to 1000 J, more frictional heat was generated at the weld interface during the USW process, and then the plastic deformation of NiTi increased since the welding energy was dispersed by shear deformation, resulting in the increasing depth of indentations on the weld material surface aided by a softening behavior of the material caused by the temperature raise.

3.2. Bonding Morphology of USW Joints

The bonding morphology at the weld interface of the welds obtained by varying the welding energies was examined in the cross section along the center of the joints parallel to the vibration direction as presented in Figure 4.

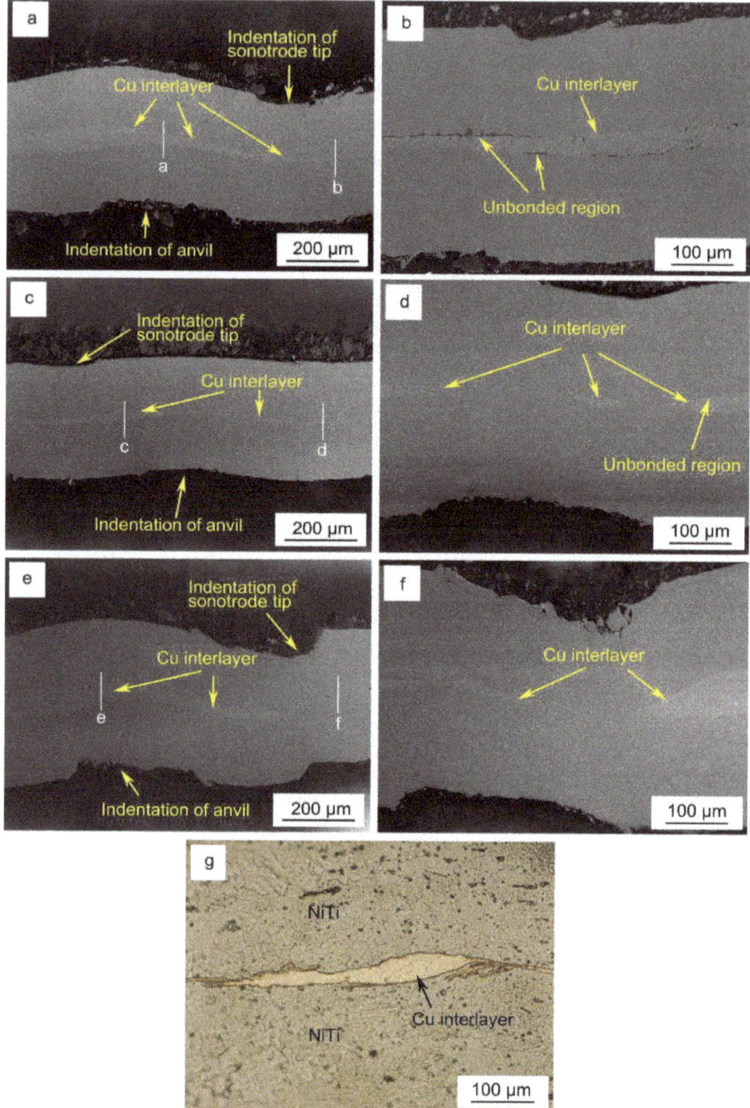

Figure 4. (**a,b**) interface morphology of the welded joint obtained at 500 J at the center (**a**) and near the periphery (**b**) of the joint; (**c,d**) interface morphology of the welded joint obtained at 700 J at the center (**c**) and near the periphery (**d**) of the joint; (**e,f**): interface morphology of the welded joint obtained at 1000 J at the center (**e**) and near the periphery (**f**) of the joint; (**g**) microstructure interface of weld of 1000 J obtained by optical microscope. The white lines in Figure 4. (**a,c,e**) represent the locals where EDS line scans were performed.

Figure 4a,b, respectively, depict the interface morphology in the center and near the periphery of the weld obtained for a welding energy of 500 J. From Figure 4a it can be observed that the weld interface showed a wave-like pattern and the thickness of the Cu interlayer was uneven due to the effect of plastic deformation on the material surface, implying the occurrence of material flow at the interface of the deformed material. Under the position of indentation of the sonotrode tip in the material, the thickness of Cu interlayer at the weld interface was thinner than that of the anvil. The weld interface was noticeably intact, resembling a friction-induced bonding characteristic [39]. Joining of NiTi with the Cu interlayer took place by the formation and growth of micro-welds at the weld interfaces owing to the close metal-to-metal contact, which produced mutual diffusion and metallurgical bonding along the interface [33,38,40]. Additionally, unbonded regions were observed near the periphery of the weld, indicating insufficient diffusion during the process. During the USW process, the highest temperature was located at the central area of the weld, thus the bonding between the NiTi and the Cu interlayer was significantly more effective close to this location that in the border of the weld [37]. In comparison, the welded joint obtained with a welding energy of 700 J, which is shown in Figure 4c,d, also presented unbonded gaps along the border of weld interface were also visible. However, these were less often observed due to increased diffusion occurring as a result of the increasing temperatures generated by the increase in the welding energy during the process [39,41].

Observations of the cross-section of the joint obtained at 1000 J showed that both at the center and periphery of the joint complete bonding was obtained, as depicted in Figure 4e,f. It can be observed that the joining interface in Figure 4f was better than that in Figure 4b,d, since no unbonded zones were observed. This can be explained by the greater plastic deformation and increase in temperature that occurred under the indent tips at both the sonotrode tip and anvil sides when 1000 J of energy were used. These results indicate that a good interfacial bonding was obtained during the NiTi with Cu foil USW process.

Figure 4g showed that NiTi and the Cu interlayer were complexly intertwined in the visible weld interface obtained for a welding energy of 1000 J, which contributes to a better bonding along the weld. During the USW process, under the clamping pressure and ultrasonic vibration, the NiTi BM adhered to the Cu foil and then shear deformation, as well as, mutual rubbing of the faying surfaces occurred. Some works [36] have shown that the strength of ultrasonic spot-welded joints is related to interfacial waves, mechanical interlocking and microbonds produced along the weld interface. From the results presented above, it can be concluded that with increased welding energy, more shear friction heat and plastic deformation energy were generated at the weld interface thus promoting joining between the NiTi BM and the Cu interlayer. Mutual extrusion and abrasion between the faying surface of NiTi and the Cu interlayer was a main source for friction heat and plastic deformation. The presence of unbonded zones in the weld interfaces obtained for lower welding energies, would promote premature fracture of joints during tensile testing, since when applying the load, the unbonded areas would become the starting points for crack initiation and propagation until failure of the joint occur.

The EDS line scan analysis at the weld interface center of different joints obtained by various welding energy conditions (the test positions for EDS analysis were depicted with white lines in Figure 4) were conducted to determine the chemical composition and to infer about the potential phases in the interface diffusion layer formed between NiTi and Cu. The results are shown in Figure 5.

It is apparent that the content distribution of Ti and Ni followed the same trend, which is opposite to the change for Cu: a decrease in either the Ni or Ti content would be compensated by an increase of Cu. The EDS line analysis results revealed a smooth and rapid change in composition of both NiTi and Cu across the weld interface border. As indicated by the ellipse in Figure 5a, changes in relative contents of NiTi and Cu elements suggests the onset of slight diffusion at the weld interface. However, the width of the diffusion zone at most weld interfaces was too small to put into evidence the formation of any IMCs layer. Therefore, the joining mechanisms of NiTi with Cu interlayer by USW can be concluded as the combination of solid-state shear plastic deformation, mechanical interlock, the formation and further expansion of micro welds, which was consistent with other previous studies [36,42,43]. Despite the element content fluctuation, it can be seen that the peaks and valleys in the element distribution profiles kept mostly stable in the rich Cu area, as shown in Figure 5a,c,e.

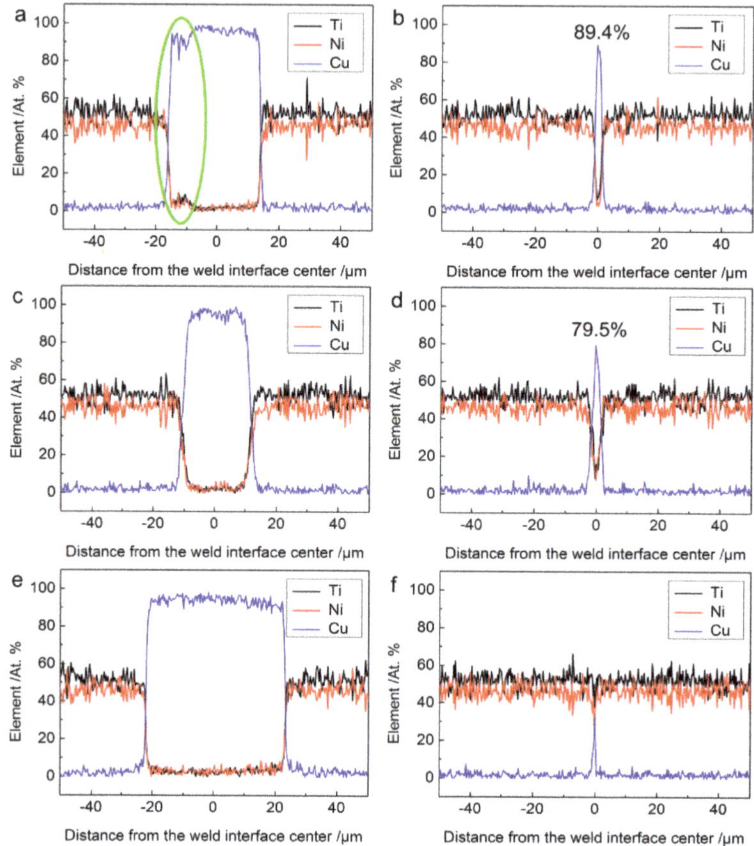

Figure 5. Chemical composition across the weld interface of NiTi joints with Cu interlayer: (**a,b**) 500 J; (**c,d**) 700 J; (**e,f**) 1000 J. The positions where the EDS line scans were performed are indicated by the white lines in Figure 4 (**a,c,e**).

The EDS analysis results also showed that the maximum content of Cu in the thinner Cu area (as depicted in Figure 4) gradually decreased with increasing welding energy, as presented in Figure 5b,d,f, with values of around 89.4%, 79.5% and 32.7%, respectively. It can be reasoned that more Cu foil was squeezed out at some positions of the weld interface with increasing welding energy. Since Cu is a soft metal with high conductivity and low yield strength, it is believed that the existence of Cu layer in the weld can compensate the thermal stress produced during the USW process [44].

3.3. Mechanical Performance and Failure Analysis

To evaluate the mechanical performance of the ultrasonic spot welded NiTi joints using Cu interlayer, tensile shear tests were conducted. Figure 6 summarizes the load-displacement curves for all welding conditions. Additionally, the load-displacement curve for a NiTi/NiTi weld obtained without Cu interlayer for a 1000 J of energy is also added for comparison.

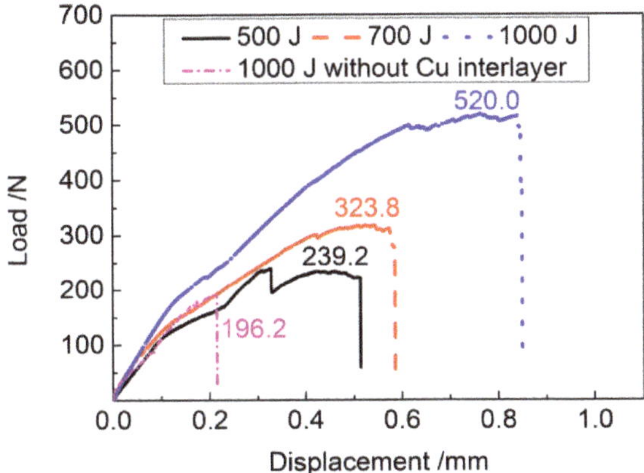

Figure 6. Load-displacement curves of NiTi joints with and without Cu interlayer at different welding energies.

The failure loads of all joints were defined as the peak tensile load on the load-displacement curve, with each joint representing its own weld strength under certain welding conditions based on the joint performance results. The failure load of the base metal with 0.15 mm thickness and 8 mm width, used in this present work is about 810 N. As shown in Figure 6, it is noticeable that a significantly higher failure load of 520 N was achieved at a welding energy of 1000 J. This higher failure strength is in accordance with the result of weld interface morphology of the 1000 J joint in Figure 4, where good bonding between NiTi and Cu was obtained due to sufficient mechanical interlocking and metallurgical adhesion as a result of a more concentrated vibratory energy. It further indicates that the weld joint with welding energy of 1000 J has the good load capacity, which can be used in distinct engineering fields. Furthermore, the use of the Cu interlayer can improve the mechanical properties of the welded joint: when no Cu interlayer is used the fracture load of the 1000 J weld is of approximately 200 N [45], which is significantly lower than the 520 N obtained when the Cu interlayer was used.

In this present work, the fracture location of all the joints occurred at the welded interface. Figure 7a,b show the schematic diagram of the fracture mode and the overall image of the fracture surface of the tensile test sample, respectively, which exhibits the interfacial fracture mode with some welded spots clearly observed. The dominant failure mode was characterized by a tearing behavior along the weld interface: it can be observed that the Cu foils present some tearing features on the fracture surface. In addition, to further understand the weld behavior and failure mode of ultrasonic spot-welded joints, the micro morphology of the corresponding fracture surface of 1000 J joint after tensile shear tests were observed by SEM. The EDS point scan analyses were performed on different areas of fracture surface, with these results presented in Figure 7e,f,h.

The judging criteria for a good weld quality could can be determined by the mechanical properties of the weld but also by the observation of developed weld spots on the fracture surfaces. Three distinct regions were observed on the fracture surface, as exhibited in Figure 7c: weld spots, scratched regions and tearing regions of Cu. The magnified images of the inserts indicated in Figure 7c suggested that on the ultrasonic spot welded NiTi joints with Cu interlayer, more significant deformation was observed under indentations of the sonotrode tip and anvil. As can be seen from Figure 7d, the fracture surface near the periphery of the weld spots was much smother and presented plastically deformed zones, while tear ridges and dimples can be observed at the center, as shown in Figure 7e,f, suggesting that the interface bonding strength near the periphery of weld spot was lower than that at the center of the weld interface.

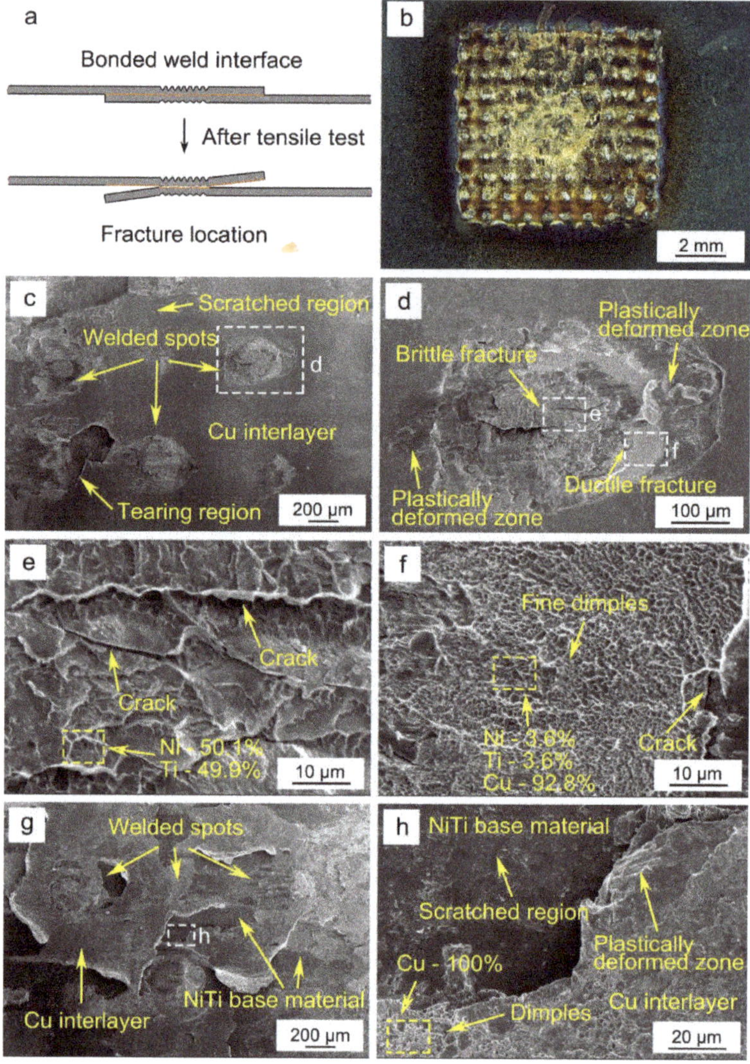

Figure 7. (**a**) Schematic diagram of fracture mode; (**b–h**) fracture morphologies of the tensile failed sample made at a welding energy of 1000 J; (**b**) overall view of fracture surface; (**c**) overall view of fracture surface by SEM; (**d**) magnified image of a weld spot depicted in (**c**); (**e,f**) magnified image of box in (**d**); (**g**) overall view of Cu interlayer on the fracture surface; (**h**) magnified image of box in (**g**).

Brittle fracture characteristics can be observed in the weld spot zone as the high-magnification SEM image presented in Figure 7e reveals. Furthermore, the features in the fracture region of the NiTi weld spot exhibited smooth step patterns and river marks with cleavage-like characteristics. The EDS point analysis showed that this region had a composition of 50.1 at.% Ni and 49.9 at.% Ti, which is good agreement with the expected composition for the as-received NiTi BM. At the weld interface, some intergranular cracking can also be observed due to the strain incompatibility between different grains, which is similar to that observed in NiTi-Cu dissimilar laser welds [15]. In addition, fine dimples also existed in the ductile fracture zone of the weld spot, as shown in Figure 7f. The EDS result shows the composition of dimples in weld spot zone was 3.6 at.% Ni, 3.6 at.% Ti and 92.8 at.%

Cu, indicating that ductile failure occurred mainly in the position of softer Cu foil. This can justify the increasing load capacity of the joints obtained with the Cu interlayer: when Cu is used as an interlayer part of the deformation is accommodated by it which provides better mechanical properties than when no interlayer is used.

An overall view of the Cu interlayer on the fracture surface is presented in Figure 7g, and obvious tearing characteristics of Cu can be observed. The magnified image of the insert presented in Figure 7h consisted of both scratched regions and plastically deformed zones. Compared with the NiTi surface, compact dimples existed in the position of Cu interlayer, suggesting that shear fracture occurred through void nucleation, growth and coalescence. During the tensile shear process, a higher tensile load can be transmitted to the 1000 J joint due to a good interface combination, resulting in a higher ultimate tensile load, which is in accordance with the tensile results presented in Figure 6. Additionally, in this study, considering the mechanical performance of joints previously discussed, it is believed that the addition of Cu interlayer into the faying surfaces is critical to enhance the bonding strength due to the increase of friction coefficient [46,47].

The SEM images of the fracture surfaces also verified that the joint was composed of micro welds between NiTi and Cu interlayer. It is possible that the fracture originated from the weak regions in the weld spot border, since this is a suitable location for nucleation and propagation of a crack, and when the crack tip reaches at the boundaries of weld spot, it can easily propagate through it, resulting in the fluctuation of tensile loads due to the existence of the multiple spots connection mechanism of ultrasonic spot-welded NiTi with Cu interlayer, as shown in Figure 6. Meanwhile, the Cu foil between NiTi began to tear, contributing to the process of tensile test until the fracture failure, with the joint being able to deform even more.

X-ray diffraction analysis was carried out at room temperature to evaluate the influence of USW on the phases composition of NiTi weld. The diffraction patterns of both BM and of the fracture surface in the center of the 1000 J weld are depicted in Figure 8.

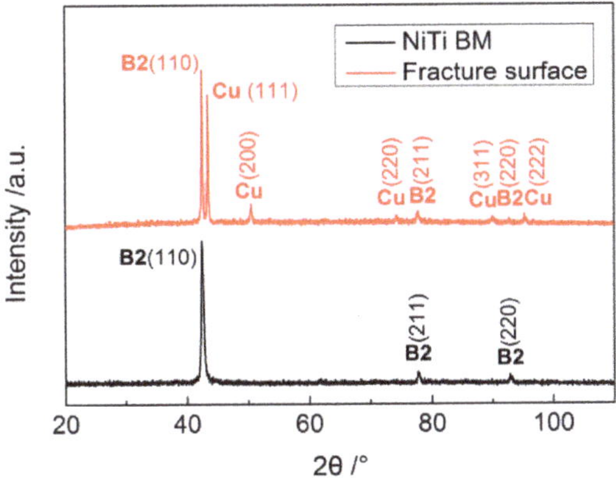

Figure 8. XRD patterns of NiTi BM and fracture surface of 1000 J weld.

The indexed pattern of the NiTi BM only consisted of B2 cubic austenite, without traces of B19' monoclinic martensitic phase, indicating the NiTi BM would be fully austenitic at room temperature. Comparing to the NiTi BM, the fracture surface of ultrasonic spot welded NiTi weld exhibited an additional pure Cu phase due to the addition of the Cu interlayer into the NiTi interface. It is noteworthy that no intermetallic phases, such as Ti_2Ni or Ni_3Ti which are usually formed during some fusion welding processes of NiTi [17–20], were detected in the weld region. Furthermore, the XRD results are consistent with the EDS line scan in Figure 5 where no intermetallic compound layer was found. It is believed that restricting the formation of the brittle phases at the weld interface is beneficial to reduce weld embrittlement, which can further prevent the formation of cold cracking [44].

In addition, the formation Cu-based intermetallics was not observed in this study either, although in fusion-based welding of NiTi to Cu they were already reported [13,15].

4. Conclusions

Ultrasonic spot welding of NiTi SMAs with Cu interlayer was performed varying the welding energy levels. The weld morphology, interface characteristics and failure behavior of the joints were analyzed in detail. The following main conclusions can be drawn:

1. With the welding energy increasing from 500 J to 1000 J, more frictional heat was generated at the weld interface and the plastic deformation of NiTi BM increased, leading to the increasing depth of indentations on the weld surface.

2. The ultrasonic spot welded NiTi weld interface presented a wave-like pattern with uneven thickness of Cu interlayer. The occurrence of shear deformation and mutual rubbing of the faying surfaces contributed to the formation and growth of microwelds at the weld interfaces. NiTi and Cu interlayer were intertwined together at the welding energy of 1000 J.

3. EDS line scan and XRD analyses revealed no intermetallic layer formation at the joint interface for all the samples although slight diffusion occurred.

4. The ultimate tensile shear load increased with increasing welding energy due to better mechanical interlocking and metallurgical adhesion at the weld interface.

5. The micro-morphologies of the fracture surface consisted of weld spots, scratched regions and tearing regions of Cu. Both brittle fracture characteristics of NiTi and ductile fracture characteristics of Cu interlayer were observed in the weld spot zone.

Author Contributions: W.Z., S.A. and Z.Z. conceived, designed the work; W.Z. and S.A. conducted the experiments; W.Z. and Y.H. analyzed the data; W.Z. and S.A. contributed to the writing of the original manuscript; J.P.O. and Z.Z. did the revision; Z.Z. and Z.L. supervised the whole procedure.

Funding: This research was funded by National Key R&D Program of China under Grant 2018YFB1107900, the National Natural Science Foundation of China under Grant 51775091 and 51575383; Natural Science Foundation of Tianjin City under Grant 18JCQNJC04100; Science and Technology Project of Sichuan Province under Grant 2018GZ0284; Fundação para a Ciência e a Tecnologia (FCT-MCTES) under Grant PEst-OE/EME/UI0667/2014.

Conflicts of Interest: The authors declare no conflict of interest.

References

1. Otsuka, K.; Ren, X. Physical metallurgy of Ti-Ni-based shape memory alloys. *Prog. Mater. Sci.* **2005**, *50*, 511–678. [CrossRef]

2. Miyazaki, S.; Igo, Y.; Otsuka, K. Effect of thermal cycling on the transformation temperatures of NiTi alloys. *Acta Metall.* **1986**, *34*, 2045–2051. [CrossRef]

3. Cisse, C.; Zaki, W.; Zineb, T.B. A review of constitutive models and modeling techniques for shape memory alloys. *Int. J. Plast.* **2016**, *76*, 244–284. [CrossRef]

4. Chowdhury, P.; Sehitoglu, H. A revisit to atomistic rationale for slip in shape memory alloys. *Prog. Mater. Sci.* **2017**, *85*, 1–42. [CrossRef]

5. Dolce, M.; Cardone, D. Mechanical behaviour of shape memory alloys for seismic applications 2. Austenite NiTi wires subjected to tension. *Int. J. Mech. Sci.* **2001**, *43*, 2657–2677. [CrossRef]

6. Cho, H.; Kim, H.Y.; Miyazaki, S. Fabrication and characterization of Ti-Ni shape memory thin film using Ti/Ni multilayer technique. *Sci. Technol. Adv. Mater.* **2005**, *6*, 678–683. [CrossRef]

7. Jani, J.M.; Leary, M.; Subic, A.; Gibson, M.A. A review of shape memory alloy research, applications and opportunities. *Mater. Des.* **2014**, *56*, 1078–1113. [CrossRef]

8. Chen, Q.A.; Thouas, G.A. Metallic implant biomaterials. *Mater. Sci. Eng. R Rep.* **2015**, *87*, 1–57. [CrossRef]

9. Zeng, Z.; Yang, M.; Oliveira, J.P.; Song, A.D.; Peng, B. Laser welding of NiTi shape memory alloy wires and tubes for multi-functional design applications. *Smart Mater. Struct.* **2016**, *25*, 085001. [CrossRef]

10. Li, H.M.; Sun, D.Q.; Cai, X.L.; Dong, P.; Wang, W.Q. Laser welding of TiNi shape memory alloy and stainless steel using Ni interlayer. *Mater. Des.* **2012**, *39*, 285–293. [CrossRef]

11. Li, Q.; Zhu, Y.X.; Guo, J.L. Microstructure and mechanical properties of resistance-welded NiTi/stainless steel joints. *J. Mater. Process. Technol.* **2017**, *249*, 538–548. [CrossRef]

12. Oliveira, J.P.; Panton, B.; Zeng, Z.; Andrei, C.M.; Zhou, Y.; Miranda, R.M.; Braz Fernandes, F.M. Laser joining of NiTi to Ti6Al4V using a niobium interlayer. *Acta Mater.* **2016**, *105*, 9–15. [CrossRef]

13. Shojaei Zoeram, A.; Akbari Mousavi, S.A.A. Laser welding of Ti-6Al-4V to Nitinol. *Mater. Des.* **2014**, *61*, 185–190. [CrossRef]

14. Shojaei Zoeram, A.; Akbari Mousavi, S.A.A. Effect of interlayer thickness on microstructure and mechanical properties of as welded Ti6Al4V/Cu/NiTi joints. *Mater. Lett.* **2014**, *133*, 5–8. [CrossRef]

15. Zeng, Z.; Oliveira, J.P.; Yang, M.; Song, D.; Peng, B. Functional fatigue behavior of NiTi-Cu dissimilar laser welds. *Mater. Des.* **2017**, *114*, 282–287. [CrossRef]

16. Tam, B.; Pequegnat, A.; Khan, M.; Zhou, Y. Resistance microwelding of Ti-55.8 wt pct Ni nitional wires and the effects of pseudoelasticity. *Metall. Mater. Trans. A* **2012**, *43*, 2969–2978. [CrossRef]

17. Fox, G.; Hahnlen, R.; Dapino, M.J. Fusion welding of nickel-titanium and 304 stainless steel tubes: Part II: Tungsten inert gas welding. *J. Intell. Mater. Syst. Struct.* **2013**, *24*, 962–972. [CrossRef]

18. Oliveira, J.P.; Miranda, R.M.; Braz Fernandes, F.M. Welding and joining of NiTi shape memory alloys: A review. *Prog. Mater. Sci.* **2017**, *88*, 412–466. [CrossRef]

19. Yang, D.; Jiang, H.C.; Zhao, M.J.; Rong, L.J. Microstructure and mechanical behaviors of electron beam welded NiTi shape memory alloys. *Mater. Des.* **2014**, *57*, 21–25. [CrossRef]

20. Tam, B.; Khan, M.I.; Zhou, Y. Mechanical and functional properties of laser-welded Ti-55.8 wt pct Ni nitinol wires. *Metall. Mater. Trans. A* **2011**, *42*, 2166–2175. [CrossRef]

21. Chan, C.W.; Man, H.C.; Yue, T.M. Effects of process parameters upon the shape memory and pseudo-elastic behaviors of laser-welded NiTi thin foil. *Metall. Mater. Trans. A* **2011**, *42*, 2264–2270. [CrossRef]

22. Ng, C.H.; Mok, E.S.H.; Man, H.C. Effect of Ta interlayer on laser welding of NiTi to AISI 316L stainless steel. *J. Mater. Process. Technol.* **2015**, *226*, 69–77. [CrossRef]

23. Zeng, Z.; Panton, B.; Oliveira, J.P.; Han, A.; Zhou, Y.N. Dissimilar laser welding of NiTi shape memory alloy and copper. *Smart Mater. Struct.* **2015**, *24*, 125036. [CrossRef]

24. Bricknell, R.H.; Melton, K.N.; Mercier, O. The structure of NiTiCu shape memory alloys. *Metall. Trans. A* **1979**, *10*, 693–697. [CrossRef]

25. Gugel, H.; Schuermann, A.; Theisen, W. Laser welding of NiTi wires. *Mater. Sci. Eng. A* **2008**, *481–482*, 668–671. [CrossRef]

26. Frick, C.P.; Ortega, A.M.; Tyber, J.; Maksound, A.E.M.; Maier, H.J.; Liu, Y.; Gall, K. Thermal processing of polycrystalline NiTi shape memory alloys. *Mater. Sci. Eng. A* **2005**, *405*, 34–49. [CrossRef]

27. Li, H.M.; Sun, D.Q.; Gu, X.Y.; Dong, P.; Lv, Z.P. Effects of the thickness of Cu filler metal on the microstructure and properties of laser-welded TiNi alloy and stainless steel joint. *Mater. Des.* **2013**, *50*, 342–350. [CrossRef]

28. Crăciunescu, C.; Ercuta, A. Modulated interaction in double-layer shape memory-based micro-designed actuators. *Sci. Technol. Adv. Mater.* **2015**, *16*, 065003. [CrossRef] [PubMed]

29. Barcellona, A.; Fratini, L.; Palmeri, D.; Maletta, C.; Brandizzi, M. Friction stir processing of Niti shape memory alloy: Microstructural characterization. *Int. J. Mater. Forming* **2010**, *3*, 1047–1050. [CrossRef]

30. Fukumoto, S.; Inoue, T.; Mizuno, S.; Okita, K.; Tomita, T.; Yamamoto, A. Friction welding of TiNi alloy to stainless steel using Ni interlayer. *Sci. Technol. Weld. Join.* **2010**, *15*, 124–130. [CrossRef]

31. Mani Prabu, S.S.; Madhu, H.C.; Perugu, C.S.; Akash, K.; Kumar, P.A.; Kailas, S.V.; Anbarasu, M.; Palani, I.A. Microstructure, mechanical properties and shape memory behaviour of friction stir welded nitinol. *Mater. Sci. Eng. A* **2017**, *693*, 233–236. [CrossRef]

32. Ni, Z.L.; Zhao, H.J.; Mi, P.B.; Ye, F.X. Microstructure and mechanical performances of ultrasonic spot welded Al/Cu joints with Al 2219 alloy particle interlayer. *Mater. Des.* **2016**, *92*, 779–786. [CrossRef]

33. Bakavos, D.; Prangnell, P.B. Mechanisms of joint and microstructure formation in high power ultrasonic spot welding 6111 aluminium automotive sheet. *Mater. Sci. Eng. A* **2010**, *527*, 6320–6334. [CrossRef]

34. Yang, J.W.; Cao, B.; He, X.C.; Luo, H.S. Microstructure evolution and mechanical properties of Cu-Al joints by ultrasonic welding. *Sci. Technol. Weld. Join.* **2014**, *19*, 500–504. [CrossRef]

35. Yang, J.W.; Cao, B. Investigation of resistance heat assisted ultrasonic welding of 6061 aluminum alloys to pure cooper. *Mater. Des.* **2015**, *74*, 19–24. [CrossRef]

36. Ni, Z.L.; Ye, F.X. Effect of lap configuration on the microstructure and mechanical properties of dissimilar ultrasonic metal welded copper-aluminum joints. *J. Mater. Process. Technol.* **2017**, *245*, 180–192. [CrossRef]
37. Macwan, A.; Kumar, A.; Chen, D.L. Ultrasonic spot welded 6111-T4 aluminum alloy to galvanized high-strength low-alloy steel: Microstructure and mechanical properties. *Mater. Des.* **2017**, *113*, 284–296. [CrossRef]
38. Shakil, M.; Tariq, N.H.; Ahmad, M.; Choudhary, M.A.; Akhter, J.I.; Babu, S.S. Effect of ultrasonic welding parameters on microstructure and mechanical properties of dissimilar joints. *Mater. Des.* **2014**, *55*, 263–273. [CrossRef]
39. Ren, D.X.; Zhao, K.M.; Pan, M.; Chang, Y.; Gang, S.; Zhao, D.W. Ultrasonic spot welding of magnesium alloy to titanium alloy. *Scripta Mater.* **2017**, *126*, 58–62. [CrossRef]
40. Wu, X.; Liu, T.; Cai, W. Microstructure, welding mechanism, and failure of Al/Cu ultrasonic welds. *J. Manuf. Process.* **2015**, *20*, 321–331. [CrossRef]
41. Elangovan, S.; Semeer, S.; Prakasan, K. Temperature and stress distribution in ultrasonic metal welding-an FEA-based study. *J. Mater. Process. Technol.* **2009**, *209*, 1143–1150. [CrossRef]
42. Chen, Y.C.; Bakavos, D.; Gholinia, A.; Prangnell, P.B. HAZ development and accelerated post-weld natural ageing in ultrasonic spot welding aluminium 6111-T4 automotive sheet. *Acta Mater.* **2012**, *60*, 2816–2828. [CrossRef]
43. Panteli, A.; Robson, J.D.; Brough, I.; Prangnell, P.B. The effect of high strain rate deformation on intermetallic reaction during ultrasonic welding aluminium to magnesium. *Mater. Sci. Eng. A* **2012**, *556*, 31–42. [CrossRef]
44. Shojaei Zoeram, A.; Rahmani, A.; Akbari Mousavi, S.A.A. Microstructure and properties analysis of laser-welded Ni-Ti and 316l sheets using copper interlayer. *J. Manuf. Process.* **2017**, *26*, 355–363. [CrossRef]
45. Zhang, W.; Ao, S.S.; Oliveira, J.P.; Zeng, Z.; Luo, Z.; Hao, Z.Z. Effect of ultrasonic spot welding on the mechanical behaviour of NiTi shape memory alloys. *Smart Mater. Struct.* **2018**, *27*, 085020. [CrossRef]
46. Panteli, A.; Robson, J.D.; Chen, Y.C.; Prangnell, P.B. The effectiveness of surface coatings on preventing interfacial reaction during ultrasonic welding of aluminum to magnesium. *Metall. Mater. Trans. A* **2013**, *44*, 5773–5781. [CrossRef]
47. Xu, L.; Wang, L.; Chen, Y.C.; Robson, J.D.; Prangnell, P.B. Effect of interfacial reaction on mechanical performance of steel to aluminum dissimilar ultrasonic spot welds. *Metall. Mater. Trans. A* **2016**, *47*, 334–346. [CrossRef]

Article

Effect Range of the Material Constraint-I. Center Crack

Jie Yang * and Lei Wang

School of Energy and Power Engineering, University of Shanghai for Science and Technology, Shanghai 200093, China; wl_21th@126.com
* Correspondence: yangjie@usst.edu.cn; Tel.: +86-21-5527-2320

Received: 30 November 2018; Accepted: 24 December 2018; Published: 25 December 2018

Abstract: Material constraints are important factor effects on the fracture behavior of welded joints. The effect range of the material constraint is an important and interesting issue which needs to be clarified, including whether the effect range of a material constraint exists or not, who will affect it, and whether the material constraint is affected by the no adjacent area or not. In this study, different basic models which reflect different single metallic welded joints, bimetallic welded joints and dissimilar metal welded joints were designed, and the fracture resistance curves and crack tip strain fields of the different models with various material constraints were calculated. Based on the results, the questions above were answered. This study has significance for developing solid mechanics, optimizing joint design, structure integrity assessment, and so on.

Keywords: material constraint; effect range; fracture resistance curve; strain field; center crack

---

1. Introduction

Constraint is the resistance of a structure or specimen against plastic deformation [1]. In recent years, the constraint effect due to structure or specimen geometry have been investigated as an important factor affecting the stress distribution around a crack. Some constraint parameters, such as T [2], Q [3,4], A_2 [5], T_Z [6–8], have been established to represent the stress fields at the crack tip under different geometry constraint conditions. In addition, the constraint effect due to material strength mismatch, which be called the material constraint, is also an important factor effects on the fracture behavior of material.

The material constraint was firstly demonstrated by Joch et al. [9] and Burstow et al. [10] to show how the slip-line fields were changed by altering the yield strength of the base material. Then, Zhang et al. [11] analyzed a two-material problem where the crack was located in the interface of two dissimilar materials and established a material constraint parameter M to consider the effect of strength mismatch on crack tip stress fields, as follows:

$$M = \frac{\sigma_{Yw}}{\sigma_{Yb}},\tag{1}$$

where the σ_{Yw} is the yield stress of the weld material and σ_{Yb} is the yield stress of the base material. They also proposed that the stress fields of an interface crack in a mismatched problem could be obtained using the J-Q-M formulation, which was derived by extending the J-Q theory. Betegón et al. [12] defined a procedure similar to the J-T, also by establishing an additional parameter β_m that quantifies the material constraint, and a total constraint parameter β_T was defined as follows:

$$\beta_T = \beta_m \cdot \sqrt{\frac{a}{h}} + \beta_g,\tag{2}$$

where β_m is a constraint parameter defined for the overmatched welded joints to quantify the material constraint effect on the crack tip stress fields, β_g is a geometry parameter by means of the T-stress to quantify the geometry constraint, a is crack length and h is weld semi-width. Recently, the author [13–15] defined a unified constraint parameter A_p based on the areas surrounded by the equivalent plastic strain (ε_p) isolines ahead of the crack tip to characterize both geometry and material constraint. The unified constraint parameter A_p was defined as follows:

$$A_p = \frac{A_{PEEQ}}{A_{ref}},\tag{3}$$

where A_{PEEQ} is the areas surrounded by the ε_p isolines ahead of the crack tip and A_{ref} is the reference areas surrounded by the ε_p isolines in a standard test.

Furthermore, many scholars focused their studies on the fracture behavior of bi-material affected by the material constraint. Negre et al. [16] and Samal et al. [17] investigated the altering of fracture resistance and crack path deviation in the bi-material interface region affected by the material constraint. Fan et al. [18–20] studied the J-resistance curves, fracture toughness, crack growth paths and stress triaxiality of bi-materials under different work-hardening mismatches. Besides, some scholars focused their studies on the fracture behavior of dissimilar metal-welded joints affected by the material constraint. Rakin et al. [21] investigated the fracture behaviors of the over-matched and under-matched high-strength low-alloyed steel weld joints. Wang et al. [22,23] studied the local fracture resistances and crack growth paths of a dissimilar metal-welded joint at different crack positions with different material constraints. Xue et al. [24] investigated the stress and strain of a micro region influenced by material yield strength mismatch at the crack tip of a dissimilar metal-welded joint. Zhu et al. [25] studied the stress fields of a crack tip affected by material constraints in a nuclear pressure steel A508-III dissimilar metal welded joint.

These studies clarified the effect of a material constraint on the fracture behaviors of welded joints, and laid the foundation for the building of accurate structure integrity assessment. Nevertheless, most studies focus their attention on the strength mismatch of both sides of the crack, such as over-match, under-match, and so on. There is another interesting and important issue, the effect range of the material constraint, which also needs to be clarified. This includes whether exists the effect zone or not, who effects it, whether the material constraint affected by the no adjacent area or not, and so on. Solving this issue is of significance in developing solid mechanics, optimizing joint design and structure integrity assessment.

Thus, in this study, different basic models which represent different single metallic-welded joints, bimetallic-welded joints and dissimilar metal-welded joints were designed. Then, the fracture resistance curves and crack tip strain fields of different models under different material constraints were calculated. Based on the results, the questions above were answered, and the effect range of the material constraint was investigated.

2. Materials and Models Design

2.1. Materials

Four different materials (A508, 52Mb, 52Mw and 316L) selected from a dissimilar metal-welded joint (DMWJ) in nuclear power plant were used in this study, as shown in Figure 1. The DMWJ was fabricated by Shanghai Company of Nuclear Power Equipment (Shanghai, China). The DMWJ was a full scale mock up of the DMWJ in a nuclear power plants. During the manufacturing process, the base metal A508 was pre-heated to 125 °C before buttering to prevent weld cracking. The buttering layer was deposited using a 1.2 mm diameter Alloy52M welding wire using automatic gas-tungsten arc welding (GTAW) on the ferritic nozzle face. The welding current, voltage and speed were 200 A, 11.5 V and 1.85 mm/s, respectively. A total of 478 weld passes were deposited, and the buttering layer with an average width of 20 mm was formed. Then, heat treatment (annealing at 610 °C for 15 h, with subsequent furnace cooling to 300 °C) was conducted on the buttering to relieve the residual stress.

Thereafter, 100% non-destructive testing was performed on the buttering. This buttering layer material is denoted as buttering 52Mb.

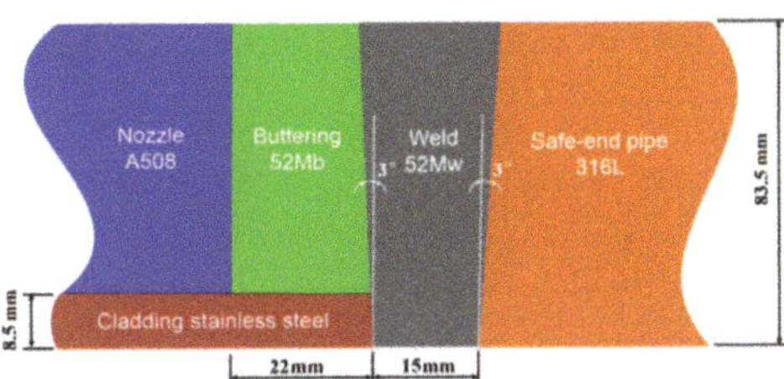

Figure 1. A dissimilar metal welded joint in a nuclear power plant.

After buttering, welding was carried out between the buttering layer and the austenitic safe-end pipe using GTAW and a 0.9 mm diameter Alloy52M welding wire. The welding current, voltage and speed were 180 A, 10 V and 1.75 mm/s, respectively. A total of 439 weld passes were deposited, and the weld with an average width of 19 mm was formed. After welding, 100% nondestructive testing was performed again on the weld. This weld metal material is denoted as weld 52Mw.

The true stress-strain curves of the four different materials at room temperature have been obtained [22], as shown in Figure 2. The measured Young's modulus E of A508, 52Mb, 52Mw and 316L are 202,410 MPa, 178,130 MPa, 178,130 MPa and 156,150 MPa, respectively, and the Poisson's ratios ν of them are 0.3 [22].

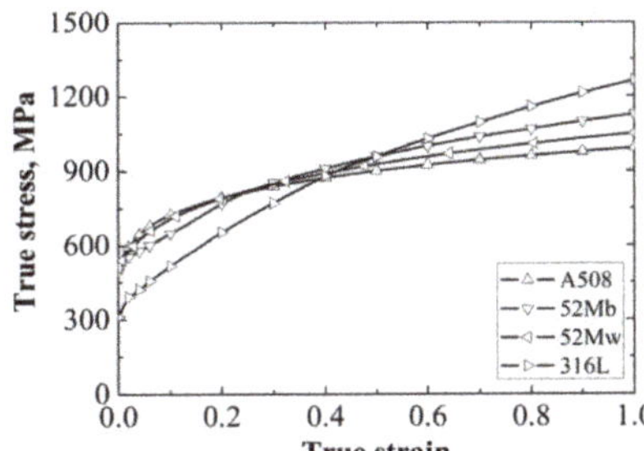

Figure 2. The true stress-strain curves of four materials. Reprinted from Materials Science and Engineering A, 568, Wang, H.T.; Wang, G.Z.; Xuan, F.Z.; Liu, C.J.; Tu, S.T.; Local mechanical properties of a dissimilar metal welded joint in nuclear power systems, 108, Copyright 2013, with permission from Elsevier. [22].

2.2. Models Design

Single-edge-notched bend (SENB) specimens were used in this study, and four different basic models were designed, which contain the "121" model, "123" model, "12321" model and "12324" model, as shown in Figure 3.

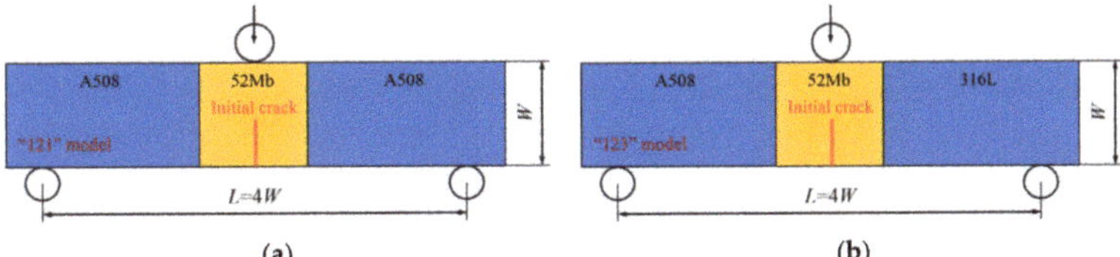

Figure 3. *Cont.*

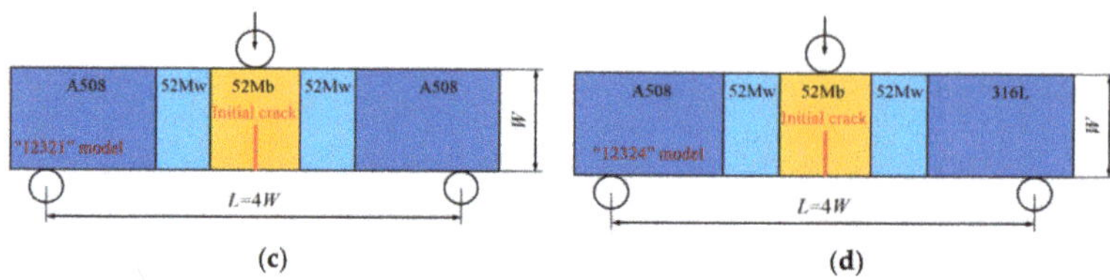

(c)

(d)

Figure 3. Four different basic models, (**a**) "121" model, (**b**) "123" model, (**c**) "12321" model and (**d**) "12324" model.

These models are similar to sandwiches. The "121" model means the model contains "1" and "2" two kinds of materials, the materials in the model from left to right are "1", "2" and "1". The "123" model means the model contains "1", "2" and "3" three kinds of materials, the materials in the model from left to right are "1", "2" and "3". The "12321" model means the model contains "1", "2" and "3" three kinds of materials, the materials in the model from left to right are "1", "2", "3", "2" and "1". The "12324" model means the model contains "1", "2", "3" and "4" four kinds of materials, the materials in the model from left to right are "1", "2", "3", "2" and "4". Furthermore, the "121" model represents the single metallic welded joint, the "123" model represents the bimetallic welded joint, the "12321" and "12324" models represent the dissimilar metal welded joint.

For all the models, a load roll is applied at the top and center of the SENB specimen, and two back-up rolls are applied at the bottom of the SENB specimen. The loading is applied at the load roll by prescribing a displacement of 6 mm, and the two back-up rolls are fixed by control displacement and rotation. The initial crack is located in the middle of specimen. All the specimen widths are 14.4 mm (W = 14.4 mm), the loading spans are 57.6 mm (L = 4W), the specimen thicknesses B are 12 mm, and the initial crack lengths are 7.2 mm (a/W = 0.5).

Different material constraints were obtained by changing the width (the direction perpendicular to the initial crack) of 52Mb or 52Mw in each model. For the "121" and "123" models, changing the width of 52Mb from 0 to 80 mm. When the width of 52Mb is 0 mm, the welded joint is the same as the A508 and 316L homogeneous material models, respectively; when the width of 52Mb is 80 mm, the welded joint is same as the 52Mb homogeneous material model. For the "12321" and "12324" models, the widths change to 52Mb and 52Mw, respectively. When changing the width of 52Mb individually, the width of 52Mw is fixed to 1 mm, and changing the width of 52Mb from 0 to 80 mm; when changing the width of 52Mw individually, the width of 52Mb is fixed to 1 mm, and changing the width of 52Mw from 0–39.5 mm.

2.3. Gurson-Tvergaard-Needleman (GTN) Damage Model

Ductile crack growth in metals is a result of nucleation, growth and coalescence of micro voids. In order to obtain the fracture resistance curves of different models, the finite element method (FEM) simulation based on Gurson-Tvergaard-Needleman (GTN) damage model was used in this study. There are nine parameters in the GTN damage model: the constitutive parameters q_1, q_2 and q_3, the void nucleation parameters ε_N, S_N and f_N, the initial void volume fraction f_0, the critical void volume fraction f_C and the final failure parameter f_F. The void coalescence occurs when the void volume fraction reaches the critical value f_C, and the fracture occurs when the void volume fraction reaches the final value f_F. These parameters have been obtained and listed in Table 1 [26].

Table 1. The GTN parameters of different materials. Reprinted by permission from Springer, Copyright 2017 [26].

Material	A508	52Mb	52Mw	316L
q_1	1.5	1.5	1.5	1.5
q_2	1	variable	1	variable
q_3	2.25	2.25	2.25	2.25

Table 1. *Cont.*

Material	A508	52Mb	52Mw	316L
ε_N	0.3	0.3	0.3	0.3
S_N	0.1	0.1	0.1	0.1
f_N	0.002	0.002	0.002	0.002
f_0	0.00008	0.000001	0.00015	0.000001
f_C	0.04	0.04	0.04	0.04
f_F	0.25	0.25	0.25	0.25

This GTN damage model has been implemented in the ABAQUS code (6.14, Dassault Systèmes group company, Shanghai, China), and is widely used to simulate the crack propagation process and calculate the *J*-resistance curve. During the finite element analysis, the 3D eight-node isoperimetric element with reduced integration (C3D8R) are used [13,14]. The typical finite element mesh for the "121" model with W_{52Mb} = 16 mm is illustrated in Figure 4a, the minimum size of mesh in the crack growth region is 0.1 mm × 0.1 mm [27], as shown in Figure 4b. This typical model contains 75,872 elements and 87,849 nodes. In addition, the surface-to-surface contact (explicit) interaction type was used in the model. Moreover, the sliding formulation is finite sliding, the mechanical constraint formulation is kinematic contact method.

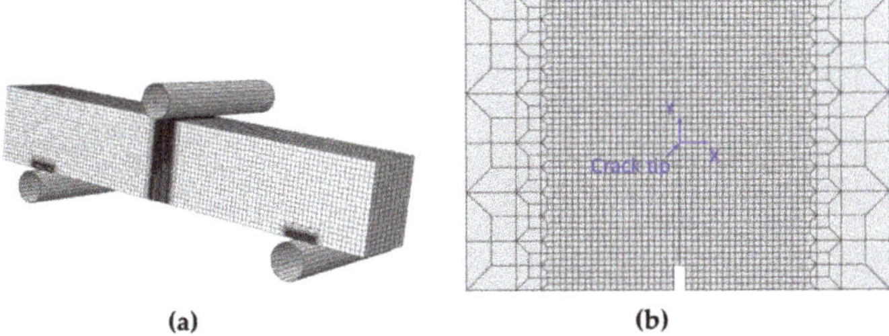

Figure 4. The whole mesh of the typical model (**a**) and the mesh in the crack growth region (**b**).

The load versus load-line displacement curve can be obtained from the FEM simulation. With instantaneous crack lengths obtained at each loading point, a crack growth resistance curve can be determined, as specified in ASTM (American Society for Testing and Materials) E1820 [28].

3. Results and Discussion

3.1. "121" Model

The *J*-resistance curves of different "121" models under different material constraints are shown in Figure 5. It can be found that increasing of the width of 52Mb from 0 to 8 mm, the *J*-resistance curves of the "121" models increase. When the width of 52Mb is up to 8 mm, the *J*-resistance curves remain steady and will not change with the increasing of the 52Mb's width.

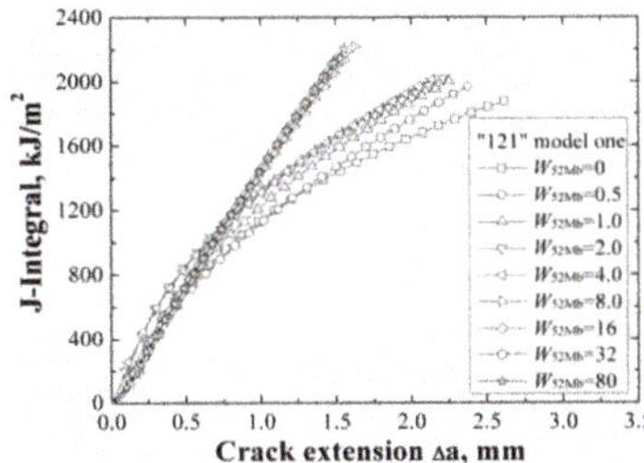

Figure 5. The *J*-resistance curves of different "121" models.

Because material A508 has lower strength than material 52Mb, the "121" model is an over-matched joint. In this model, when the W_{52Mb} = 0 mm, it is the same with the homogeneous material A508; when the W_{52Mb} = 80 mm, it is the same with the homogeneous material 52Mb. Thus, the results in Figure 5 show that for an over-matched joint, the *J*-resistance curve of the joint is higher than the base material. In addition, a notable phenomenon is that when the width of 52Mb is up to 8 mm, the *J*-resistance curves of the "121" models are same with the *J*-resistance curve of homogeneous material 52Mb. It means that the crack is out of the effect range of the material constraint induced by the A508/52Mb interface. In this condition, it does not matter even if the material on the outside is soft or hard. That is, when the crack locates out of the effect range of material constraint, the fracture resistance curve of the weld joint no longer influenced by the material constraint anymore. Of course, the effect range is also related to different materials and models.

Figure 6 shows the distributions of equivalent plastic strain ε_p = 0.1 isoline at crack tip at the same *J*-integral (J = 1600 kJ/m^2) for different "121" models. It can be found that though the distributions of equivalent plastic strain are different for different models, but the equivalent plastic strains surrounded by ε_p = 0.1 isoline are within the scope of 8 mm for all the models. When the interface is located within this scope, the *J*-resistance curve will be affected by the material constraint; when the interface is located outside this scope, the *J*-resistance curve will not be affected by material constraint. This scope is the effect range of the material constraint.

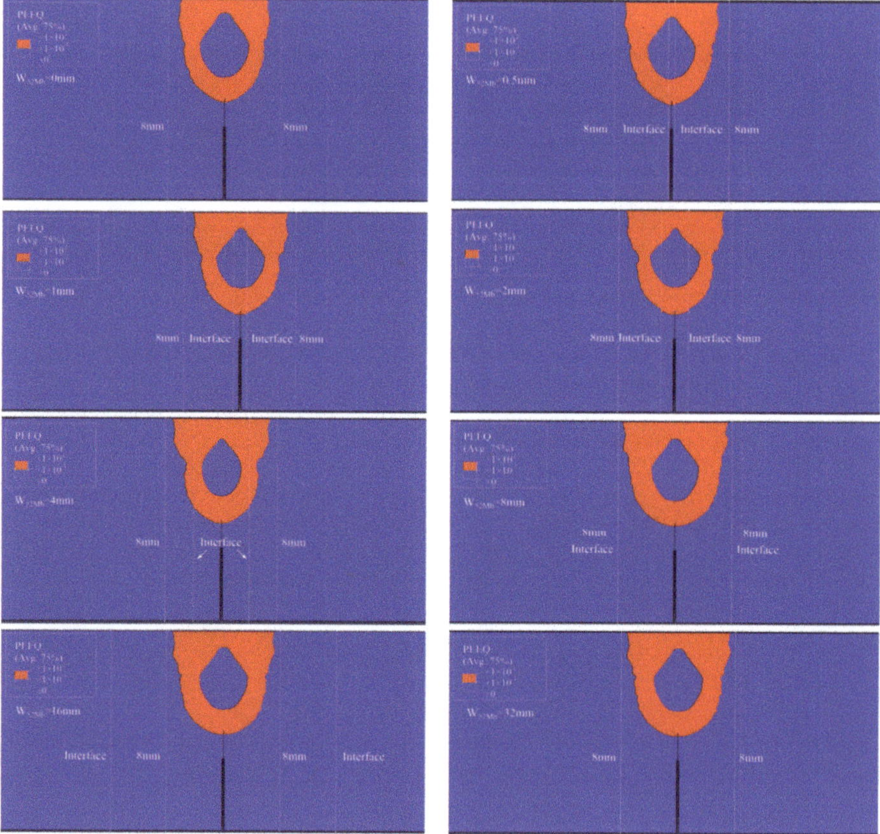

Figure 6. *Cont.*

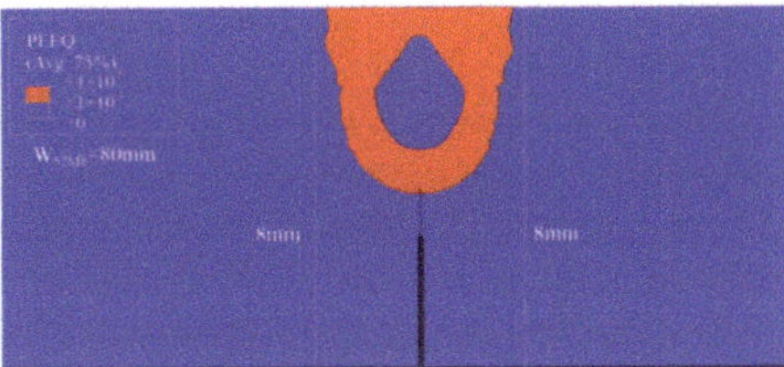

Figure 6. The distributions of $\varepsilon_p = 0.1$ isoline at crack tip at $J = 1600$ kJ/m^2.

In addition, the areas surround by the $\varepsilon_p = 0.1$ isoline reflect the same change rule with the *J*-resistance curves, as shown in Figure 7. Because the constraint is the resistance of a structure against plastic deformation, at the same *J*-integral (driving force) a lower plastic deformation reflects a higher constraint and a lower *J*-resistance curve, and vice versa. The same change rules between *J*-resistance curves and areas can prove each other and also reflect the change rules are related to the constraint.

Furthermore, Figure 7 also shows that the *J*-resistance curve of material is controlled by the strain fields at crack tip rather than stress fields. It should be noted that the $\varepsilon_p = 0.1$ isoline was selected here, when a small ε_p value was selected, the scope will beyond the 8 mm. Therefore, there may exist a main control value or control zone. For this study, the main control value is $\varepsilon_p = 0.1$.

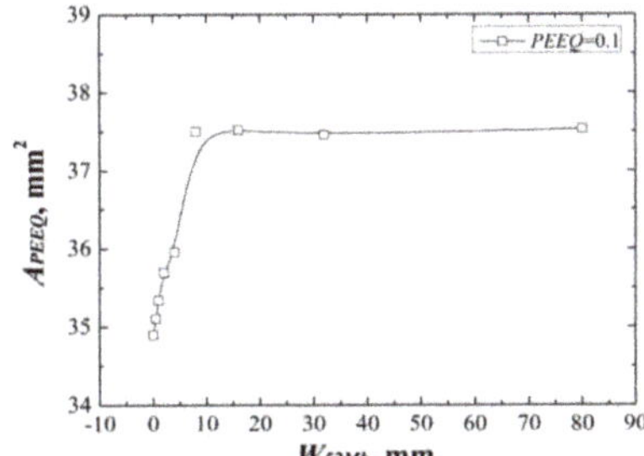

Figure 7. The areas surround by the $\varepsilon_p = 0.1$ isoline at the same *J*-integral for "121" model.

3.2. "123" Model

The *J*-resistance curves of different "123" models under different material constraints are shown in Figure 8. It can be found that increasing of the width of 52Mb, the *J*-resistance curves of the models increase firstly then decrease, and finally remain steady. The model with $W_{52Mb} = 0$ mm has the lowest *J*-resistance curve and the model with $W_{52Mb} = 4$ mm has the highest *J*-resistance curve. When the width of 52Mb is up to 16 mm, the *J*-resistance curve will not change with increasing of the 52Mb's width.

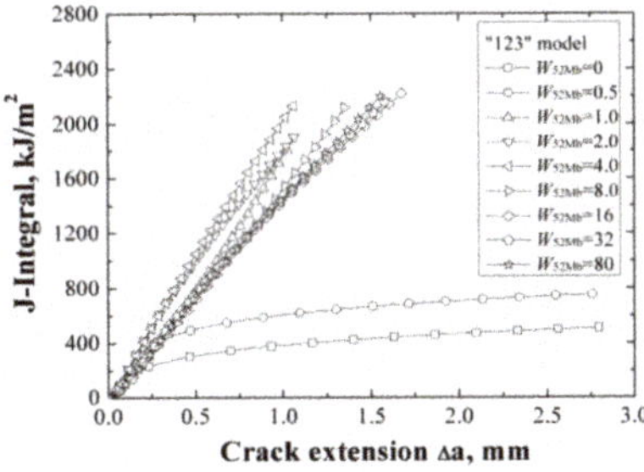

Figure 8. The *J*-resistance curves of different "123" models.

When the W_{52Mb} = 0 mm, the model is the same with the bimetallic welded joint with an interface crack. In this condition, the model has the lowest *J*-resistance curve, which shows that the interface crack in bimetallic welded joint is very dangerous. With increasing of the width of 52Mb, the *J*-resistance curve of the model increases. When the W_{52Mb} = 4 mm, there exists an optimal width and the model has the highest *J*-resistance curve. Then, the *J*-resistance curves of the models decrease and remain steady at last.

The same with the "121" model, when the width of 52Mb up to a value, the *J*-resistance curve of the model is same with the *J*-resistance curve of homogeneous material 52Mb. That is, an effect range also exists. By contrast with the "121" model, the steady value is different and is related to the materials on both sides of the crack.

Figure 9 shows the areas surround by the ε_p = 0.1 isoline at crack tip at the same *J*-integral (J = 1600 kJ/m^2) for different "123" models. It reflects the same change rule with the *J*-resistance curves. The same change rules can prove each other also.

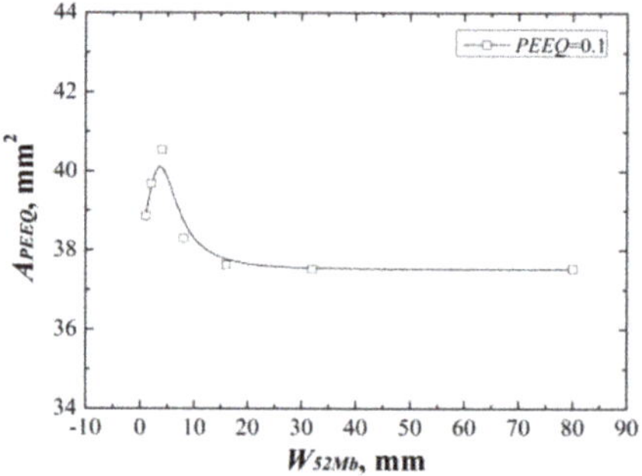

Figure 9. The areas surround by the ε_p = 0.1 isoline at the same *J*-integral for "123" model.

3.3. "12321" Model

When changing the width of 52Mb individually, the *J*-resistance curves of different "12321" models under different material constraints are shown in Figure 10a. It can be found that increasing the width of 52Mb, the *J*-resistance curves of different "12321" models increase and remain steady when the width of 52Mb is up to 16mm. It is similar with the "121" model; the strength of the material 52Mb is higher than the materials A508 and 52Mw, with increasing of the width of 52Mb, the *J*-resistance curve of the model increases. When the width of 52Mb over the effect range of the material constraint, the *J*-resistance curve of the "12321" model is unchanged.

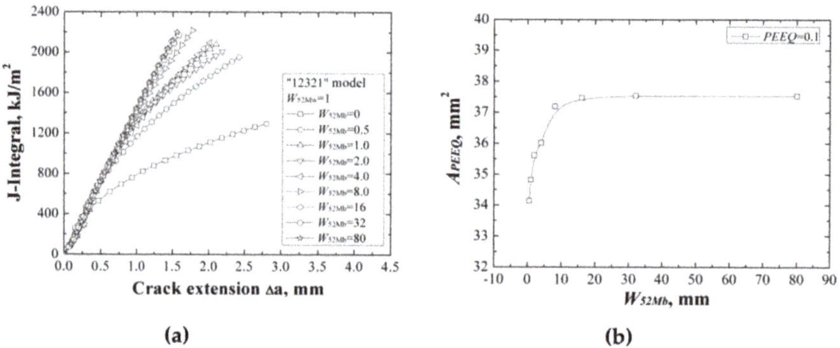

(a)   (b)

Figure 10. The *J*-resistance curves (**a**) and the areas surround by the ε_p = 0.1 isoline (**b**) of different "12321" models with the same W_{52Mw}.

The areas surrounded by the $\varepsilon_p = 0.1$ isoline at crack tip at the same J-integral ($J = 1600$ kJ/m^2) for different "12321" models are shown in Figure 10b, which also reflects the same change rule with the J-resistance curves.

When changing the width of 52Mw individually, the J-resistance curves of different "12321" models under different material constraints are shown in Figure 11a. It can be found that increasing of the width of 52Mw, the J-resistance curves of different "12321" models increase firstly then decrease, and finally remain steady. The model with $W_{52Mw} = 0$ mm has the lowest J-resistance curve and the model with $W_{52Mw} = 16$ mm has the highest J-resistance curve. When the width of 52Mw up to 32 mm, the J-resistance curve of the model will not change by increasing the 52Mw's width.

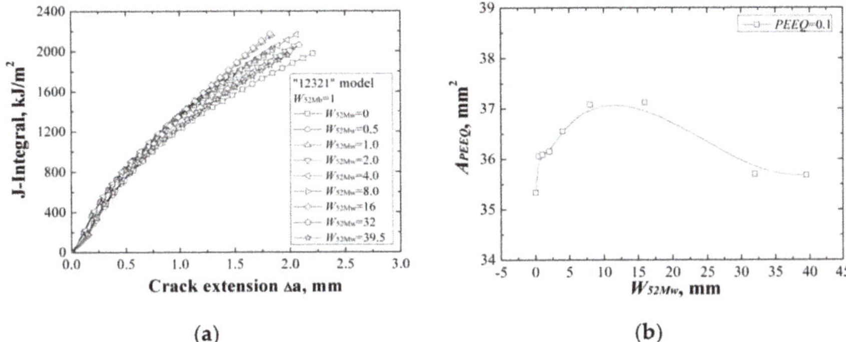

Figure 11. The J-resistance curves (**a**) and the areas surround by the $\varepsilon_p = 0.1$ isoline (**b**) of different "12321" models with the same W_{52Mb}.

Because the strength of the material 52Mw is higher than the material A508, thus, increasing of the width of 52Mw, the J-resistance curve of the model increases firstly. When the $W_{52Mw} = 16$ mm, there exists an optimal width and the model has the highest J-resistance curve. Then, the J-resistance curves of the models decrease and remain steady at last when the total width of 52Mb and 52Mw over the effect range of the material constraint.

The areas surrounded by the $\varepsilon_p = 0.1$ isoline at crack tip at the same J-integral ($J = 1600$ kJ/m^2) for different "12321" models are shown in Figure 11b, which also reflects the same change rule with the J-resistance curves.

3.4. "12324" Model

When changing the width of 52Mb individually, the J-resistance curves of different "12324" models under different material constraints are shown in Figure 12a. It can be found that increasing the width of 52Mb, the J-resistance curves of the models increase firstly then decrease, and finally remain steady. The models with $W_{52Mb} = 0$ mm and $W_{52Mb} = 0.5$ mm have the lowest J-resistance curves and the model with $W_{52Mb} = 2$ mm has the highest J-resistance curve. When the width of 52Mb up to 8 mm, the J-resistance curve will not change by increasing the 52Mb's width.

It is the same with "12321" model, because the strength of material 52Mb is higher than the material 52Mw, and increasing the width of 52Mb, the J-resistance curves of the models increase firstly. However, the strength of material 52Mb is lower than the material 316L, the J-resistance curves of the models do not always increase. When the $W_{52Mb} = 2$ mm, there exists an optimal width and the model has the highest J-resistance curve. Then, the J-resistance curves of the models decrease and remain steady at last when the width of 52Mb over the effect range of the material constraint.

The areas surrounded by the $\varepsilon_p = 0.1$ isoline at crack tip at the same J-integral ($J = 1600$ kJ/m^2) for different "12324" models are shown in Figure 12b, which also reflects the same change rule with the J-resistance curves.

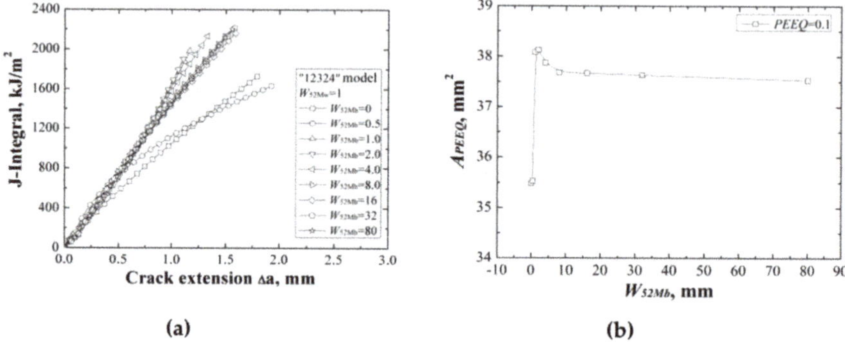

(a)   **(b)**

Figure 12. The *J*-resistance curves (**a**) and the areas surround by the ε_p = 0.1 isoline (**b**) of different "12324" models with the same W_{52Mw}.

When changing the width of 52Mw individually, the *J*-resistance curves of different "12324" models under different material constraints are shown in Figure 13a. It can be found that increasing of the width of 52Mw, the *J*-resistance curves of the "12324" models decrease. When the width of 52Mw is up to 32 mm, the *J*-resistance curves of the models remain steady and will not change with increasing the 52Mw's width. This is because although the strength of material 52Mw is higher than the material A508, it is much lower than the materials 316L and 52Mb. Increasing of the width of 52Mw, the *J*-resistance curves of the models decrease until the total width of 52Mb and 52Mw over the effect range of the material constraint.

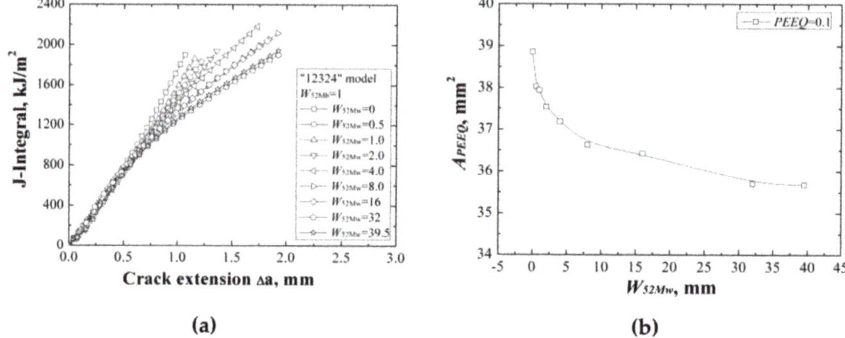

(a)   **(b)**

Figure 13. The *J*-resistance curves (**a**) and the areas surround by the ε_p = 0.1 isoline (**b**) of different "12324" models with the same W_{52Mb}.

The areas surrounded by the ε_p = 0.1 isoline at crack tip at the same *J*-integral (*J* = 1600 kJ/m^2) for different "12324" models are shown in Figure 13b, which also reflects the same change rule with the *J*-resistance curves.

In general, the above results show the effect range of the material constraint. Comparing all the models, it can be found that the effect range of the material constraint is real. The effect range relates to the materials on both sides of the crack. It should be pointed that the effect range is not only related to the adjacent material, but also non-adjacent material. It can be proved by comparing the *J*-resistance curves of the "12321" model and "12324" model. When the two cracks have the same adjacent material and dimensions but different non-adjacent materials, the *J*-resistance curves of the two models are different. That is, the *J*-resistance curves influenced by all the materials within the effect range, no matter whether they are adjacent or not.

4. Conclusions

In this study, the center crack was selected, different models that reflect different welded joints with various material constraints were designed, and the effect range of the material constraint was studied. The main results obtained are summarized as follows:

(1) For all the weld joints, the effect ranges of the material constraints are real. When the crack locates without the effect range of material constraint, the fracture resistance curves of the weld joints are no longer influenced by the material constraint any more.

(2) The *J*-resistance curves of the weld joints are influenced by all the materials within the effect range, no matter whether the material is adjacent to the crack or not.

(3) The areas surrounded by the ε_p isoline reflect the same change rule with the *J*-resistance curves. The *J*-resistance curves of the materials are controlled by the strain fields rather than the stress fields, and there may exist a main control value or control zone.

Author Contributions: Conceptualization, J.Y.; Methodology, J.Y.; Software, L.W.; Validation, J.Y. and L.W.; Writing-Original Draft Preparation, J.Y. and L.W.; Writing-Review and Editing, J.Y.

Funding: This research was funded by [National Natural Science Foundation of China] grant number [51605292].

Conflicts of Interest: The authors declare no conflict of interest.

References

1. Brocks, W.; Schmitt, W. The second parameter in J-R curves: Constraint or triaxiality. *ASTM Int.* **1995**. [CrossRef]
2. Larsson, S.G.; Carlsson, A.J. Influence of non-singular stress terms and specimen geometry on small-scale yielding at crack tips in elastic-plastic materials. *J. Mech. Phys. Solids* **1973**, *21*, 263–277. [CrossRef]
3. O'Dowd, N.P.; Shih, C.F. Family of crack-tip fields characterized by a triaxiality parameter. I. Structure of fields. *J. Mech. Phys. Solids* **1991**, *39*, 989–1015. [CrossRef]
4. O'Dowd, N.P.; Shih, C.F. Family of crack-tip fields characterized by a triaxiality parameter. II. Fracture applications. *J. Mech. Phys. Solids* **1992**, *40*, 939–963. [CrossRef]
5. Chao, Y.J.; Yang, S.; Sutton, M.A. On the fracture of solids characterized by one or two parameters: Theory and practice. *J. Mech. Phys. Solids* **1994**, *42*, 629–647. [CrossRef]
6. Guo, W. Elastoplastic three dimensional crack border field-I. Singular structure of the field. *Eng. Fract. Mech.* **1993**, *46*, 93–104. [CrossRef]
7. Guo, W. Elastoplastic three dimensional crack border field-II. Asymptotic solution for the field. *Eng. Fract. Mech.* **1993**, *46*, 105–113. [CrossRef]
8. Guo, W. Elasto-plastic three-dimensional crack border field-III. Fracture parameters. *Eng. Fract. Mech.* **1995**, *51*, 51–71. [CrossRef]
9. Joch, J.; Ainsworth, R.A.; Hyde, T.H. Limit load and J-estimates for idealised problems of deeply cracked welded joints in plane-strain bending and tension. *Fatigue Fract. Eng. Mater. Struct.* **1993**, *16*, 1061–1079. [CrossRef]
10. Burstow, M.C.; Ainsworth, R.A. Comparison of analytical, numerical and experimental solutions to problems of deeply cracked welded joints in bending. *Fatigue Fract. Eng. Mater. Struct.* **1995**, *18*, 221–234. [CrossRef]
11. Zhang, Z.L.; Hauge, M.; Thaulow, C. Two-parameter characterization of the near-tip stress fields for a bi-material elastic-plastic interface crack. *Int. J. Fract.* **1996**, *79*, 65–83. [CrossRef]
12. Betegón, C.; Peñuelas, I. A constraint based parameter for quantifying the crack tip stress fields in welded joints. *Eng. Fract. Mech.* **2006**, *73*, 1865–1877. [CrossRef]
13. Yang, J.; Wang, G.Z.; Xuan, F.Z.; Tu, S.T. Unified characterisation of in-plane and out-of-plane constraint based on crack-tip equivalent plastic strain. *Fatigue Fract. Eng. Mater. Struct.* **2013**, *36*, 504–514. [CrossRef]
14. Yang, J.; Wang, G.Z.; Xuan, F.Z.; Tu, S.T. Unified correlation of in-plane and out-of-plane constraints with fracture toughness. *Fatigue Fract. Eng. Mater. Struct.* **2014**, *37*, 132–145. [CrossRef]
15. Yang, J.; Wang, G.Z.; Xuan, F.Z.; Tu, S.T. Unified correlation of in-plane and out-of-plane constraint with fracture resistance of a dissimilar metal welded joint. *Eng. Fract. Mech.* **2014**, *115*, 296–307. [CrossRef]

16. Negre, P.; Steglich, D.; Brocks, W. Crack extension at an interface: prediction of fracture toughness and simulation of crack path deviation. *Int. J. Fract.* **2005**, *134*, 209–229. [CrossRef]

17. Samal, M.K.; Balani, K.; Seidenfuss, M.; Roos, E. An experimental and numerical investigation of fracture resistance behaviour of a dissimilar metal welded joint. *Proc. Inst. Mech. Eng. C J. Mech. Eng. Sci.* **2009**, *223*, 1507–1523. [CrossRef]

18. Fan, K.; Wang, G.Z.; Tu, S.T.; Xuan, F.Z. Geometry and material constraint effects on fracture resistance behavior of bi-material interfaces. *Int. J. Fract.* **2016**, *201*, 143–155. [CrossRef]

19. Fan, K.; Wang, G.Z.; Xuan, F.Z.; Tu, S.T. Effects of work hardening mismatch on fracture resistance behavior of bi-material interface regions. *Mater. Des.* **2015**, *68*, 186–194. [CrossRef]

20. Fan, K.; Wang, G.Z.; Xuan, F.Z.; Tu, S.T. Local fracture resistance behavior of interface regions in a dissimilar metal welded joint. *Eng. Fract. Mech.* **2015**, *136*, 279–291. [CrossRef]

21. Rakin, M.; Medjo, B.; Gubeljak, N.; Sedmak, A. Micromechanical assessment of mismatch effects on fracture of high-strength low alloyed steel welded joints. *Eng. Fract. Mech.* **2013**, *109*, 221–235. [CrossRef]

22. Wang, H.T.; Wang, G.Z.; Xuan, F.Z.; Liu, C.J.; Tu, S.T. Local mechanical properties of a dissimilar metal welded joint in nuclear power systems. *Mater. Sci. Eng. A* **2013**, *568*, 108–117. [CrossRef]

23. Wang, H.T.; Wang, G.Z.; Xuan, F.Z.; Tu, S.T. An experimental investigation of local fracture resistance and crack growth paths in a dissimilar metal welded joint. *Mater. Des.* **2013**, *44*, 179–189. [CrossRef]

24. Xue, H.; Sun, J. Study on micro region of crack tip of welded joints under different matches of yield stress. *Hot Work. Technol.* **2016**. [CrossRef]

25. Zhu, Z.Q.; Jing, H.Y.; Ge, J.G.; Chen, L.G. Effects of strength mis-matching on the fracture behavior of nuclear pressure steel A508-III welded joint. *Materi. Sci. Eng. A* **2005**, *390*, 113–117. [CrossRef]

26. Yang, J. Micromechanical analysis of in-plane constraint effect on local fracture behavior of cracks in the weakest locations of dissimilar metal welded joint. *Acta Metall. Sin.* **2017**, *30*, 840–850. [CrossRef]

27. Østby, E.; Thaulow, C.; Zhang, Z.L. Numerical simulations of specimen size and mismatch effects in ductile crack growth-Part I: Tearing resistance and crack growth paths. *Eng. Fract. Mech.* **2007**, *74*, 1770–1792. [CrossRef]

28. *Standard Test Method for Measurement of Fracture Toughness*; ASTM E1820-08a; ASTM International: Philadelphia, PA, USA, 2008.

Article

Study of the Effect of Current Pulse Frequency on Ti-6Al-4V Alloy Coating Formation by Micro Arc Oxidation

Alexander Sobolev, Alexey Kossenko and Konstantin Borodianskiy *

Department of Chemical Engineering, Ariel University, Ariel 40700, Israel; sobolev@ariel.ac.il (A.S.); kossenkoa@ariel.ac.il (A.K.)
* Correspondence: konstantinb@ariel.ac.il; Tel.: +972-3-9143085

Received: 6 November 2019; Accepted: 27 November 2019; Published: 1 December 2019

Abstract: The micro arc oxidation (MAO) process has been applied to produce ceramic oxide coating on Ti-6Al-4V alloy. The MAO process was carried out at the symmetric bipolar square pulse in electrolyte containing Na_2CO_3 and Na_2SiO_3. The effect of current frequency on the surface morphology, the chemical and the phase compositions as well as the corrosion resistance was examined. Morphology and cross-sectional investigation by electron microscopy evaluated more compacted and less porous coating produced by high current frequency (1000 Hz). This alloy also exhibited a high corrosion resistance in comparison with the untreated alloy. Additionally, the alloy subjected to MAO treatment by a current frequency of 1000 Hz showed a higher corrosion resistance in comparison with alloys obtained by lower current frequencies. This behavior was attributed to more compacted and less porous morphology of the coating.

Keywords: micro arc oxidation; current pulse frequency; Ti-6Al-4V alloy; titanium oxide; corrosion resistance

---

1. Introduction

Titanium alloys are widely applicable in marine, aerospace, chemical and biomedical industries due to their high specific strength, corrosion resistance and excellent biocompatibility [1–4]. However, their advanced properties are limited due to the lack of the protective surface properties. One of the most promising environmentally friendly processes of coating formation is micro arc oxidation (MAO).

MAO is an electrochemical processing of ceramic oxide coating formation on the metallic substrate by the appearance of dielectric breakdown and plasma discharges. The basic principle of the MAO technology is an application of a high voltage between the treated metal and the electrode. During the process, micro-arc discharge migration points appear on the surface lead to the oxide coating formation [5–8]. Usually, MAO treatment is suitable for the enhanced mechanical properties and corrosion resistance of the alloys' surface [9–11].

Ti-6Al-4V is one of the most commonly applicable titanium alloys because of its outstanding properties [12]. This Ti alloy contains both aluminum and vanadium; Al stabilizes α phase and has the HCP lattice, and V stabilizes β phase and has the BCC lattice. Consequently, this alloy exhibits advanced mechanical properties as α phase provides a high strength while β phase provides high ductility. Moreover, Ti-6Al-4V alloy is one of the main metallic biomaterials due to its high biocompatibility [13–15]. Additional treatment of this alloy by MAO process will lead to the production of an oxide coating with excellent corrosion resistance for applications in wide range of industries.

The majority of research works have been devoted to studying the electrolyte chemical composition effect on the structure and the properties of the created coatings [16–19]. However, the influence of the process electrical parameters affects also the formation of the coating, resulting in different surface properties [20,21].

Therefore, the goal of the present work is an investigation of the current pulse frequency effect of the MAO process on the coating formation on Ti-6Al-4V alloy. The obtained coating morphology, and the chemical and the phase compositions were also examined in details in the work. Furthermore, the influence of the current frequency on the corrosion resistance of the produced alloys' surface was also examined.

2. Materials and Methods

2.1. Coating Production

Ti-6Al-4V alloy samples (Grade V, max. wt.%: 5.5–6.7 Al, 3.5–4.0 V, 0.25 Fe, 0.2 O, 0.05 N, 0.08 C, bal. Ti) (Scope Metals Group Ltd., Israel) with a surface area of 62 cm^2 were grounded up to abrasive paper grits #1200 and subjected to the ultrasonic cleaning in acetone.

The MAO process was carried out by MP2-AS 35 power supply (Magpulls, Sinzheim, Germany) with the follow electrical parameters: I_{max} = 5 A, U_{max} = 1000 V. Electrical parameters were pulsed (symmetric bipolar square pulse) at a frequency from 200 to 1000 Hz. Names of the experimental samples and their current pulse parameters listed in Table 1. The process was carried out in electrolyte contained sodium carbonate (Na$_2$CO$_3$, Sigma Aldrich, St. Louis, MO, USA) = 10 g/L and sodium metasilicate pentahydrate (Na$_2$SiO$_3$·5H$_2$O, Sigma Aldrich, St. Louis, MO, USA) = 2 g/L. The process was conducted at 30 °C; in a water-cooled glass vial for 15 min. The examined alloy served as a working electrode and a stainless-steel sheet served as a counter electrode. The anodic current density was set at 0.05 A/cm^2 and the applied voltage was 250 V, with the duty cycle 50% and symmetrical anode/cathode pulse. Electrical parameters of the process were recorded by Scope Meter 199C, 200 MHz, 2.5 GS s^{-1} (Fluke, Everett, WA, USA).

Table 1. Names of examined samples in the research work.

Name of the Sample	Frequency Applied (Hz)
Sam–200 Hz	200 Hz
Sam–400 Hz	400 Hz
Sam–600 Hz	600 Hz
Sam–800 Hz	800 Hz
Sam–1000 Hz	1000 Hz

2.2. Coating Characterization

Scanning electron microscope (SEM) MAIA3 (TESCAN, Brno, Czech Republic) was used for surface morphology and cross-sectional examinations. EDS system by (Oxford instruments, Abingdon, UK) with an X-Max^N detector was applied for elemental analysis detection. The phase analysis of the coating was detected using the X'Pert Pro diffractometer (PANalytical B.V., Almelo, the Netherlands) with Cu$_\alpha$ radiation (λ = 1.542Å) at the grazing incidence mode (grazing angle of 3°) with a 2θ range from 20 to 70° (step size of 0.03°) at 40 kV and 40 mA.

Coating elemental analysis was also examined by particle induced X-ray emission (PIXE) method by 1.7 MV Pelletron accelerator equipped with SuperSilicon Drift Detector, X-123 SDD (Amptek Inc., Bedford, MA, USA) positioned at 45° to the beam (I~4.3 nA with a nominal diameter of 1.5 mm). 2.017 MeV 4He$^+$ ± 1 KeV beam was used to collect spectra. The samples were coated by carbon in order to prevent charge effect during irradiation. GUPIX software (University of Guelph, ON, Canada) was used for the PIXE data analysis.

Light scattering measurements were performed using XploRA ONE™ Raman confocal microscope (HORIBA Scientific, Piscataway, NJ, USA) using 532 nm laser excitation line and a constant power of 20 mW.

2.3. Coating Corrosion Resistance Test

The corrosion resistance examination was done on PARSTAT 4000A potentiostat/galvanostat (Princeton Applied Research, Oak Ridge, TN, USA) using potentiodynamic polarization test in a 3.5 wt.% NaCl (Sigma-Aldrich, St. Louis, MO, USA) solution, using a three- electrode cell configuration. Pt sheet served as a counter electrode and saturated Ag/AgCl (Metrohm Autolab B.V., Almelo, the Netherlands) served as a reference electrode. The polarization resistance was detected in the range of ± 0.25 V with respect to the recorded corrosion potential at a scan rate of 0.1 mV/s. Prior to the test, the samples were placed in the 3.5 wt.% NaCl solution for 1 h in order to reach the steady state of a working electrode.

3. Results and Discussion

3.1. Process Characterization

The recorded electric characteristics in the process are illustrated in Figure 1.

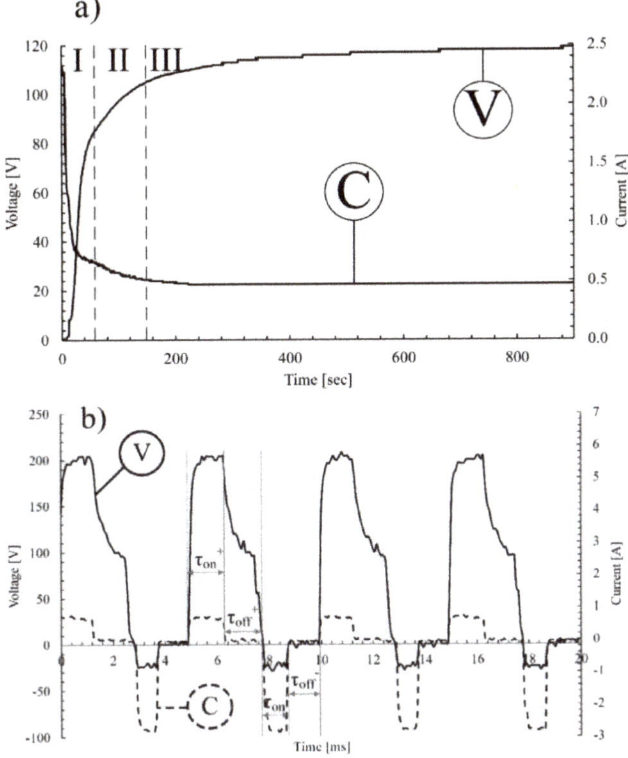

Figure 1. Typical plots of: (**a**) voltage- and current-time behavior; (**b**) the waveform of the MAO process applied on alloy Ti-6Al-4V. V—voltage-time curve; C—current-time curve.

Figure 1 presents a typical voltage- and current-time behavior for examined samples; the obtained plots are the same for all five examined samples. The voltage-time plot contains three main stages, which are clearly detected on the plot (Figure 1a). In the initial stage, the voltage increases rapidly up to 80 V. At this stage the amorphous layer is formed (Figure 1a, area I) and the process is accompanied by the gas bubbles appearance on the materials' surface, which was also shown previously in the work of Snizhko et al. [22]. In the second stage of the process, the voltage has reached 105 V and dielectric breakdowns came out with the uniformly appeared electric sparks on the alloys' surface (Figure 1a, area II). At this stage, an amorphous to crystalline structure transition of the oxide coating occurred, as described in work of Hussein et al. [23]. In the last stage of the MAO process, the voltage was near the plateau (Figure 1a, area III). Here, sparks became larger and more intense, which corresponds with the increase in the charge transfer resistance. This behavior is also confirmed by work of Mortazavi et al. [24].

On the current-time plot, the rapid drop of the current at the initial stage is detected (Figure 1a, area I). However, the process current begins at 0 A, and it extremely quickly reaches the maximum value as a result of the quasi short circuit. Simultaneously, the amorphous layer is rapidly growing. Its growth terminates as sufficient layer thickness is reached. In the second stage of the processing, the transformation of the oxide layer occurs resulting in the formation of low conductive coating (Figure 1a, area II), and finally stabilizes (Figure 1a, area III). The same behavior was also observed by Ahounbar et al. [25].

The obtained waveform that was determined by the nature of the process is shown in Figure 1b. τ_{off} refers to the period when the current is not supplied. At this period of time, titanium ions may release to electrolyte or may form a thin amorphous layer on the alloys' surface. When the voltage value is τ_{on} positive (anodic polarization), the current is supplied, and micro discharges appear over the dielectric breakdown potential. At this period of time, the MAO process is accompanied by a strong electric field and a high temperature and pressure [6]. Thus, several components of the electrolyte, the coating, or the substrate may ionize and decompose resulting in titanium cations release. This strong electric field also affects the diffusion of the oxygen anions towards the Ti substrate. Oxygen ions interact with Ti ions, producing titanium oxide amorphous structure that later transforms into the crystalline phase. Newly formed TiO_2 coating is subjected to the compaction during subsequent dielectric breakdowns. When the value is τ_{on} negative (cathodic polarization), the dielectric breakdowns do not occur due to the high electrical conductivity of the created oxide coating [9].

3.2. Surface Characterization

Coating morphology and cross-sections images of the produced surfaces by MAO processing are shown in Figures 2 and 3.

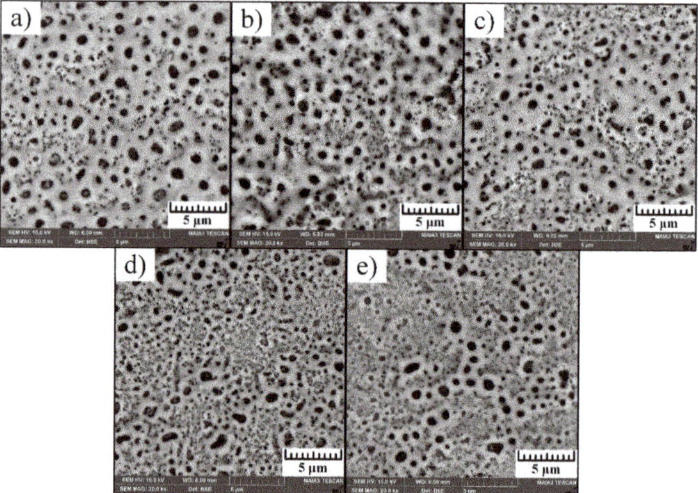

Figure 2. SEM images of surface morphology of the alloy Ti-6Al-4V treated by MAO with various current pulses: (**a**) Sam–200 Hz; (**b**) Sam–400 Hz; (**c**) Sam–600 Hz; (**d**) Sam–800 Hz; (**e**) Sam–1000 Hz.

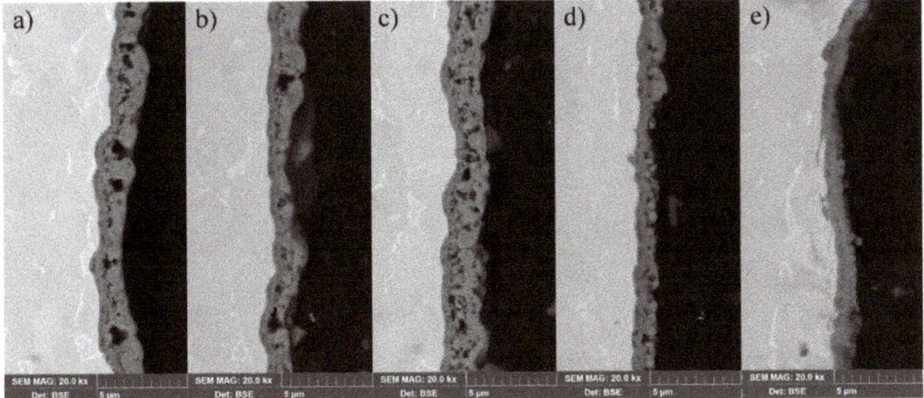

Figure 3. SEM images of cross-sections of the alloy Ti-6Al-4V treated by MAO with various current pulses: (**a**) Sam–200 Hz; (**b**) Sam–400 Hz; (**c**) Sam–600 Hz; (**d**) Sam–800 Hz; (**e**) Sam–1000 Hz.

Investigation of the surface morphology presented by SEM in Figure 2 revealed numerous micro-sized and sub-micron pores. These pores are formed in the process as the result of dielectric breakdowns at the extreme high temperature. Surface porosity change was also detected on the morphology images where sub-micron pores tend to disappear with the frequency increase. This was also clearly detected on the cross-section images presented in Figure 3.

MAO treatment is a high-energy process associated with a high current and a high voltage that affect surface morphology formation. The distribution of this current and voltage over the surface is not uniform and strongly depends on the localized dielectric properties of the developed coating. Thus, appearance of pores on the surface is spontaneous during the dielectric breakdowns, which are also influenced by their size and form. Therefore, control of the coating porosity may be done by the variety of the current frequency.

Additionally, to reduce porosity, the cross-sectional images also evaluated compaction of the coating with the growth of current pulses in MAO process. Thickness of the coatings was determined on the SEM cross-sectional images. It is clearly seen that the thickness of the produced coating reduced with the current pulse increase. Sam–200 Hz has a coating thickness of 2 µm, which was reduced up to 1 µm for the Sam–1000 Hz. This can be attributed to the reduction of current peak time duration with the increase of the frequency as the result of τ_{on} and τ_{off} reduction.

Observation of Figure 3a–c revealed almost the same coating thickness with different pores size and their number. The large pores detected in Sam–200 Hz became smaller—mostly sub-micron—in Sam–600 Hz due to the dielectric breakdowns caused by localized high temperature. This phenomenon may be attributed to the decrease of the localized temperature as the result of a specific power density reduction. With the increase of the current frequency (Figure 3d,e), the local re-melt of the initial porous coating occurs, resulting in compaction and reduction of its porosity.

Examination of the cross-sectional images showed in Figure 3 also revealed thickness of all coatings. The thickness was found to be almost the same over the surface depth of each sample.

3.3. Chemical and Phase Analysis

The elemental composition of the developed coatings detected by EDS is presented in Table 2.

Table 2. Elemental composition of the alloy Ti-6Al-4V treated by MAO with various current pulses detected by EDS investigation (at. %).

Sample	Ti	Al	V	Si	O
Sam–200 Hz	24.2 ± 1.3	2.3 ± 0.5	0.9 ± 0.1	5.1 ± 0.9	Balance
Sam–400 Hz	24.1 ± 1.5	2.5 ± 0.3	0.9 ± 0.1	3.5 ± 0.7	Balance
Sam–600 Hz	24.4 ± 2.1	2.8 ± 0.3	1.0 ± 0.2	3.7 ± 0.7	Balance
Sam–800 Hz	27.4 ± 2.3	3.1 ± 0.7	1.1 ± 0.3	2.2 ± 0.6	Balance
Sam–1000Hz	24.0 ± 1.2	2.5 ± 0.5	0.9 ± 0.2	4.4 ± 1.0	Balance

EDS study revealed that the appearance of Ti, Al and V originated from the Ti-6Al-4V alloy and Si originated from the electrolyte decomposition. The chemical composition of all samples is almost the same and is not affected by the applied current pulse.

Figure 4 shows PIXE spectra of the formed coatings.

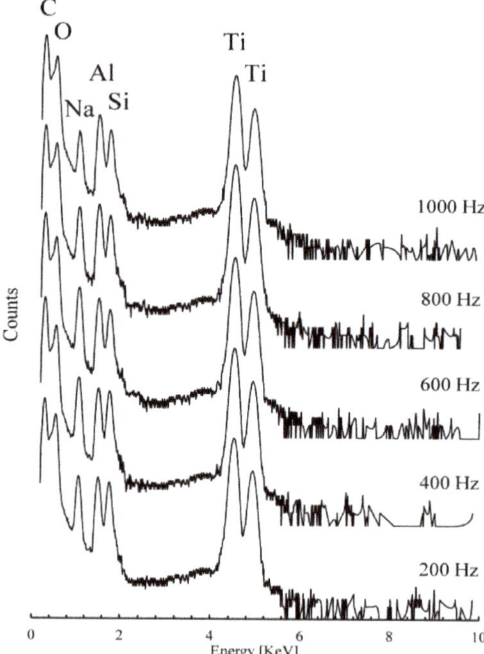

Figure 4. PIXE spectra of the alloy Ti-6Al-4V coatings subjected to MAO treatment with various current pulses.

PIXE analysis clearly detected the following chemical elements: C, O, Al, Na, Si and Ti. Titanium and oxygen are main elements of the developed titanium oxide coating, aluminum is the major alloying element in Ti-6Al-4V alloy, and carbon peak was detected because the surface was coated in order to eliminate charge effect. Sodium and silicon were detected because of their appearance in the coating as the result of electrolyte decomposition during MAO processing. As it is clearly seen in Figure 4, there are no changes in the detected peaks were found in all examined samples.

The phase composition of the obtained coatings was revealed by the XRD measurements and are shown in Figure 5.

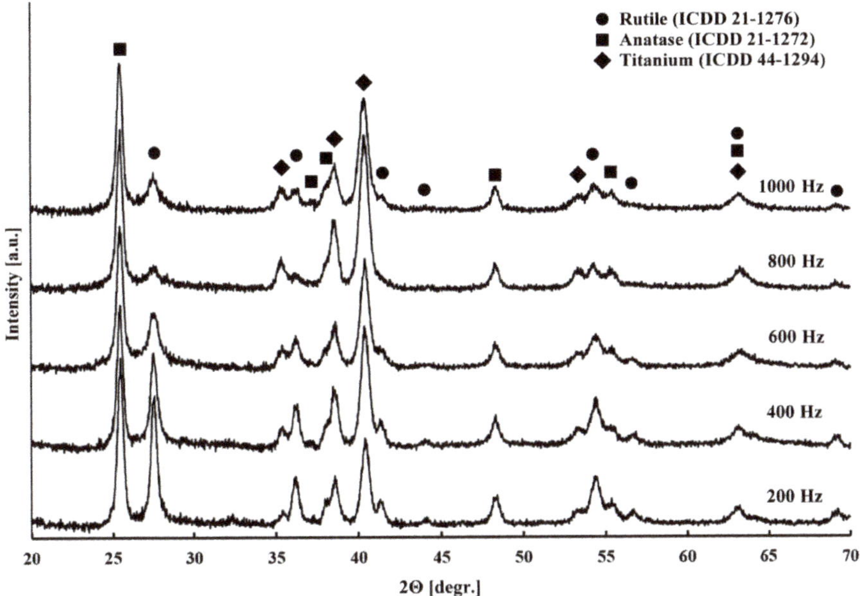

Figure 5. X-ray diffraction patterns of the alloy Ti-6Al-4V surfaces after MAO treatment with various current pulses.

XRD measurement evaluated the presence of the titanium phase (ICDD 44-1294) along with two titanium oxide phases, rutile (ICDD 21-1276) and anatase (ICDD 21-1272). It is detected that the content of the rutile phase was reduced with the growth of the processing frequency. It is clearly seen on the rutile analytical peak intensity change at 27.6°, that the peak is relatively low for Sam–1000 Hz as compared to the same peak for Sam–200 Hz alloy. Additional peaks of rutile at 36.2 and 41.2° were also clearly detected in the pattern obtained in the Sam–200 Hz alloy, and their intensity was significantly reduced with the current frequency increase. Simultaneously, the analytical peak of anatase at 25.5° and its additional peak at 55.5° intensity were detected to be increased with the current frequency increase. This behavior may be attributed to the anatase-to-rutile transformation that occurred when the process reached a dielectric breakdown temperature, as described in by Hanaor et al. [26]. A metastable anatase phase is irreversible and transforms into rutile phase at a low frequency where the contribution of the current is more significant. No more phases were detected by XRD analysis. However, additional phases originated from the electrolyte decomposition may appear in the coating in amorphous structure or even in crystalline with a content that is below the detection limit of XRD.

Apart from XRD analysis, the TiO_2 coatings obtained by MAO process were also investigated by Raman spectroscopy (Figure 6).

Raman spectra illustrated in Figure 6 are almost the same for all examined samples and show that the TiO_2–anatase phase, which was detected with the active vibrations at: 147 (Eg1), 198 (Eg2), 398 (B1g), 515 (A1g + B1g doublet band), 640 (Eg3), and 796 cm^{-1} (B1g[F], first overtone of B1g at 398 cm^{-1}) [27]. The intensity of the Raman spectra peaks increased with the current frequency increase as the result of the produced surface porosity variation.

Due to the low penetration, the rutile phase was not detected by Raman spectroscopy. However, the anatase-to-rutile phase transformation was found by XRD and described above. Consequently, rutile was formed as the result of the transformation in the inner layer of the coating.

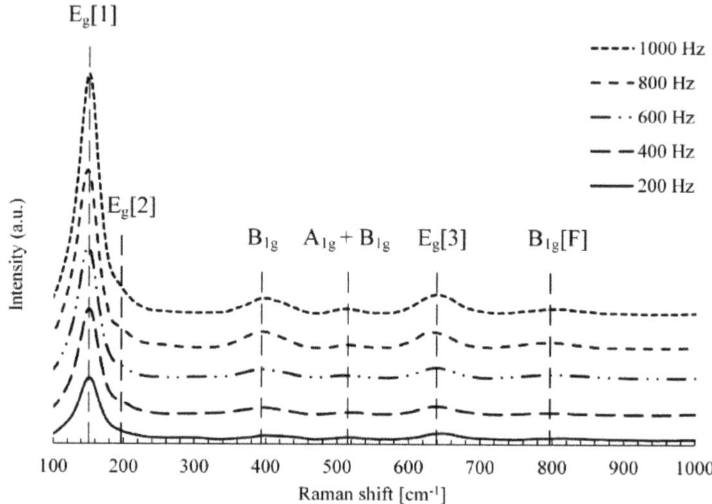

Figure 6. Raman spectra of the alloy Ti-6Al-4V surfaces after MAO treatment with various current pulses.

3.4. Corrosion Resistance Investigation

The corrosion properties of the treated samples were determined by a potentiodynamic polarization method. The obtained curves are illustrated in Figure 7.

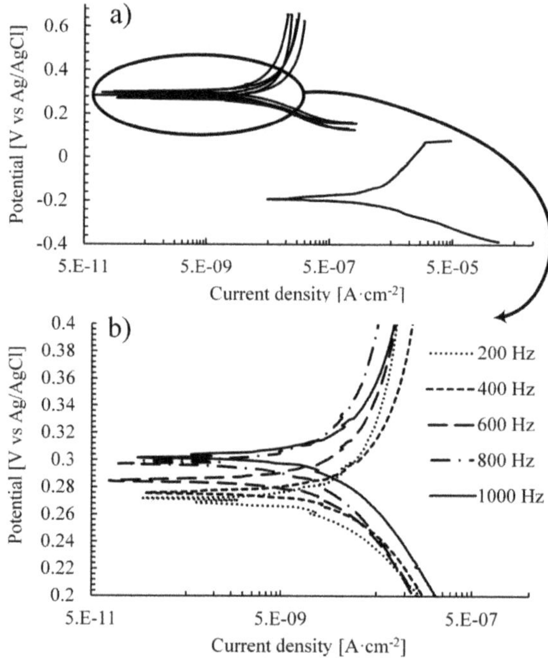

Figure 7. (**a**). potentiodynamic polarization curves for alloy Ti-6Al-4V after MAO treatment with various current pulses (upper curves) and curve for base Ti-6Al-4V (lower curve). (**b**). enlarged curves for treated alloys. Examination was carried out in 3.5 wt.% NaCl.

According to theory, the higher the corrosion potential, the higher the corrosion resistance. Potentiodynamic polarization curves on Figure 7a clearly indicated a shift of the coated samples to the more positive potential. This may be attributed to the reduction of the anodic and the cathodic processes due to the oxide coating appearance.

The polarization resistance (R_p) was calculated according to Equation (1):

$$R_p = \frac{\beta_a \times \beta_c}{2.3 \times i_{corr}(\beta_a + \beta_c)}. \tag{1}$$

The Tafel slopes β_a and β_c were calculated from the anodic and the cathodic curves on the plot (Figure 7b). The corrosion potentials (E_{corr}), the corrosion current densities (i_{corr}) and the polarization resistance (R_p) are listed in Table 3.

Table 3. Calculated corrosion test results for alloy Ti-6Al-4V after MAO treatment with various current pulses and the base alloy.

Samples	E_{corr} (mV)	i_{corr} (nA/cm^2)	β_a (mV/decade)	β_c (mV/decade)	R_p (kΩ cm^2)	Corrosion Rate (mm/year)
Untreated alloy	−198.386	2141.5	310.368	88.89	14	0.018644
Sam–200Hz	270.387	49.319	333.266	130.897	832	0.000429
Sam–400Hz	269.917	33.000	304.585	92.802	937	0.000287
Sam–600Hz	303.337	30.534	225.418	117.102	1097	0.000266
Sam–800Hz	282.228	33.364	308.202	129.618	1189	0.000290
Sam–1000Hz	300.085	17.260	216.575	113.596	1877	0.000150

Calculation results in Table 3 revealed that the corrosion rate of coated alloy was at least 40 times lower compared to the untreated alloy. The lowest value was obtained on Sam–1000 Hz, which exhibited 125 times higher corrosion resistance in comparison with untreated alloy. Thus, the rate of the corrosion was 0.000150 mm/year and 0.018644 mm/year for the Sam–1000 Hz and untreated alloy, respectively. Furthermore, the higher the applied current pulse in MAO process, the higher the corrosion resistance. It is well correlated with the morphology investigation of the coatings where Sam–1000 Hz was found to be more compact with lower porosity compared to other examined samples.

4. Conclusions

In the current work, ceramic oxide coating on the Ti-6Al-4V alloy using MAO processing has been produced. The effect of different current pulse frequency on the coating morphology, phase composition and corrosion resistance has been studied. It was found that the application of different current frequency affected the porosity and compaction of the formed oxide coating.

Surface morphology and cross-sectional investigation by electron microscopy showed that the higher the applied current pulse frequency, the more compact and less porous the obtained coating is. After processing with a pulse current of 1000 Hz, the coating with a thickness of 1 μm and with a low porosity has been obtained. PIXE analysis revealed the appearance of titanium oxide coating on treated alloys. Two phases of titanium oxide, anatase and rutile were identified by XRD. Raman spectroscopy investigation detected the appearance of anatase phase on the surface of the examined samples. Rutile appeared in inner layer of the coating as the result of the partial transformation of the anatase phase.

Potentiodynamic polarization curves have evaluated that the corrosion resistance of the produced coating 40 times higher compared to the untreated alloy. Moreover, it was found that the higher the current pulse frequency, the higher the corrosion resistance of the obtained coating resulting in the formation of compacted and low porous oxide coating.

Author Contributions: Conceptualization, A.S., A.K. and K.B.; formal analysis A.S., A.K. and K.B.; investigation, A.S.; writing—original draft preparation A.S. and K.B.; writing—review and editing, A.S. and K.B.; supervision K.B.

Funding: This research received no external funding.

Conflicts of Interest: The authors declare no conflict of interest.

References

1. Gu, Y.; Ma, A.; Jiang, J.; Li, H.; Song, D.; Wu, H.; Yuan, Y. Simultaneously improving mechanical properties and corrosion resistance of pure Ti by continuous ECAP plus short-duration annealing. *Mater. Charact.* **2018**, *138*, 38–47. [CrossRef]
2. Gurrappa, I. Characterization of titanium alloy Ti-6Al-4V for chemical, marine and industrial applications. *Mater. Charact.* **2003**, *51*, 131–139. [CrossRef]
3. Wang, H.-Y.; Zhu, R.-F.; Lu, Y.-P.; Xiao, G.-Y.; Zhao, X.-C.; He, K.; Ma, X.-N. Preparation and properties of plasma electrolytic oxidation coating on sandblasted pure titanium by a combination treatment. *Mater. Sci. Eng. C* **2014**, *42*, 657–664. [CrossRef] [PubMed]
4. Kaur, M.; Singh, K. Review on titanium and titanium based alloys as biomaterials for orthopaedic applications. *Mater. Sci. Eng. C* **2019**, *102*, 844–862. [CrossRef] [PubMed]
5. Sobolev, A.; Wolicki, I.; Kossenko, A.; Zinigrad, M.; Borodianskiy, K. Coating formation on Ti-6Al-4V alloy by micro arc oxidation in molten salt. *Materials* **2018**, *11*, 1611. [CrossRef]
6. Yerokhin, A.L.; Nie, X.; Leyland, A.; Matthews, A.; Dowey, S.J. Plasma electrolysis for surface engineering. *Surf. Coat. Technol.* **1999**, *122*, 73–93. [CrossRef]
7. Yerokhin, A.L.; Nie, X.; Leyland, A.; Matthews, A. Characterization of oxide films produced by plasma electrolytic oxidation of a Ti–6Al–4V alloy. *Surf. Coat. Technol.* **2000**, *130*, 195–206. [CrossRef]
8. Nominé, A.; Martin, J.; Henrion, G.; Belmonte, T. Effect of cathodic micro-discharges on oxide growth during plasma electrolytic oxidation (PEO). *Surf. Coat. Technol.* **2015**, *269*, 131–137. [CrossRef]
9. Cheng, Y.; Wu, X.-Q.; Xue, Z.; Matykina, E.; Skeldon, P.; Thompson, G.E. Microstructure, corrosion and wear performance of plasma electrolytic oxidation coatings formed on Ti–6Al–4V alloy in silicate-hexametaphosphate electrolyte. *Surf. Coat. Technol.* **2013**, *217*, 129–139. [CrossRef]
10. Zhu, L.; Petrova, R.S.; Gashinski, J.P.; Yang, Z. The effect of surface roughness on PEO-treated Ti-6Al-4V alloy and corrosion resistance. *Surf. Coat. Technol.* **2017**, *325*, 22–29. [CrossRef]
11. Sobolev, A.; Valkov, A.; Kossenko, A.; Wolicki, I.; Zinigrad, M.; Borodianskiy, K. Bioactive coating on titanium alloy with high osseointegration and antibacterial Ag nanoparticles. *ACS Appl. Mater. Interfaces* **2019**, *11*, 39534–39544. [CrossRef] [PubMed]
12. Mierzejewska, Ż.; Hudák, R.; Sidun, J. Mechanical Properties and Microstructure of DMLS Ti6Al4V Alloy Dedicated to Biomedical Applications. *Materials* **2019**, *12*, 176. [CrossRef] [PubMed]
13. Santos-Coquillat, A.; Gonzalez Tenorio, R.; Mohedano, M.; Martinez-Campos, E.; Arrabal, R.; Matykina, E. Tailoring of antibacterial and osteogenic properties of Ti6Al4V by plasma electrolytic oxidation. *Appl. Surf. Sci.* **2018**, *454*, 157–172. [CrossRef]
14. Shah, F.A.; Trobos, M.; Thomsen, P.; Palmquist, A. Commercially pure titanium (cp-Ti) versus titanium alloy (Ti6Al4V) materials as bone anchored implants—Is one truly better than the other? *Mater. Sci. Eng. C* **2016**, *62*, 960–966. [CrossRef]
15. Durdu, S.; Usta, M.; Berkem, A.S. Bioactive coatings on Ti6Al4V alloy formed by plasma electrolytic oxidation. *Surf. Coat. Technol.* **2016**, *301*, 85–93. [CrossRef]
16. Wheeler, J.M.; Collier, C.A.; Paillard, J.M.; Curran, J.A. Evaluation of micromechanical behaviour of plasma electrolytic oxidation (PEO) coatings on Ti–6Al–4V. *Surf. Coat. Technol.* **2010**, *204*, 3399–3409. [CrossRef]
17. Aliasghari, S.; Němcová, A.; Čížek, J.; Gholinia, A.; Skeldon, P.; Thompson, G.E. Effects of reagent purity on plasma electrolytic oxidation of titanium in an aluminate—Phosphate electrolyte. *Int. J. Surf. Eng. Coat.* **2016**, *94*, 32–42. [CrossRef]
18. Stojadinović, S.; Vasilić, R.; Petković, M.; Kasalica, B.; Belča, I.; Žekić, A.; Zeković, L. Characterization of the plasma electrolytic oxidation of titanium in sodium metasilicate. *Appl. Surf. Sci.* **2013**, *265*, 226–233. [CrossRef]
19. Shokouhfar, M.; Dehghanian, C.; Baradaran, A. Preparation of ceramic coating on Ti substrate by Plasma electrolytic oxidation in different electrolytes and evaluation of its corrosion resistance: Part 2. *Appl. Surf. Sci.* **2012**, *258*, 2416–2423. [CrossRef]

20. Songur, F.; Dikici, B.; Niinomi, M.; Arslan, E. The plasma electrolytic oxidation (PEO) coatings to enhance in-vitro corrosion resistance of Ti–29Nb–13Ta–4.6Zr alloys: The combined effect of duty cycle and the deposition frequency. *Surf. Coat. Technol.* **2019**, *374*, 345–354. [CrossRef]

21. Torres-Ceron, D.A.; Restrepo-Parra, E.; Acosta-Medina, C.D.; Escobar-Rincon, D.; Ospina-Ospina, R. Study of duty cycle influence on the band gap energy of TiO2/P coatings obtained by PEO process. *Surf. Coat. Technol.* **2019**, *375*, 221–228. [CrossRef]

22. Snizhko, L.O.; Yerokhin, A.L.; Pilkington, A.; Gurevina, N.L.; Misnyankin, D.O.; Leyland, A.; Matthews, A. Anodic processes in plasma electrolytic oxidation of aluminium in alkaline solutions. *Electrochem. Acta* **2004**, *49*, 2085–2095. [CrossRef]

23. Hussein, R.O.; Nie, X.; Northwood, D.O.; Yerokhin, A.; Matthews, A. Spectroscopic study of electrolytic plasma and discharging behaviour during the plasma electrolytic oxidation (PEO) process. *J. Phys. D Appl. Phys.* **2010**, *43*, 13. [CrossRef]

24. Mortazavi, G.; Jiang, J.; Meletis, E.I. Investigation of the plasma electrolytic oxidation mechanism of titanium. *Appl. Surf. Sci.* **2019**, *488*, 370–382. [CrossRef]

25. Ahounbar, E.; Khoei, S.M.M.; Omidvar, H. Characteristics of in-situ synthesized hydroxyapatite on TiO$_2$ ceramic via plasma electrolytic oxidation. *Ceram. Int.* **2018**, *45*, 3118–3125. [CrossRef]

26. Hanaor, D.A.H.; Sorrell, C.C. Review of the anatase to rutile phase transformation. *J. Mater. Sci.* **2011**, *46*, 855–874. [CrossRef]

27. Balachandran, U.; Eror, N.G. Raman spectra of titanium dioxide. *J. Solid State Chem.* **1982**, *42*, 276–282. [CrossRef]

Article

Coating Formation on Ti-6Al-4V Alloy by Micro Arc Oxidation in Molten Salt

Alexander Sobolev, Israel Wolicki, Alexey Kossenko, Michael Zinigrad and Konstantin Borodianskiy *

Zimin Advanced Materials Laboratory, Department of Chemical Engineering, Biotechnology and Materials, Ariel University, Ariel 40700, Israel; sobolev@ariel.ac.il (A.S.); imwolicki@gmail.com (I.W.); kossenkoa@ariel.ac.il (A.K.); zinigrad@ariel.ac.il (M.Z.)
* Correspondence: konstantinb@ariel.ac.il; Tel.: +972-3-9143085

Received: 12 August 2018; Accepted: 3 September 2018; Published: 4 September 2018

Abstract: Micro Arc Oxidation (MAO) is an electrochemical surface treatment process to produce oxide protective coatings on some metals. MAO is usually conducted in an aqueous electrolyte, which requires an intensive bath cooling and leads to the formation of a coating containing impurities that originate in the electrolyte. In the current work, we applied an alternative ceramic coating to the Ti-6Al-4V alloy using the MAO process in molten nitrate salt at a temperature of 280 °C. The obtained coating morphology, chemical and phase composition, and corrosion resistance were investigated and described. The obtained results showed that a coating of 2.5 μm was formed after 10 min of treatment, containing titanium oxide and titanium-aluminum intermetallic phases. Morphological examination indicated that the coating is free of cracks and contains round, homogeneously distributed pores. Corrosion resistance testing indicated that the protective oxide coating on Ti alloy is 20 times more resistive than the untreated alloy.

Keywords: micro arc oxidation; titanium coating; titanium oxide; molten salt

---

1. Introduction

Micro arc oxidation (MAO) is one of the most promising methods for the surface treatment of metals and alloys, and has recently received wide acceptance from various branches of industry. MAO is generally used to produce multipurpose wear-, corrosion-, and heat-resistant dielectric and decorative coatings on valve metals, such as Al, Mg, Ti, Ta, Nb, Zr, and Be [1–5].

Currently, titanium is an appealing metal due to its high specific strength [6], corrosion resistance [7], and excellent biocompatibility [8]. The use of MAO favors adapting surface composition, crystallographic structure, and morphology to achieve a large functionality that cannot be provided by the parent metal. The versatility of coatings obtained by MAO on Ti alloys, together with the simplicity and low cost of this treatment method, stimulated numerous attempts to coat titanium by the MAO [9–12] approach for various applications, including tribological, biomedical, dielectric, and photovoltaic coatings. Ti-6Al-4V is the most widely used Ti alloy as it contains stabilizer elements for both α and β phases for good creep and strength, respectively.

The mechanism of the MAO process is based on the anodizing electrochemical reaction, which occurs on a metallic surface and is accompanied by microarc discharge to form an oxide ceramic surface layer with a particular morphology and phase composition [13,14].

The formation of the coating results in numerous difficulties that affect different factors of the layer quality. Among those factors are the chemical composition, the concentration and temperature of the electrolyte, the duration of the treatment process, the chemical composition and structure of the substrate, and the electrical parameters of the MAO process [15–18]. The following factors affecting MAO in an aqueous electrolyte are considered as undesirable: the necessity of forced cooling of the treatment bath, an increased current density, the thermal dissociation of the electrolyte, the formation of compounds in the ceramic coating, and the low growth rate. Those issues can be solved by replacing an aqueous electrolyte with molten salts. The application of molten salt as an electrolyte in the MAO process has been reported by us in an earlier work on Al alloy surface treatment [19,20].

In the present work, the formation of a ceramic coating on Ti-6Al-4V alloy in a mixture of molten nitrate salts by MAO process was obtained. The chemical and phase composition of the obtained coating as well as its morphology and corrosion resistance are investigated and illustrated.

2. Materials and Methods

2.1. MAO Process

Titanium alloy Ti-6Al-4V rectangular specimens (Scope Metals Group Ltd., Bne Ayish, Israel, chemical composition shown in Table 1) with a surface area of 0.2 dm^2 were ground using abrasive papers grits #280, #400, #600, #1000, #2400, and #4000, respectively, and then subjected to ultrasonic cleaning in acetone. The surface roughness is maintained to R_a = 3 μm after polishing.

Table 1. Chemical composition of the alloy Ti-6Al-4V.

Chemical Element, mass %			
V	Fe	Al	Ti
4	0.11	6	Base

MAO treatment was performed at 280 °C in the electrolyte with a eutectic composition of KNO_3-$NaNO_3$ (Sigma-Aldrich, St. Louis, MO, USA) with the mass % of 54.3 and 45.7, respectively. The electrolyte was held in a nickel crucible (99.95% Ni), which served as a counter-electrode. The surface ratio of anode-to-cathode was 1:30, the anodic current density was 250 mA/cm^2, and the voltage was limited by the galvanostatic mode. The applied power supply had the following parameters: I_{max} = 35 A, U_{max} = 1000 V; current and voltage were pulsed with a square-wave sweep at a frequency of 1 Hz (t_a = t_k = 0.5 s) by a Digit-EL PG-872 pulse generator (Minsk, Belarus). The duration of the MAO treatment was 10 min, with a coating growth rate of 0.25 μm/min. Finally, the obtained specimens were air-cooled, rinsed with distilled water, and dried. The behavior of current vs. time and voltage vs. time was recorded by a Fluke Scope Meter 199C (Eindhoven, The Netherlands) (200 MHz, 2.5 GS s^{-1}). A schematic of the detailed experimental setup is given in Figure 1.

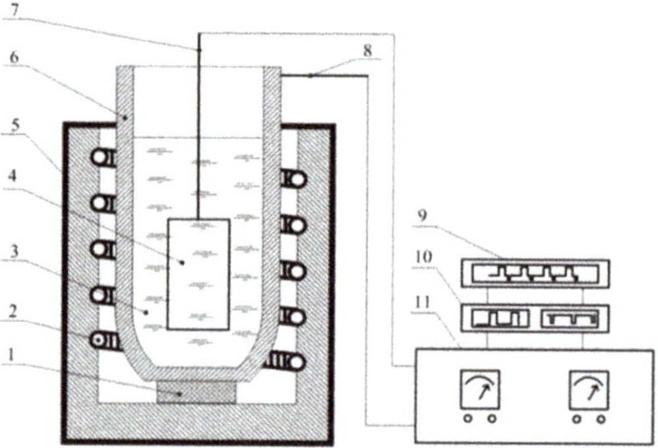

Figure 1. Schematic of experimental setup: 1—ceramic stand; 2—heating element; 3—molten salt electrolyte; 4—specimen subjected to MAO treatment; 5—furnace with automatic temperature controller; 6—nickel crucible; 7,8—current connectors; 9—data logger; 10—pulse generator; 11—power supply.

2.2. Characterization Techniques

The morphology examinations of the obtained coatings were done on the cross section of the treated specimen by TESCAN MAIA3 scanning electron microscopy (SEM) (Brno, Czech Republic) equipped with an energy dispersive X-ray spectroscopy (EDS) system by Oxford Instruments (Abingdon, UK) with an X-Max^N detector. The phase analysis of the coating was determined by the X'Pert Pro diffractometer (PANalytical B.V., Almelo, The Netherlands) with Cu_α radiation (λ = 1.542 Å) at the grazing incidence mode (angle of 3°) with a 2θ range from 30° to 80° (step size of 0.03°) at 40 kV and 40 mA.

The corrosion behavior of the treated and untreated specimens was examined by a potentiodynamic polarization test in a 3.5 wt % NaCl (Sigma-Aldrich Co.) solution by PARSTAT 4000A potentiostat/galvanostat (Princeton Applied Research, Oak Ridge, TN, USA). A three-electrode cell configuration was used for the corrosion test, wherein a Pt sheet acted as a counter-electrode and saturated Ag/AgCl (Metrohm Autolab B.V., Utrecht, The Netherlands) acted as a reference electrode. The polarization resistance of a sample was detected at the range of ±250 mV with respect to the recorded corrosion potential at a scan rate of 0.1 mV/s. Prior to the potentiodynamic polarization test, the samples were kept in the 3.5 wt % NaCl solution for 60 min in order to reach the steady state of a working electrode.

3. Results and Discussion

3.1. MAO Process Characterization

Plots of voltage and current as a function of time during the MAO process are presented in Figure 2a,b, respectively.

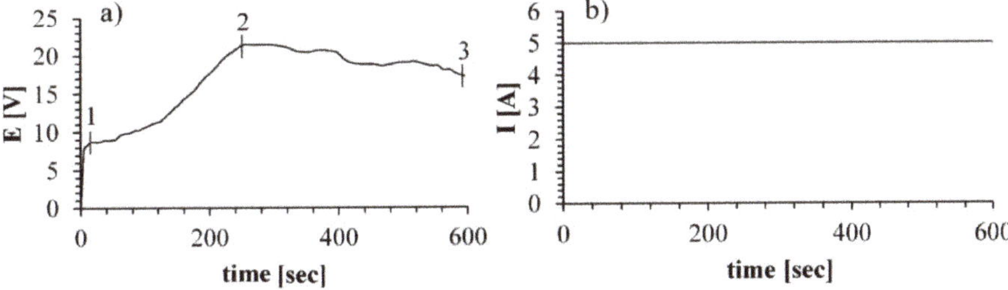

Figure 2. Plot of electric parameters of the MAO process applied on alloy Ti-6Al-4V: (**a**) voltage as a function of treatment time; (**b**) current as a function of treatment time.

Here, we sought to establish optimal conditions for plasma-mediated oxidation of Ti alloy. Briefly, the sample was immersed in a molten salt electrolyte in a nickel crucible, and the voltage was applied so that the sample served as a positive pole and the crucible as a negative pole. During the first few seconds of the process, a double electric layer was formed, followed by the charging, accompanied by the adsorption of gas bubbles and the formation of an amorphous film on the specimen surface (Figure 2a area 0–1). With the increase in the treatment time, a thicker oxide layer was formed, followed by a dielectric breakdown (Figure 2a area 1–2), which was accompanied by the formation of sparks on the specimen surface. It can be noted that during the MAO process, the voltage turns to the stationary mode after 300 s, meaning that the sparking process moves into the so-called micro arc oxidation mode (Figure 2a, area 2–3).

The process applied in molten salt is conducted at significantly lower potentials, about 22 V compared to the potentials of 300–600 V in the process conducted in aqueous electrolyte [21]. The current values of both processes are in the same range [22]. Those parameters indicate that the MAO process in molten salt is a more energy-efficient process and therefore is more economically beneficial.

3.2. Morphology and Elemental Analysis

The surface morphology and chemical compositions of the specimen treated by MAO process were investigated by SEM and EDS, respectively. The corresponding SEM and EDS images are shown in Figure 3a–c.

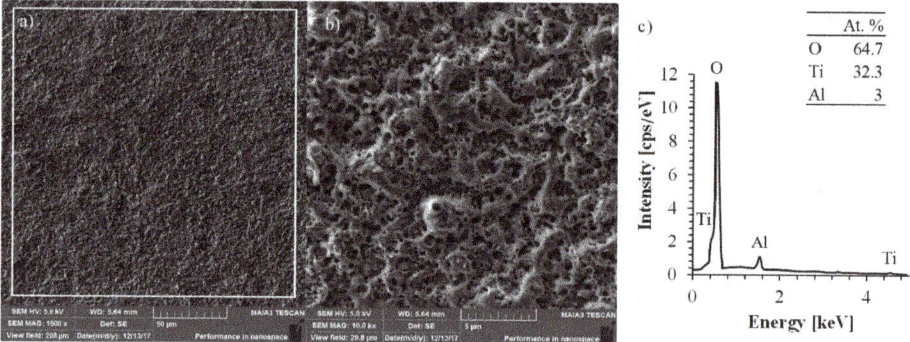

Figure 3. SEM image of surface morphology of the alloy Ti-6Al-4V treated by MAO with magnification: (**a**) 1000×; (**b**) 10,000× and (**c**) the elemental composition obtained by EDS.

The surface of the oxide coating has a typical morphology usually obtained by the MAO process upon valve metals [23]. The homogeneously distributed round-shaped pores formed on the surface in the locations where the electrical high-temperature breakdowns took place.

During the electrical breakdown, the temperature of the discharge reached several thousand degrees, as noted by estimations made by Hussein et al. [24], resulting in the creation of a newly formed oxide layer that was first melted and then recrystallized. Surface morphology investigation showed that the formed coating has no cracks on the surface, indicating a low cooling rate of the newly formed oxides. Additionally, the obtained surface consists of round pores of a diameter ranging from 0.15 µm up to 0.5 µm. These pores are significantly smaller than the pores formed in the MAO process conducted in aqueous electrolyte (usually 2.5–15 µm, depending on the applied potential and processing time [25]).

The atomic composition of titanium, aluminum, and oxygen obtained by EDS analysis were 32.3%, 3.0%, and 64.7%, respectively. These values clearly indicate that the formed coating is free of impurities. That is contrary to the coating obtained by MAO treatment in aqueous electrolyte, which usually includes additional components originating from the electrolyte [26,27].

3.3. Phase Analysis

The XRD pattern of the alloy Ti-6Al-4V surface after MAO treatment is shown in Figure 4.

XRD investigation evaluated the presence of the following phases in the obtained oxide-based coating: titanium dioxide in the form of rutile [28] and intermetallic of $Al_{0.3}Ti_{1.7}$ [29]. They were expected to be formed in the coating, rutile due to the oxidation process and Ti/Al intermetallic resulting in the noticeable presence of Al in the alloy. XRD measurements evaluated that no new phase was formed during the process and no impurities were detected in the coating.

The cross section line scan and the microstructure of the Ti-6Al-4V alloy treated by MAO are shown in Figure 5.

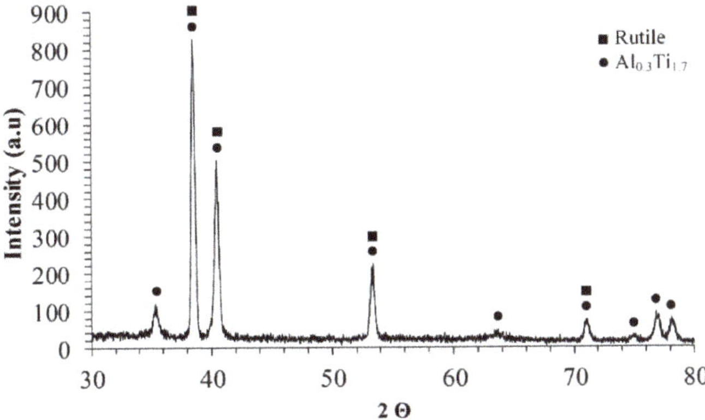

Figure 4. X-ray diffraction pattern of the alloy Ti-6Al-4V surfaces after MAO treatment.

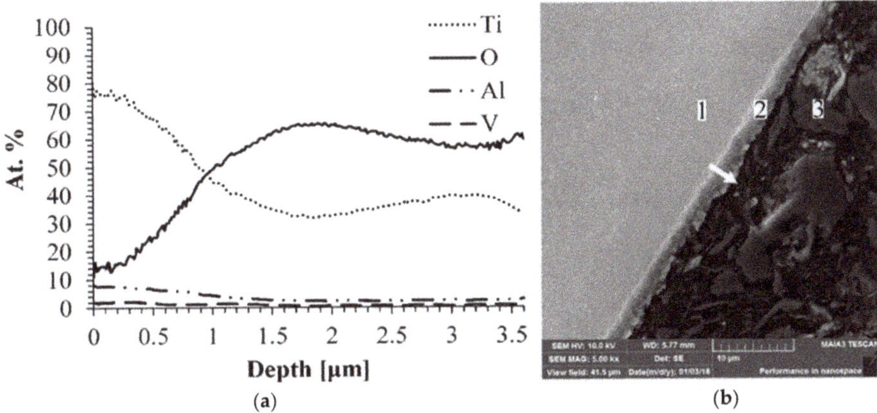

(a) (b)

Figure 5. EDS line scan of the alloy Ti-6Al-4V after MAO treatment (**a**) and its cross section microphotograph obtained by SEM (**b**). Arrow indicates the direction of the EDS line scan analysis. Points in images attributed to: 1—base alloy; 2—oxide layer; 3—resin.

SEM micrograph, jointly with the EDS line scan, indicated that the obtained oxide layer is uniform and its thickness is about 2.5 µm. Moreover, the elemental analysis detected only components that fit the composition of the expected oxide layer and no additional impurities. Usually, impurities are detected in the coating after the process that is carried out in the aqueous electrolyte. Aliasghari et al. detected the presence of phosphorous in the coating formed on Ti by the MAO process in an electrolyte containing phosphoric acid [30].

3.4. Corrosion Resistance Investigation

The corrosion properties of the treated specimen were determined by the potentiodynamic polarization method. The obtained curve on the coated specimen was compared to the curve of the untreated and both are illustrated in Figure 6.

The obtained curves in Figure 6 are presented in semi-logarithmic coordinates. A higher corrosion resistance of the specimen is obtained when the corrosion potential is higher and the corrosion current density is lower. Therefore, it is clearly seen that the corrosion potential of the coated specimen shifted to be more positive and the current density to a more negative value, providing higher corrosion protection to the alloy. That may indicate the reduction of the anodic and cathodic processes due to the presence of a newly formed protective oxide layer on the metallic surface. The movement of the corrosion potential towards the anodic area also indicates the improvement of the treated specimen's resistance.

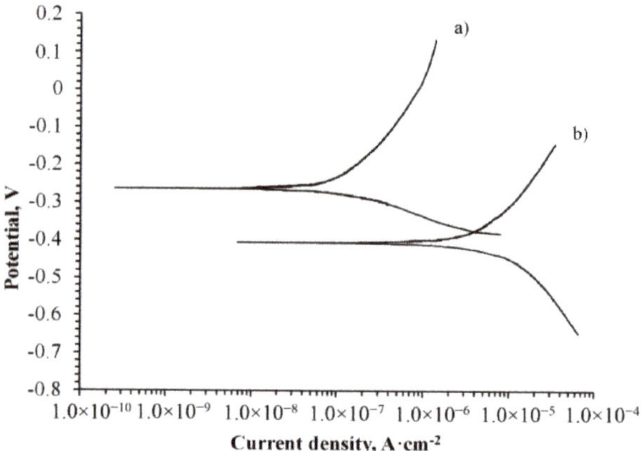

Figure 6. Potentiodynamic polarization curves for alloy Ti-6Al-4V (a) treated by MAO process and (b) untreated specimen. Both examined in 3.5 wt % NaCl.

Based on the corrosion currents and obtained slopes of cathodic and anodic curves, the polarization resistance (R_p) was calculated according to Equation (1):

$$R_p = \frac{\beta_a \times \beta_c}{2.3 \times i_{corr}(\beta_a + \beta_c)} \tag{1}$$

The Tafel slopes, β_a and β_c, were calculated from the anodic and cathodic curves on the plot. Results of calculations that present the corrosion potentials (E_{corr}), corrosion current densities (i_{corr}), and the polarization resistance (R_p) are summarized in Table 2.

Table 2. Calculated corrosion test results of untreated alloy Ti-6Al-4V and treated by MAO process specimens. Both examined in 3.5 wt % NaCl.

Samples	E_{corr} [mV]	$i_{corr} \times 10^{-6}$ [A]	β_a [mV/decade]	β_c [mV/decade]	$R_p \times 10^3$ [Ω/cm^2]
Untreated alloy Ti-6Al-4V	−398	6.95	395	316	10.98
Treated alloy Ti-6Al-4V	−260	0.16	390	97	213.76

The calculations presented in Table 2 show that the polarization resistance of the treated specimen is 213.76 kΩ/cm², while the untreated specimen has a resistance of 10.98 kΩ/cm². Those values show that the oxide protective coating on Ti alloy is almost 20 times higher than the untreated one. Our results, together with those of additional previous works [31,32], lead to the conclusion that MAO treatment can be applied to improve the corrosion resistance of metals.

The polarization resistance of an alloy treated in molten salt is higher than that of a similar alloy obtained in aqueous electrolyte [33]. This can be attributed to the lack of impurities in the coating and the presence of smaller pores, which conduct current and therefore reduce corrosion resistance.

4. Conclusions

A new approach to ceramic protective coating formation by the MAO process in molten salt was described. A TiAl6V4 alloy has been subjected to MAO treatment in a eutectic nitrate molten salt mixture. The thickness of the oxide coating reached 2.5 μm and the morphology examination evaluated the presence of a typical structure with homogeneously distributed round pores.

Phase composition analysis detected the presence of titanium oxide and titanium aluminum intermetallic ($Al_{0.3}Ti_{1.7}$). This was confirmed by EDS analysis.

The protective coating was subjected to corrosion resistance testing, and it was found that the specimen by coated MAO treatment is 20 times more resistant than the untreated specimen.

A comparison of the process conducted in molten salt with the process conducted in aqueous electrolyte showed the following benefits: the pores obtained on the surface are smaller, the coating has no impurities, the corrosion resistance is higher, and the process is economically beneficial due to the significantly lower potentials applied.

Author Contributions: Conceptualization and Methodology, A.S., A.K., M.Z. and K.B.; Experimental, A.S., I.W. and A.K.; SEM, EDS, and Micro-Hardness, A.S. and I.W.; XRD, A.K.; Data analysis A.S., M.Z. and K.B.; Writing, Reviewing and Editing A.S. and K.B.; Supervision M.Z. and K.B.

Funding: This research received no external funding.

Acknowledgments: This work was carried out with the support of the Ministry of Aliyah and Integration, the State of Israel.

Conflicts of Interest: The authors declare no conflict of interest.

References

1. Yerokhin, A.L.; Nie, X.; Leyland, A.; Matthews, A. Characterization of oxide films produced by plasma electrolytic oxidation of a Ti–6Al–4V alloy. *Surf. Coat. Technol.* **2000**, *130*, 195–206. [CrossRef]

2. Wang, Y.; Jiang, B.; Lei, T.; Guo, L. Dependence of growth features of microarc oxidation coatings of titanium alloy on control modes of alternate pulse. *Mater. Lett.* **2004**, *58*, 1907–1911. [CrossRef]

3. Fei, C.; Hai, Z.; Chen, C.; Xia, Y.J. Study on the tribological performance of ceramic coatings on titanium alloy surfaces obtained through microarc oxidation. *Prog. Org. Coat.* **2009**, *64*, 264–267. [CrossRef]

4. Huang, P.; Wang, F.; Xu, K.; Han, Y. Mechanical properties of titania prepared by plasma electrolytic oxidation at different voltages. *Surf. Coat. Technol.* **2007**, *201*, 5168–5171. [CrossRef]

5. Santos-Coquillat, A.; Gonzalez Tenorio, R.; Mohedano, M.; Martinez-Campos, E.; Arrabal, R.; Matykina, E. Tailoring of antibacterial and osteogenic properties of Ti6Al4V by plasma electrolytic oxidation. *Appl. Surf. Sci.* **2018**, *454*, 157–172. [CrossRef]

6. Niinomi, M. Mechanical properties of biomedical titanium alloys. *Mater. Sci. Eng. A* **1998**, *243*, 231–236. [CrossRef]

7. Gu, Y.; Ma, A.; Jiang, J.; Li, H.; Song, D.; Wu, H.; Yuan, Y. Simultaneously improving mechanical properties and corrosion resistance of pure Ti by continuous ECAP plus short-duration annealing. *Mater. Charact.* **2018**, *138*, 38–47. [CrossRef]

8. Geetha, M.; Singh, A.K.; Asokamani, R.; Gogia, A.K. Ti based biomaterials, the ultimate choice for orthopaedic implants—A review. *Prog. Mater. Sci.* **2009**, *54*, 397–425. [CrossRef]

9. Lederer, S.; Lutz, P.; Fürbeth, W. Surface modification of Ti 13Nb 13Zr by plasma electrolytic oxidation. *Surf. Coat. Technol.* **2018**, *335*, 62–71. [CrossRef]

10. Wheeler, J.M.; Collier, C.A.; Paillard, J.M.; Curran, J.A. Evaluation of micromechanical behaviour of plasma electrolytic oxidation (PEO) coatings on Ti–6Al–4V. *Surf. Coat. Technol.* **2010**, *204*, 3399–3409. [CrossRef]

11. Khorasanian, M.; Dehghan, A.; Shariat, M.H.; Bahrololoom, M.E.; Javadpour, S. Microstructure and wear resistance of oxide coatings on Ti–6Al–4V produced by plasma electrolytic oxidation in an inexpensive electrolyte. *Surf. Coat. Technol.* **2011**, *206*, 1495–1502. [CrossRef]

12. Fakhr Nabavi, H.; Aliofkhazraei, M.; Sabour Rouhaghdam, A. Electrical characteristics and discharge properties of hybrid plasma electrolytic oxidation on titanium. *J. Alloys Compd.* **2017**, *728*, 464–475. [CrossRef]

13. Yerokhin, A.L.; Nie, X.; Leyland, A.; Matthews, A.; Dowey, S.J. Plasma electrolysis for surface engineering. *Surf. Coat. Technol.* **1999**, *122*, 73–93. [CrossRef]

14. Yerokhin, A.L.; Lyubimov, V.V.; Ashitkov, R.V. Phase formation in ceramic coatings during plasma electrolytic oxidation of Aluminium Alloys. *Ceram. Int.* **1998**, *24*, 1–6. [CrossRef]

15. Kossenko, A.; Zinigrad, M. A universal electrolyte for the plasma electrolytic oxidation of aluminum and magnesium alloys. *Mater. Des.* **2015**, *88*, 302–309. [CrossRef]

16. Habazakia, H.; Tsunekawa, S.; Tsuji, E.; Nakayama, T. Formation and characterization of wear-resistant PEO coatings formed on β-titanium alloy at different electrolyte temperatures. *Appl. Surf. Sci.* **2012**, *259*, 711–718. [CrossRef]

17. Liu, Y.J.; Xu, J.Y.; Lin, W.; Gao, C.; Zhang, J.C.; Chen, X.H. Effects of different electrolyte systems on the formation of micro-arc oxidation ceramic coatings of 6061 aluminum alloy. *Rev. Adv. Mater. Sci.* **2013**, *33*, 126–130.

18. Al Bosta, M.M.S.; Ma, K.J. Influence of electrolyte temperature on properties and infrared emissivity of MAO ceramic coating on 6061 aluminum alloy. *Infrared Phys. Technol.* **2014**, *67*, 63–72. [CrossRef]

19. Sobolev, A.; Kossenko, A.; Zinigrad, M.; Borodianskiy, K. Comparison of plasma electrolytic oxidation coatings on Al alloy created in aqueous solution and molten salt electrolytes. *Surf. Coat. Technol.* **2018**, *344*, 590–595. [CrossRef]

20. Sobolev, A.; Kossenko, A.; Zinigrad, M.; Borodianskiy, K. An investigation of oxide coating synthesized on an aluminum alloy by plasma electrolytic oxidation in molten salt. *Appl. Sci.* **2017**, *7*, 889. [CrossRef]

21. Montazeri, M.; Dehghanian, C.; Shokouhfar, M.; Baradaran, A. Investigation of the voltage and time effects on the formation of hydroxyapatite-containing titania prepared by plasma electrolytic oxidation on Ti–6Al–4V alloy and its corrosion behavior. *Appl. Surf. Sci.* **2011**, *257*, 7268–7275. [CrossRef]

22. Li, Q.; Yang, W.; Liu, C.; Wang, D.; Liang, J. Correlations between the growth mechanism and properties of micro-arc oxidation coatings on titanium alloy: Effects of electrolytes. *Surf. Coat. Technol.* **2017**, *316*, 162–170. [CrossRef]

23. Wang, J.H.; Wang, J.; Lu, Y.; Du, M.H.; Han, F.Z. Effects of single pulse energy on the properties of ceramic coating prepared by micro-arc oxidation on Ti alloy. *Appl. Surf. Sci.* **2015**, *324*, 405–413. [CrossRef]

24. Hussein, R.O.; Northwood, D.O.; Nie, X. Coating growth behavior during the plasma electrolytic oxidation process. *J. Vac. Sci. Technol. A* **2010**, *28*, 766–773. [CrossRef]

25. Wang, Y.; Lei, T.; Jiang, B.; Guo, L. Growth, Microstructure and mechanical properties of microarc oxidation coatings on titanium alloy in phosphate-containing solution. *Appl. Surf. Sci.* **2004**, *233*, 258–267. [CrossRef]

26. Lugovskoy, A.; Zinigrad, M.; Kossenko, A.; Kazanski, B. Production of ceramic layers on aluminum alloys by plasma electrolytic oxidation in alkaline silicate electrolytes. *Appl. Surf. Sci.* **2013**, *264*, 743–747. [CrossRef]

27. Aliasghari, S.; Němcová, A.; Čížek, J.; Gholinia, A.; Skeldon, P.; Thompson, G.E. Effects of reagent purity on plasma electrolytic oxidation of titanium in an aluminate-phosphate electrolyte. *Int. J. Surf. Eng. Coat.* **2016**, *94*, 32–42. [CrossRef]

28. Tomaszewski, P.E. Structural Phase Transitions in Crystals. *Phase Transit.* **1992**, *38*, 127–220. [CrossRef]

29. Clark, D.; Jepson, K.S.; Lewis, G.I. A study of the titanium-aluminium system up to 40 al.-percent aluminum. *J. Inst. Metals* **1963**, *91*, 197–203.

30. Aliasghari, S.; Skeldon, P.; Thompson, G.E. Plasma electrolytic oxidation of titanium in a phosphate/silicate electrolyte and tribological performance of the coatings. *Appl. Surf. Sci.* **2014**, *316*, 463–476. [CrossRef]

31. Movahedi, N.; Habibolahzadeh, A. Effect of plasma electrolytic oxidation treatment on corrosion behavior of closed-cell Al-A356 alloy foam. *Mater. Lett.* **2016**, *164*, 558–561. [CrossRef]

32. Venugopal, A.; Srinath, J.; Rama Krishna, L.; Ramesh Narayanan, P.; Sharma, S.C.; Venkitakrishnan, P.V. Corrosion and nanomechanical behaviors of plasma electrolytic oxidation coated AA7020-T6 aluminum alloy. *Mater. Sci. Eng. A* **2016**, *660*, 39–46. [CrossRef]

33. Shokouhfar, M.; Dehghanian, C.; Baradaran, A. Preparation of ceramic coating on Ti substrate by Plasma electrolytic oxidation in different electrolytes and evaluation of its corrosion resistance. *Appl. Surf. Sci.* **2011**, *257*, 2617–2624. [CrossRef]

 *materials*

Article

CVD Diamond Interaction with Fe at Elevated Temperatures

Sergei Zenkin *, Aleksandr Gaydaychuk, Vitaly Okhotnikov and Stepan Linnik

Research School of High-Energy Physics, Tomsk Polytechnic University, Savinyh str. 2a, Tomsk 634050, Russia; gaydaychuk@tpu.ru (A.G.); ohotnikov@tpu.ru (V.O.); linniksa@tpu.ru (S.L.)
* Correspondence: zen@tpu.ru

Received: 8 November 2018; Accepted: 5 December 2018; Published: 10 December 2018

Abstract: Chemical vapor deposition (CVD) diamond is a prospective thin film material for cutting tools applications due to the extreme combination of hardness, chemical inertness, and thermal conductivity. However, the CVD diamond cutting ability of ferrous materials is strongly limited due to its extreme affinity to iron, cobalt, or nickel. The diamond–iron interaction and the diffusion behavior in this system are not well studied and are believed to be similar to the graphite–iron mechanism. In this article, we focus on the medium-temperature working range of 400–800 °C of a CVD diamond–Fe system and show that for these temperatures etching of diamond by Fe is not as strong as is generally accepted. The starting point of the diamond graphitization in contact with iron was found around 400 °C. Our results show that CVD diamond is applicable for the cutting of ferrous materials under medium-temperature conditions.

Keywords: CVD diamond; diffusion; Fe–C interaction

1. Introduction

Chemical vapor deposition (CVD) diamond is a prospective thin film material for cutting tools applications due to the extreme combination of hardness, chemical inertness, and thermal conductivity [1]. However, the CVD diamond cutting ability of ferrous materials is strongly limited due to its extreme affinity to iron, cobalt, or nickel [2] attributed to the phase transformation of diamond to graphite and subsequent diffusion of carbon into the metal [3]. This phenomenon is frequently used for diamond catalytic etching [4,5], patterning [6], or polishing [7]. Jin et al. [8] report the thinning of CVD diamond, caused by the reaction with iron foil with a speed up to 2 µm/h at 900 °C in argon. Ralchenko et al. [9] show that for the CVD diamond–Fe system in a hydrogen atmosphere the etching speed increases up to 8 µm/min due to the formation of gaseous hydrocarbons, primarily methane. Giménez et al. [10] show an extreme increase of the chemical wear rate of polycrystalline diamond during iron-based materials machining in the temperature range from 700 to 1300 °C. Most of these experiments were carried out at extreme temperatures up to 1000 °C, while the diamond–iron interaction and diffusion behavior in the diamond–Fe system at lower temperatures are not well studied and are believed to be similar to the graphite–iron mechanism.

In the article, we are focused on the interaction between CVD grown microcrystalline diamond and thermally evaporated Fe at elevated temperatures in the range of 400–800 °C under vacuum conditions. We show that in this temperature range, etching of CVD diamond is not as strong as for the 900–1000 °C interval [8,10], making a CVD diamond applicable for the cutting of ferrous materials under medium-temperature conditions.

2. Materials and Methods

As a substrate in this work, we used monocrystalline Si (100) with the size of 10 mm × 10 mm × 0.38 mm. The diamond–iron system preparation is schematically shown in Figure 1. The samples were ultrasonically cleaned with acetone prior to diamond layer deposition.

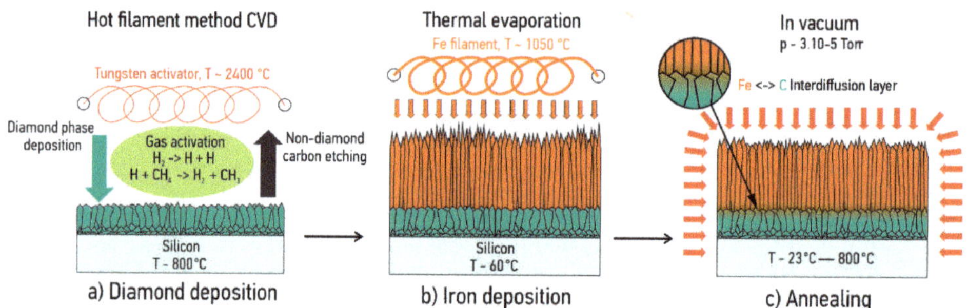

Figure 1. A schematic illustration of the experimental procedure: (**a**) Diamond deposition; (**b**) Iron deposition; (**c**) Annealing.

Polycrystalline diamond coatings were deposited in the self-made hot filament chemical vapor deposition reactor using a hydrogen–methane mixture (ratio of H_2:CH_4 = 50:1) with a total gas flow rate 106 mL/min (Bronkhorst EL-FLOW, Bronkhorst High-Tech B.V., Ruurlo, The Netherlands). The distance between tungsten filaments (Ø = 0.16 mm) and substrates was 10 ± 0.5 mm. The substrate temperature during the deposition was maintained at 800 ± 25 °C using an infrared thermal imager (ULIRvision TI170, ULIRvision Technology Co., Ltd., Zhejiang, China). The pressure in the reactor during the deposition was maintained at 20 ± 1 Torr (Pfeiffer Vacuum CMR 372, Pfeiffer Vacuum, Annecy, France) and the current at 6.5 ± 0.01 A per filament, see Figure 1a. The total diamond film thickness was 2 ± 0.1 µm.

After diamond layer synthesis, samples were placed in the vacuum chamber. The base pressure in the evacuated chamber was 8×10^{-6} Torr. Prior to Fe deposition, samples were cleaned by an Ar^+ ion source with an energy of 3.5 keV.

Fe films were deposited by the two-step thermal evaporation process, see Figure 1b:

- An initial 3 µm layer was evaporated with ion source assistance using an Ar^+ ion energy of 500 eV.
- Subsequently, Fe deposition was carried out without ion source assistance. The total Fe film thickness was 10 ± 0.5 µm.

After that, samples were annealed under vacuum conditions during 30 min at fixed temperatures in the range of 400–800 °C, see Figure 1c. In order to determine the interaction behavior between CVD diamond and evaporated Fe films, we used cross-sectional scanning electron microscopy, XRD, Raman spectroscopy, and EDX measurements. This combination of techniques allows for clarification of the interdiffusion process in the CVD diamond–Fe system at elevated temperatures and for the determination of the temperature starting point of the diamond graphitization. The coatings cross-sectional morphology was studied using a scanning electron microscope (Vega3, TESCAN, Brno, Czech Republic). Structural characteristics of the coatings were studied using X-ray diffraction (Shimadzu XRD 6000, Shimadzu, Kyoto, Japan) in the Bragg–Brentano configuration with Cu Kα (λ = 0.154 nm) radiation in the range of $2\theta = 20°$–$90°$, with a sampling pitch equal to 0.01°, and an integration time of 1 s. Raman spectra were recorded using a NanoScan Technology Centaur IHR spectrometer (NanoScan Technology, Dolgoprudny, Russia) with a 514.5 nm source laser.

3. Results and discussion

The results of XRD measurements are shown in Figure 2. The CVD diamond–Fe system that was not annealed shows only diamond reflexes at 2θ = 44.05° and 75.4° attributed to the (111) and (220) crystallographic planes, respectively, in combination with low-intensity broad peaks of α-Fe at 2θ = 44.76° (Fe (110)) and 82.43° (Fe (211)). A strong change in the microstructure of the system can be created only after annealing above 400 °C, as shown in Figure 2. For this sample, one can detect the formation of the graphite phase with the peak position at 2θ ≈ 24.8° ("G" in Figure 2), indicating the start of the carbon diffusion into the iron layer. For a higher-temperature interval, 600–800 °C, we can detect the recrystallization of the thermally evaporated Fe film, starting at 600 °C, attributed to the formation of strong α-Fe peaks at 2θ = 44.76° (Fe (110)), 65.13° (Fe (200)) and 82.43° (Fe (211)). Analysis of the full width at half maximum (FWHM) of the α-Fe peak at 2θ = 82.43° using the Scherrer equation shows the coherent scattering region broadening from 23 to 38 nm, pointing to the increase of the crystallite size in the iron film. However, due to the limited ability of the Fe_3C formation (only 25 at.% of carbon was involved in the formation of Fe_3C), a strong peak due to α-Fe overlap appears at 2θ ≈ 45° in the XRD pattern. Only low-intensity Fe_3C peaks at 2θ = 43.11° and 46.01° can be detected for a diamond–Fe system heated at 800 °C. At the same time, neither the intensity or the width of the diamond peak at 2θ = 44.05° changes as a result of heating the system, even at 800 °C. Based on the data, one can assume that only a minor amount of carbon from the CVD diamond film was involved in the formation of Fe_3C or graphite during the interdiffusion process.

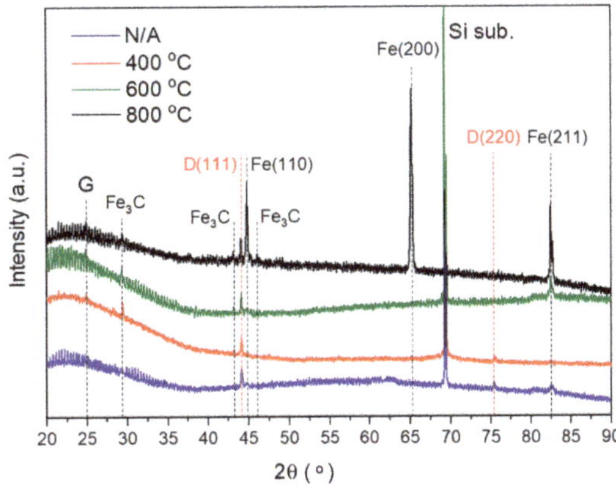

Figure 2. XRD patterns of the CVD diamond–Fe system at elevated temperatures.

In order to support our assumption, we used scanning electron microscopy measurements of the cross-section of the CVD diamond–Fe interface as a function of the annealing temperature, see Figure 3. The iron films that were not annealed are characterized by a strongly columnar void-free microstructure and exhibit homogeneous interface contact with the diamond film, as shown in Figure 3a. For diamond–Fe systems heated in the range 400–500 °C, see Figure 3b,c, there is no visible change in the system structure and the formation of an Fe–C transition layer is not clearly observed, despite the fact that the formation of both graphite and Fe_3C is well-evidenced in the XRD spectrum at 400 °C, see the red curve of Figure 2. Further increasing the annealing temperature, see Figure 2d–f, leads to:

- Recrystallization of the thermally evaporated Fe film, starting at 600 °C, in agreement with XRD measurements.
- Formation of an Fe–C transition layer due to the interdiffusion that takes place in the diamond–Fe system in the temperature range 600–800 °C. The thickness of this transition layer is relatively low (approx. 0.5 μm) due to the limited ability of Fe_3C to form. Diffusion of carbon is undetectable on the SEM cross-sections.

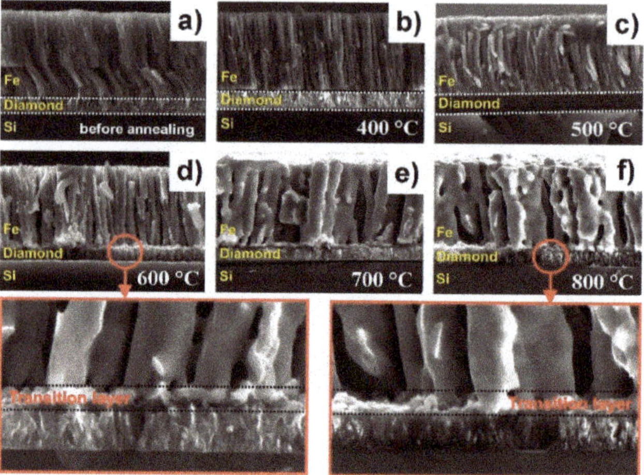

Figure 3. Cross-sectional scanning electron microscopy image of a CVD diamond–Fe system (**a**) as deposited, heated at (**b**) 400 °C; (**c**) 500 °C; (**d**) 600 °C; (**e**) 700 °C; and (**f**) 800 °C. The transition layer in the CVD diamond–Fe interface annealed at 600 °C (bottom left) and 800 °C (bottom right) during 30 min under vacuum conditions.

To clarify the Fe–C interdiffusion, we measured the Raman spectra of the CVD diamond–Fe interface, highlighted in Figure 3 (marked as transition layer). To carry out the measurements, we chemically etched the top Fe layer using concentrated nitric acid. After that, we measured the Raman spectra from the top of the diamond film. The results of these measurements are shown in Figure 4. For samples annealed at 400 °C (blue curve) and 600 °C (red curve), one can detect three strong peaks at 1333 cm^{-1}, 1471 cm^{-1}, and 1578 cm^{-1}, attributed to the diamond (pure sp^3–hybridization of carbon), trans-polyacetylene, and G–band (graphitic carbon structure), respectively. It can be seen that, as the annealing temperature increases, the Raman spectra of 600 °C-heated samples clearly exhibit a decrease of the diamond peak intensity at 1333 cm^{-1} and an increase of the non–diamond peak centered at 1578 cm^{-1}.

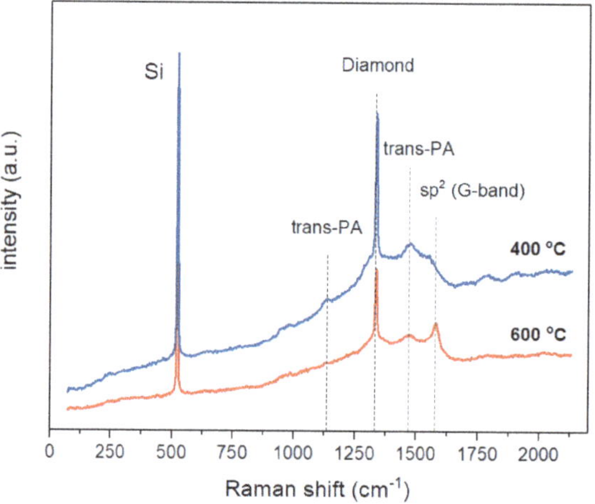

Figure 4. Raman spectra of the CVD diamond–Fe interface for films heated at 400 °C (blue) and 600 °C (red) indicating the formation of a graphite phase due to the carbon–iron interdiffusion.

In order to make a quantitative evaluation of the changes in the quality of the diamond film, we investigated a change of the diamond Raman peak width. The obtained data are presented in Figure 5.

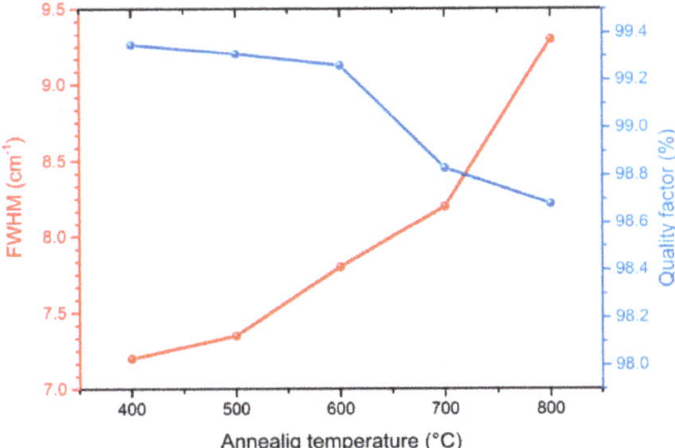

Figure 5. FWHM of the Raman diamond peak at 1333 cm^{-1} and the corresponding diamond quality factor as a function of the annealing temperature.

As can be seen, increasing the annealing temperature of the diamond–iron system from 400 to 800 °C does not lead to a significant deterioration in the quality of the diamond coating. The diamond peak FWHM for 400 °C is 7.2 cm^{-1}, while for 800 °C, the annealed film is around 9.3 cm^{-1} (FWHM = 2 cm^{-1} for the highest available quality type IIa diamond [11]). The diamond phase purity was also quantified by the calculation of the diamond quality factor described as:

$$Q = \frac{I_{\text{diamond}}}{\left(I_{\text{diamond}} + \frac{I_{\text{a-carbon}}}{233}\right)} \times 100\% \tag{1}$$

here I_{diamond} is the diamond peak intensity at 1333 cm^{-1}, $I_{\text{a-carbon}}$ is the sum of the intensities of the observed nondiamond carbon lines [12].

As it is shown in Figure 5, the diamond quality factor is in the range of 98.7%–99.4% for the temperature interval 400–800 °C, confirming the high phase purity of the diamond film and a low interaction ability of the CVD diamond–Fe system in the investigated temperature range.

For a deeper understanding of the element distribution during the interdiffusion process, we measured concentrations of carbon and iron using the EDX method on the cross-section of the sintered diamond–Fe system. The results of these measurements are shown in Figure 6. We select three representative samples annealed at 400, 600, and 800 °C. The strong diffusion of carbon into the Fe volume starts even at temperatures lower than 400 °C due to its extreme diffusion coefficient [13], Figure 6a 25% of carbon was involved in the Fe$_3$C formation while its residual part diffuses into the Fe bulk with the formation of graphite, in agreement with the XRD measurements shown in Figure 2. One can conclude, that at temperatures above 400 °C only the diffusion of carbon into the iron occurs. However, at higher temperatures (from 600 °C) diffusion of Fe into the diamond layer becomes substantial, see Figure 6b,c. The calculated effective diffusion length (x) of the Fe into the diamond film is represented in Figure 7. It was obtained from the distance between the top of the diamond layer and the point where the Fe concentration plot reaches 0%. EDS measurements were performed at five different points for each distance. The statistical deviation of the concentration measurement was $\pm$ 5 at.%. Due to the fact that we cannot detect a major change of the diamond layer using SEM measurements while Raman spectroscopy measurements show a minor reduction of the diamond quality parameter, we assume that diffusion occurs along the grain boundaries of the diamond layer. For the determination of the diffusion coefficient and diffusion activation energy of iron, we used Fick's second law:

$$D = \frac{x^2}{4t} \tag{2}$$

here x is the effective diffusion length of the Fe, D is the diffusion coefficient, and t is the diffusion time.

Diffusion coefficients of the Fe at different temperatures, Equation (2), were used to prepare the Arrhenius plot, as shown in Figure 7. From the intersection of the linear fits with the ordinate, we obtained the pre-exponential factor, D_0, and derived the activation energy, Q, from their slope using the Arrhenius equation for the diffusion coefficient, D:

$$D = D_0 \exp\left(-\frac{Q}{RT}\right) \tag{3}$$

Here D is the Fe diffusion coefficient into the diamond film, D_0 is the pre-exponential factor, Q is the diffusion activation energy, R is the gas constant, and T is the absolute temperature.

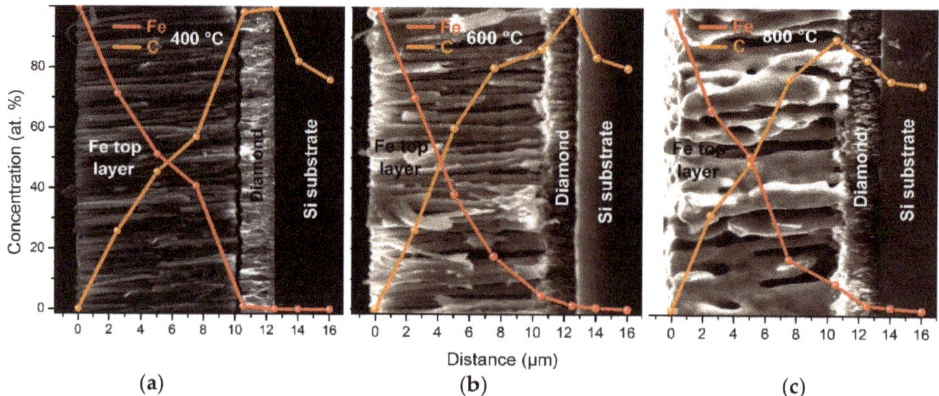

Figure 6. EDX plots of carbon and iron for the CVD diamond–Fe system annealed at (**a**) 400 °C; (**b**) 600 °C and (**c**) 800 °C.

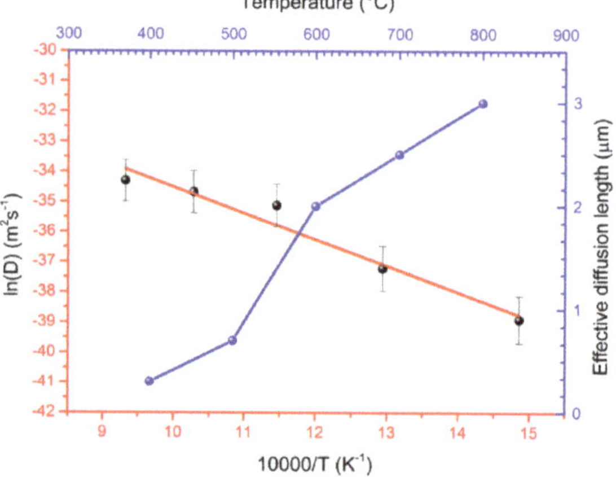

Figure 7. Arrhenius plot for Fe diffusion into the diamond layer and the corresponding diffusion length of the Fe as a function of the annealing temperature.

It can be seen that the diffusion coefficient of the Fe increases upon increasing the annealing temperature and changes from the 1.25×10^{-17} m^2 s^{-1} at 400 °C up to 1.25×10^{-15} m^2 s^{-1} at 800 °C.

The pre-exponential factor and the activation energy obtained from the graph were 5.6×10^{-12} m^2 s^{-1} and 69.1 kJ/mol, respectively.

4. Conclusions

To summarize, we can conclude that:

- The CVD diamond–Fe interaction in the range of 400–800 °C is significantly lower in comparison with previously reported data for the range of temperatures 900–1300 °C. Diamond etching and graphitization in the range of 400–800 °C are undetectable during microscopic measurements and can be detected only by XRD and Raman measurements.

- Annealing of the diamond in contact with Fe slightly reduces the quality parameter in the range of 98.7–99.4%, confirming a slight interaction of diamond–Fe at the selected temperature range.

- Strong diffusion of carbon into the Fe occurs even at low-temperature annealing conditions of 400 °C.

- Formation of an Fe–C transition layer due to the interdiffusion process in the diamond–Fe system can be detected in the temperature range 600–800 °C. The thickness of this transition layer is relatively low (approx. 0.5 μm) due to the limited ability of the Fe$_3$C to form.

- When annealing at 600 °C, one can detect a diffusion of the Fe into the diamond film. The Fe diffusion coefficient was estimated from 1.25×10^{-17} m^2 s^{-1} at 400 °C up to 1.25×10^{-15} m^2 s^{-1} at 800 °C. The diffusion activation energy of the Fe was determined to be 69.1 kJ/mol.

Author Contributions: Methodology, S.Z., A.G. and S.L.; Formal Analysis, S.Z., A.G., V.O. and S.L.; Resources S.Z., and S.L; Writing–Original Draft Preparation, S.Z. and A.G.; Writing–Review and Editing, V.O. and. S.L.

Funding: The research was funded by the Tomsk Polytechnic University within the framework of Tomsk Polytechnic University Competitiveness Enhancement Program grant.

Conflicts of Interest: The authors declare no conflict of interest.

References

1. Gracio, J.J.; Fan, Q.H.; Madaleno, J.C. Diamond growth by chemical vapour deposition. *J. Phys. D Appl. Phys.* **2010**, *43*, 374017. [CrossRef]
2. Shimada, S.; Tanaka, H.; Higuchi, M.; Yamaguchi, T.; Honda, S.; Obata, K. Thermo-chemical wear mechanism of diamond tool in machining of ferrous metals. *CIRP Ann.* **2004**, *53*, 57–60. [CrossRef]
3. Nakamura, E.; Hirakuri, K.K.; Ohyama, M.; Friedbacher, G.; Mutsukura, N. High quality chemical vapor deposition diamond growth on iron and stainless steel substrates. *J. Appl. Phys.* **2002**, *92*, 3393. [CrossRef]
4. Mehedi, H.; Arnault, J.-C.; Eon, D.; Hébert, C.; Carole, D.; Omnes, F.; Gheeraert, E. Etching mechanism of diamond by Ni nanoparticles for fabrication of nanopores. *Carbon* **2013**, *59*, 448–456. [CrossRef]
5. Zaitsev, A.M.; Kosaca, G.; Richarz, B.; Raiko, V.; Job, R.; Fries, T.; Fahrner, W.R. Thermochemical polishing of CVD diamond films. *Diam. Relat. Mater.* **1998**, *7*, 1108–1117. [CrossRef]
6. Jin, S.; Graebner, J.E.; Tiefel, T.H.; Kammlott, G.W. Thinning and patterning of CVD diamond films by diffusional reaction. *Diam. Relat. Mater.* **1993**, *2*, 1038–1042. [CrossRef]
7. Tokura, H.; Yang, C.; Yoshikawa, M. Study on the polishing of chemically vapour deposited diamond film. *Thin Solid Films* **1992**, *212*, 49–55. [CrossRef]
8. Jin, S.; Graebner, J.E.; Kammlott, G.W.; Tiefel, T.H.; Kosinski, S.G.; Chen, L.H.; Fastnacht, R.A. Massive thinning of diamond films by a diffusion process. *Appl. Phys. Lett.* **1992**, *60*, 1948. [CrossRef]
9. Ralchenko, V.G.; Kononenko, T.V.; Pimenov, S.M.; Chernenko, N.V.; Loubnin, E.N.; Armeyev, V.Y.; Zlobin, A.Y. Catalytic interaction of Fe, Ni and Pt with diamond films: Patterning applications. *Diam. Relat. Mater.* **1993**, *2*, 904–909. [CrossRef]
10. Giménez, S.; Biest, O.v.d.; Vleugels, J. The role of chemical wear in machining iron based materials by PCD and PCBN super-hard tool materials. *Diam. Relat. Mater.* **2007**, *16*, 435–445. [CrossRef]
11. Girolami, M.; Bellucci, A.; Calvani, P.; Orlando, S.; Valentini, V.; Trucchi, D.M. Raman investigation of femtosecond laser-induced graphitic columns in single-crystal diamond. *Appl. Phys. A* **2014**, *117*, 143–147. [CrossRef]

12. Guillemet, T.; Xie, Z.Q.; Zhou, Y.S.; Park, J.B.; Veillere, A.; Xiong, W.; Heintz, J.M.; Silvain, J.F.; Chandra, N.; Lu, Y.F. Stress and phase purity analyses of diamond films deposited through laser-assisted combustion synthesis. *ACS Appl. Mater. Interfaces* **2011**, *3*, 4120–4125. [CrossRef]
13. Verein Deutscher Eisenhüttenleute. *Werkstoffkunde Stahl. Band 1: Grundlagen*; Springer-Verlag GmbH: Berlin/Heidelberg, German, 1984. (In Germany)

Article

Properties and Structure of Deposited Nanocrystalline Coatings in Relation to Selected Construction Materials Resistant to Abrasive Wear

Jacek Górka [1,*], Artur Czupryński [1], Marcin Żuk [1], Marcin Adamiak [2] and Adam Kopyść [2]

[1]   Department of Welding Engineering, Silesian University of Technology, Konarskiego 18A,
      44-100 Gliwice, Poland; artur.czuprynski@polsl.pl (A.C.); marcin.zuk@polsl.pl (M.Ż.)
[2]   Institute of Engineering Materials and Biomaterials, Silesian University of Technology, Konarskiego 18A,
      44-100 Gliwice, Poland; marcin.adamiak@polsl.pl (M.A.); adam.kopysc@gmail.com (A.K.)
*   Correspondence: jacek.gorka@polsl.pl; Tel.: +48-32-237-1445

Received: 6 June 2018; Accepted: 9 July 2018; Published: 10 July 2018

Abstract: Presented in this work are the properties and structure characteristics of MMA (Manual Metal Arc) deposited nanocrystalline coatings (Fe-Cr-Nb-B) applied to an iron nanoalloy matrix on an S355N steel substrate in relation to selected construction materials resistant to abrasive wear currently used in industry. The obtained overlay welds were subjected to macro and microscopic metallographic examinations; grain size was determined by X-ray diffraction (XRD), and chemical composition of precipitates was determined by energy-dispersive X-ray spectroscopy (EDS) during scanning electron microscopy (SEM). The size of the crystalline grains of the Fe-Cr-Nb-B nanocrystalline microstructure was analyzed using an Xpert PRO X-ray diffractometer. Analysis of the test results of the obtained layers of arc-welded Fe-Cr-Nb-B-type alloy confirmed that the obtained layers are made of crystallites with a size of 20 nm, which classifies these layers as nanocrystalline. The obtained nanocrystalline coatings were assessed by hardness and with the use of metal-mineral abrasion testing. The results of the coatings' properties tests were compared to HARDOX 400 alloy steel.

Keywords: abrasive wear; nanocrystalline layers; abrasion resistant plate; deposit weld

1. Introduction

Wearing of machine parts poses an important problem in terms of scientific, technical and economic potential. In most cases, the mechanisms of wear are highly complex and include many interrelated factors whose impact depends on the environment and working conditions of the part in question. The variety of the types of wear modes leads to the specialization of materials used in the construction of parts subject to abrasive wear in order to ensure the highest resistance to wearing of surface layers under specific operating conditions. One of the types of such specialized materials is Abrasion Resistant (AR) plates [1–7]. The structure and properties of the surface layer are, to a large extent, the main factor when considering materials for machine parts in terms of durability. In recent years, there has been dynamic research in the development of new abrasion-resistant materials containing layers of unique properties and structures differing significantly from previous work. In particular, new approaches such as improved hardness, resistance to impact loads and low coefficients of friction are the main concerns when developing new materials [8–19]. The rapid development of nano-structurally modified materials is foreshadowing an increase in their application in novel welding technologies. The wide variety of properties inherent to nano-structually modified materials are a paradigm shift, bringing new possibilities through the use of nanomaterial-based surfacing technologies. Nanomaterials are classified as single or multiphase polycrystals characterized by a grain size on the order of 1×10^{-9} m to 250×10^{-9} m in diameter. At the upper limit of this range, the term "very fine" is used with respect to grain size on the order of 250–1000 nm in diameter [20–27]. Nanocrystalline materials are structurally characterized by a high volume fraction at the grain boundaries, which significantly changes their physical, chemical and mechanical properties in comparison to conventional coarse grains, whose grain size is usually on the order of 10–300 μm. Until now, existing nanomaterials used for nanostructure coatings and layers have shown a significantly higher (many times) wear resistance compared to traditional steel alloy-based materials. Due to the high cost and continuous development of their production technology, nanostructural materials have not been widely applied [28–33].

2. Experimental Section

The purpose of the performed experiments was to compare the structure and properties of Fe-Cr-Nb-B nanocrystalline surface layers deposited by manual metal arc (MMA) welding with a covered electrode 3.2 mm in diameter to previously used abrasion-resistant materials. HARDOX 400 steel was used as a reference material in the assessment of resistance to metal-mineral abrasion. The following materials were tested in the experiments:

(1) Nanocrystalline layer deposited with a covered electrode—NANO (Fe-Cr-Nb-B)
(2) Abrasion-resistant plate—ABRECOPLATE
(3) Abrasion-resistant plate—CDP
(4) Abrasion-resistant layer deposited by covered electrode—ABRADUR 64
(5) Abrasion-resistant layer deposited by GMA with a ceramo-metallic wire containing 50% WC (tungsten carbide)
(6) Abrasion-resistant sheet—HARDOX 400.

Metallographic examinations of the deposited coatings were carried out on a Zeiss SteREO Discovery, and LEICA MEF4A optical microscope (Leica Microsystems, Cambridge, UK) in addition to a Zeiss Supra 35 (SEM) scanning electron microscope (Carl Zeiss GmbH, Jena, Germany). Metallographic images were taken of transverse sections with respect to weld deposition direction. The specimens were prepared by standard metallographic techniques and etched in Nital solution. Phase compositions of the investigated materials in addition to the size of the crystallographic grains of nanocrystalline microstructure were determined using an Xpert PRO X-ray diffractometer with step data logging, employing the filtered K@ X-rays. The chemical composition of precipitates was determined by energy-dispersive X-ray spectroscopy (EDS) with an EDAX detector. Hardness tests were conducted on a Zwick Roell ZHR hardness tester based on the Rockwell method, while cross-sectional hardness was measured using the Vickers method with a Future-Tech FM-700 tester (Future-Tech Corp., Kawasaki, Kanagawa Prefecture, Japan).

2.1. Nanocrystalline Layer Deposited with Covered Fe-Cr-Nb-B Electrode

The Fe-Cr-Nb-B nanocrystalline layer, shown in Table 1, deposited by MMA with a 3.2 mm covered electrode on an S355N steel alloy substrate is shown in Figure 1. MMA deposit welding was performed with a constant direct current of 100 A in the flat position (PA). During welding, the electrode was set at an angle of 90° with respect to the surface of the welding base. The sheet surface was ground and pre-heated with a gas torch to a temperature of 80 °C.

Table 1. Chemical composition and hardness of the tested deposit weld.

Chemical Composition, wt %							HRC
C	Cr	B	Nb	Mn	Si	Fe	
1.4	15.2	4.0	3.4	0.4	0.4	remainder	68–70

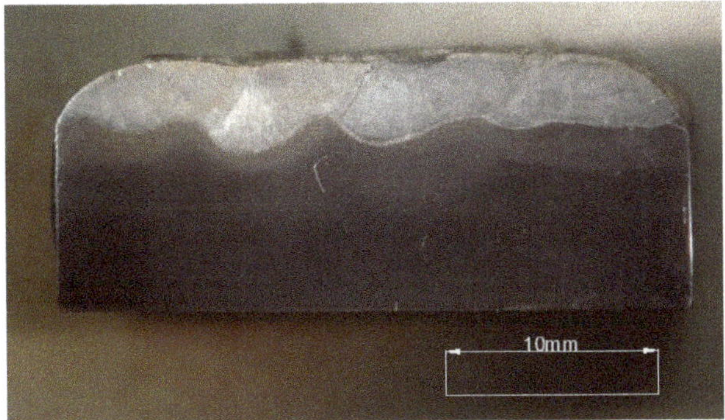

Figure 1. Macrostructure of the cross-section of a nanocrystalline layer, the average width of the overlay layer is 3 mm.

According to the manufacturer's data, the Fe-Cr-Nb-B nanocrystalline alloy electrodes mostly consist of very hard boron carbide fractions evenly distributed in a semi-amorphous iron alloy matrix, yielding a hardness of 67–70 HRC. The deposit weld should have a high abrasion resistance and an increased resistance to dynamic loads due to these characteristics. The electrodes can be used for both direct- and alternating-current welding.

2.2. Abrasion Resistant Plate—ABRECOPLATE

ABRECOPLATE abrasion-resistant materials are produced in the following formats: plates (straight or beveled, by special order), bars, buttons (in the shape of a dome, octagonal, protecting screws). ABRECOPLATE is a layered material composed of chromium-molybdenum white cast iron, metallurgically connected to a soft-structural steel underlying plate, as detailed in Table 2.

Table 2. Chemical composition and physical characteristics of the abrasion-resistant layer of ABRECOPLATE.

Chemical Composition, wt %						
C	Cr	Mo	Mn	Si	Ni	Fe
2.8–3.6	14.0–18.0	2.3–3.5	0.5–1.5	1.0 max.	5.0 max.	remainder
Mechanical Properties						
HRC		Heat Resistance (°C)		Creep Resistance (°C)		
64		540		595		

ABRECOPLATE's high abrasion-resistance properties are due to the structure of the surface layer. The special heat treatment (hardening throughout) of cast iron allows for the obtainment of a microstructure consisting of chromium-molybdenum carbides in an almost completely martensitic matrix. The substrate of these abrasive plates is a soft-structural steel. Undercoated cast iron is joined by soldering with a soft copper-based binder which ensures adequate stress transfer, shown in Figure 2.

An important advantage of ABRECOPLATE abrasive discs is the content of abrasive material in relation to the primer, which is 3:1.

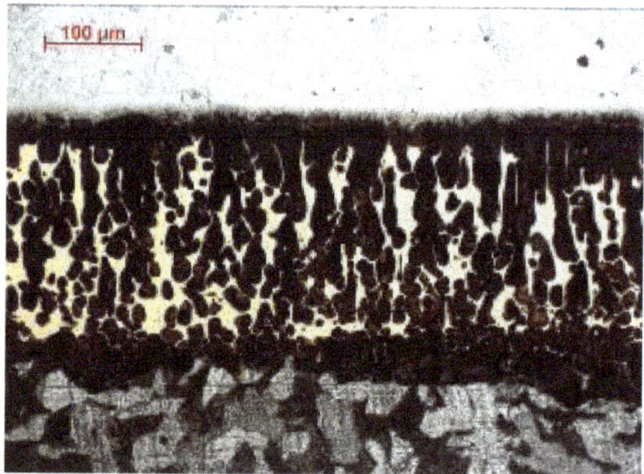

Figure 2. Microstructure of the intermediate layer of ABRECOPLATE cross-section.

2.3. CDP Plate

Abrasion-resistant plates are manufactured by welding a sheet of non-alloy, low-alloy or high-alloy steel with shielded powder wire or self-shielded wire, as shown in Table 3 and Figure 3.

Table 3. Chemical composition of the surface deposit weld.

Chemical Composition, wt %							HRC
C	Cr	B	Nb	Mn	Si	Fe	
5.2	22.0	1.8	7.0	0.4	0.4	remainder	57–62

Figure 3. Macrostructure of a cross-section of CDP plate.

The deposited layer exhibits a very high abrasion resistance and has a standard thickness of 3–18 mm. Typical dimensions of abrasion plates are: 1000 × 2000 mm, 1500 × 3000 mm, and 2000 × 3000 mm. It is possible to cut flat elements of any shape from abrasion plates and further shape them by bending and rolling. They are joined to the regenerated substrate with fillet welds, continuous or intermittent, depending on the type of abrasive plate load. The high content of carbon, chromium, and niobium allows for the obtainment of a structure similar to that of cast iron with the inclusion of very hard chromium borides, niobium carbides, and iron carbides.

2.4. Abrasion Resistant Layer Deposited by Covered ABRADUR 64 Electrode

The abrasion-resistant layer was deposited by manual welding of an S335JR steel with a covered electrode, DIN 8555: E 10-UM-65-GR with a diameter of 5.0 mm and a welding current of 270 A. The welding process was carried out using a buffer layer made with a covered electrode, ERWS 19-12-3 L with a diameter of 3.25 mm and a welding current of 110 A, as shown in Figure 4. The purpose of the buffer layer was to transfer the stresses between the base material and hardfacing. The chemical composition and properties of the weld metal, ABRADUR 64 are given in Table 4.

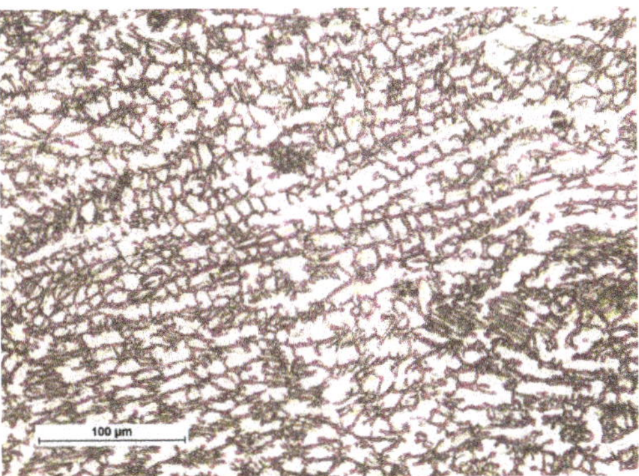

Figure 4. Microstructure of austenitic intermediate layer.

Table 4. Chemical composition and properties of the abrasion-resistant layer of ABRADUR 64-covered electrodes.

Chemical Composition, wt %				HRC
C	Cr	Nb	Fe	
7.0	24.0	7.0	remainder	64

2.5. Abrasion Resistant Layer Formed by GMA with Ceramo-Metallic Wire

The abrasion-resistant layer was made by single-layer GMA welding of 15 HM steel using a ceramo-metallic wire with a 50% nickel carbide content of tungsten carbide (WC). The analysis of the chemical composition is shown in Table 5.

Table 5. Chemical composition of the layer formed by GMA (Gas Metal Active) with a ceramo-metallic wire layer.

Mass Percent of Elements in Abrasion Resistant, wt %					
Ni	C	Si	Cr	B	WC
Remainder	0.4	2.5	3.0	1.5	50

2.6. HARDOX 400 Steel Sheet

HARDOX type steels are defined as "high-quality abrasion-resistant steels". This group of materials is derived from low-alloy steels destined for heat treatment and belongs to a new generation of machinable and weldable structural steels. Materials made of HARDOX steel are used where resistance to abrasion is required such as in the presence of variable loads, e.g., feeders, crushers, sieves, shafts, elements of incline lift, conveyors, blades, gears and chains, dumpers, loaders, trucks, motor-carriages, dozers, loading buckets, and screw conveyors. All types of HARDOX steel are delivered in the hardened condition (water-hardened). In the case of specifically required hardnesses, tempering is also performed. These steels can be bent, cut, drilled, machined, or turned under strictly defined conditions. HARDOX sheets can be machined using high-speed steel (HSS) or tools made of sintered carbides. The chemical composition and properties of HARDOX 400 steel are given in Table 6.

Table 6. Chemical composition and mechanical properties of HARDOX 400.

Chemical Composition, wt %					
C	**Mn**	**Mo**	**Cr**	**Si**	**Ni**
0.14–0.32	1.60	0.25–0.60	0.30–1.40	0.70	0.25–1.50
Mechanical Properties					
HBW		**Tensile Strength (MPa)**		**Yield Strength (MPa)**	
370–430		1250		1000	

3. Results and Discussion

3.1. ASTM G65-00 Metal–Mineral Abrasion Resistance Test

The tests of abrasive wear resistance for the selected materials were carried out in accordance with ASTM G 65-00. Procedure A, which is the most demanding examination of abrasion resistance, was used for the tests. During the test, the sample was mounted in a special fixture in which it was clamped to a rubber wheel 228.6 mm in diameter. The test sample was pressed against the rubber wheel with a force of 130 N. Abrasive in the form of granular sand (gradation 250 μm) was delivered through the nozzle at the location of sample contact with the rubber wheel. The abrasive flow rate was 300–400 g/min. The wheel rotated in the direction corresponding to the abrasive flow, at a speed of 200 rpm through 6000 revolutions. The tested samples were 25 × 75 × thickness of the sample, mm, as shown in Figure 5. The mass loss was determined using an analytical balance precise to 0.0001 g (measurement accuracy up to 5 decimal places). To compare the results of abrasion resistance, the density of plates and abrasive layers was determined. As a measure of abrasion, the volume loss of the sample (mm^3) was determined, and is shown in Equation (1), Table 7, and Figure 6.

$$Volume\ loss\ (\text{mm}^3) = mass\ loss\ (\text{g})\ :\ density\ (\frac{\text{g}}{\text{cm}^3}) \times 1000 \tag{1}$$

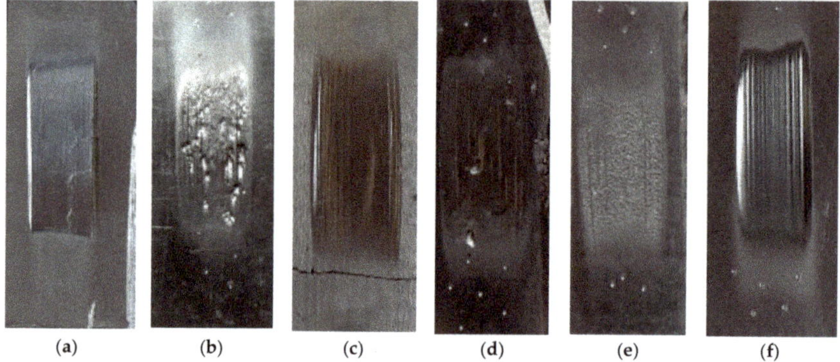

(a) (b) (c) (d) (e) (f)

Figure 5. Abrasion tests samples following the abrasion test: (**a**) NANO; (**b**) ABRECOPLATE; (**c**) CDP; (**d**) ABRADUR 64; (**e**) WC; (**f**) HARDOX 400.

The tests have shown that the relative resistance to metal-to-metal abrasive wear of the nanocrystalline layer is 11 times higher when compared to the Hardox400 reference metal sheet.

Table 7. ASTM G65-00 abrasion resistance test results.

Material/Density (g/cm^3)	Sample Number	Sample Weight before Test (g)	Sample Weight after Test (g)	Mass Loss (g)	Average Mass Loss (g)	Average Volume Loss (mm^3)	Relative Abrasion Resistance *
NANO/8.78	1	102.9477	102.8393	0.1084	0.1113	12.6765	10.95
	2	101.7964	101.6821	0.1143			
ABRECOPLATE/7.5961	1	173.7335	173.6133	0.1202	0.11635	15.3170	9.07
	2	173.6714	173.5589	0.1125			
CDP/7.1724	1	128.6154	128.4378	0.1776	0.1697	23.3881	5.94
	2	128.9438	128.7821	0.1617			
ABRADUR 64/7.1544	1	136.2893	136.0933	0.1960	0.19825	27.7102	5.01
	2	139.6675	139.4670	0.2005			
WC/10.6808	1	179.6026	179.3009	0.3017	0.32360	30.2974	4.58
	2	181.8750	181.5295	0.3455			
HARDOX 400/7.7115	1	62.1029	61.0320	1.0709	1.0691	138.8705	1.00
	2	62.5591	61.4918	1.0673			

* Relative abrasion resistance compared to HARDOX 400.

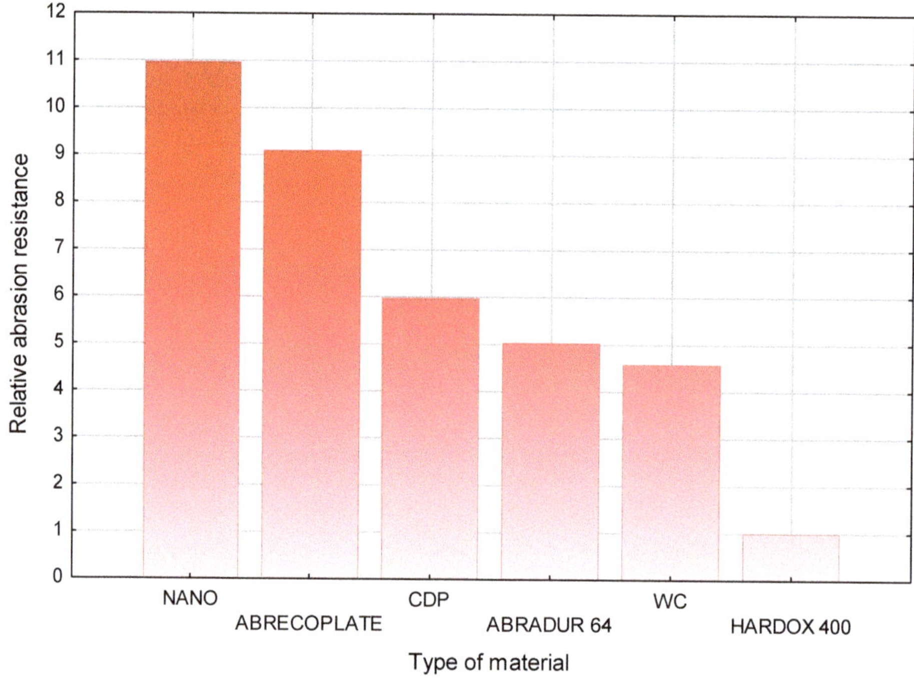

Figure 6. Comparison of the results of relative abrasive wear resistance of selected construction materials.

3.2. Metallographic Testing

The microscopic observations made it possible to determine the microscopic structure of the materials studied. The observations, carried out using a light microscope, did not show internal defects in the layers due to welding methods and material defects in the case of HARDOX 400 sheet, ABRECOPLATE, and CDP plates, Figure 7.

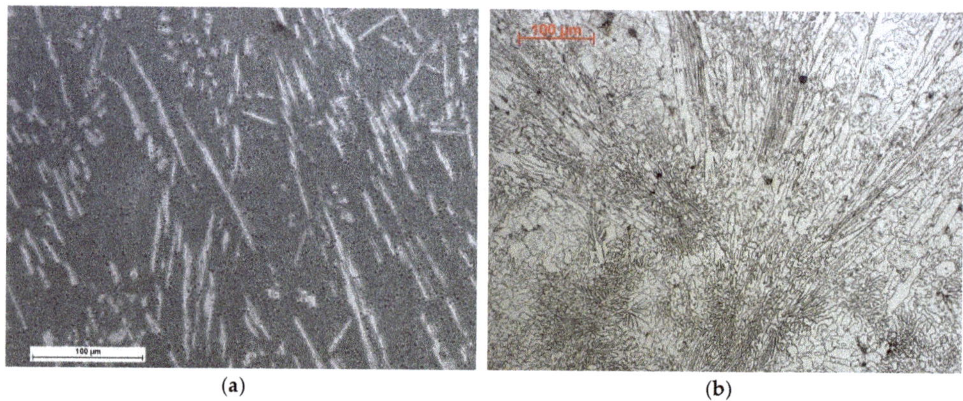

(a)   (b)

Figure 7. *Cont.*

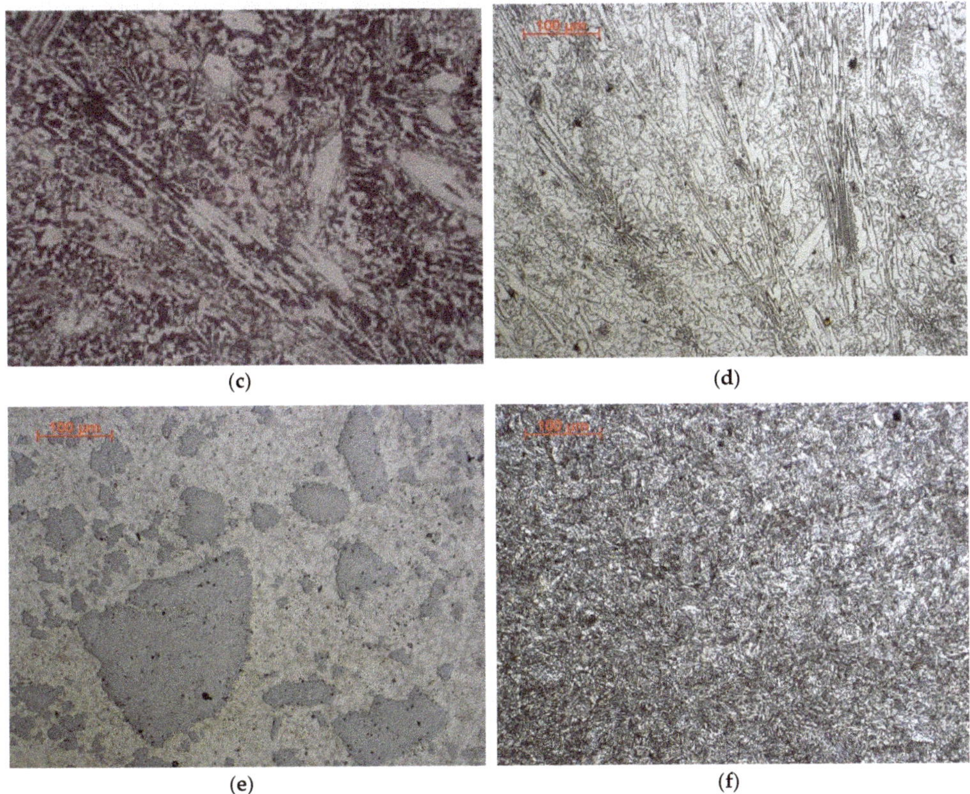

Figure 7. Microstructure of layers resistant to abrasive wear. (**a**) NANO; (**b**) ABRECOPLATE; (**c**) CDP; (**d**) ABRADUR 64; (**e**) WC; (**f**) HARDOX 400; Reagents selected for abrasion resistant material.

Metallographic examinations carried out on a scanning microscope technique (SE) revealed large amounts of primary carbide precipitates highly dispersed in the zone of the nanocrystalline matrix, as shown in Figures 8 and 9.

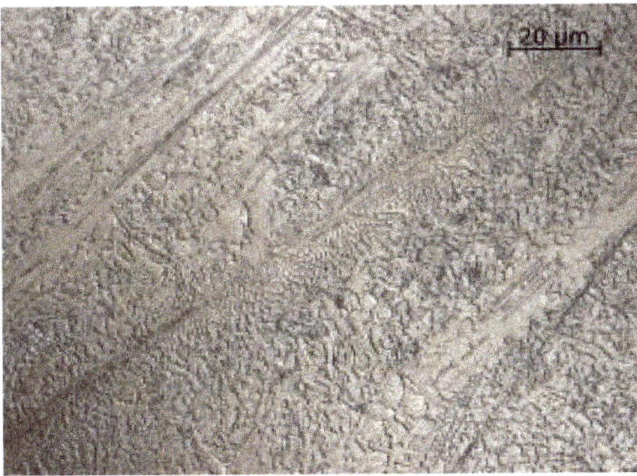

Figure 8. Microstructure of a nanocrystalline layer with eutectic like carbides of NANO (Fe-Cr-Nb-B).

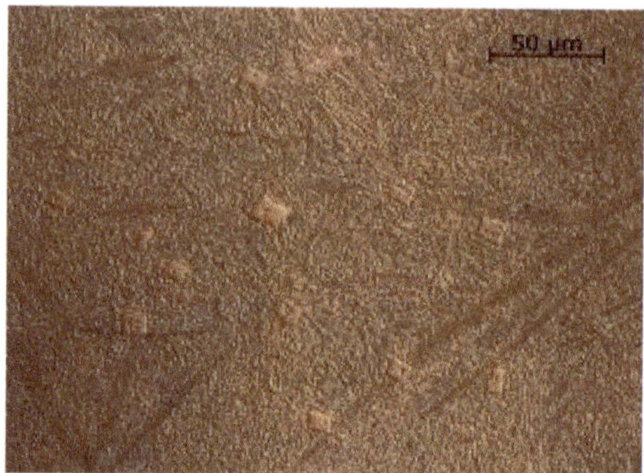

Figure 9. Microstructure of a nanocrystalline layer with carbide precipitates of NANO (Fe-Cr-Nb-B).

SEM analysis showed that the observed carbide precipitates which appeared larger in size were niobium and chromium carbides, shown in Figures 10 and 11.

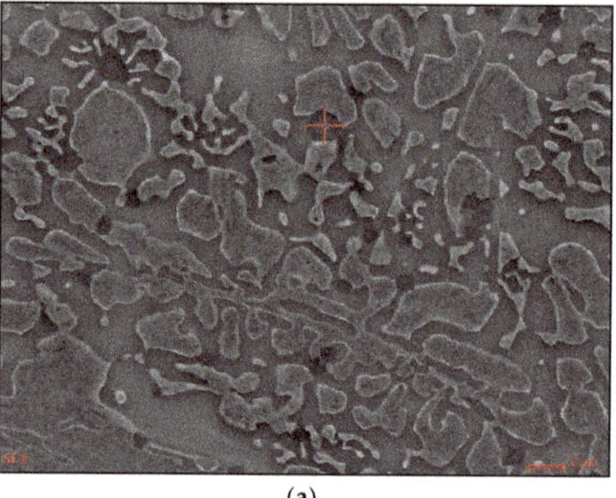

(a)

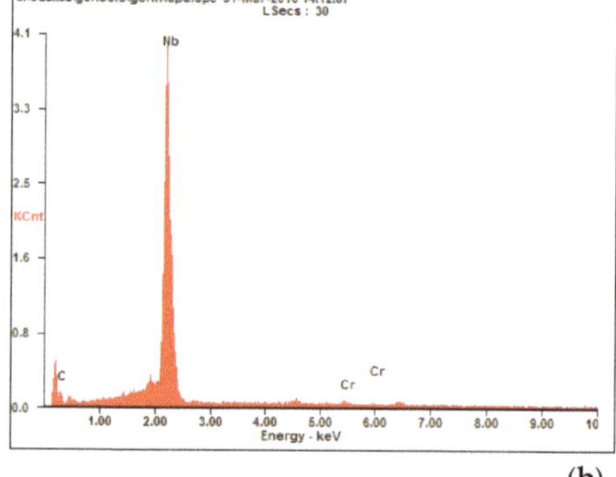

Element	Wt (%)	At (%)
CK	32.90	75.36
AlK	00.99	1.01
NbL	57.52	17.84
CrK	1.99	1.05
FeK	7.60	3.74
Matrix	Correction	ZAF

(b)

Figure 10. (**a**) SEM image of the nanostructural layered microstructure over the EDS analysis area—a precipitation of niobium carbide; (**b**) the EDS spectrum with the above indicated spot-separation of niobium carbide and results of the quantitative elemental analysis.

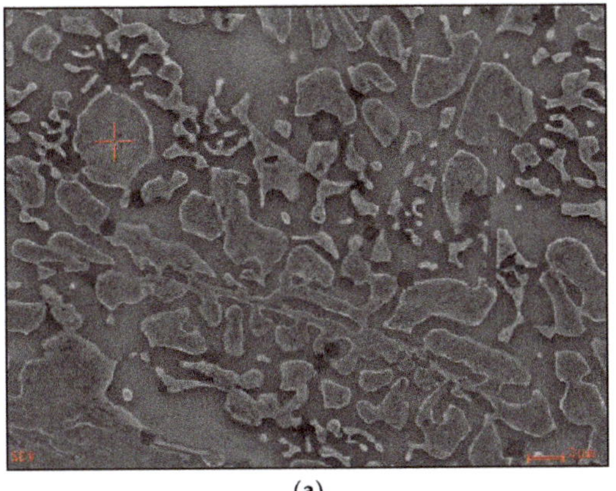

(a)

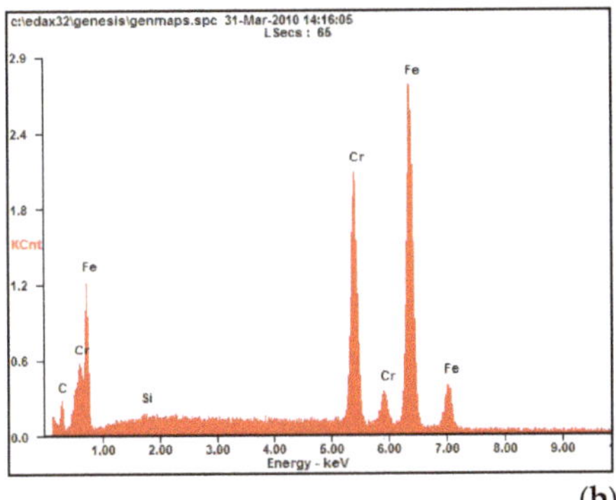

Element	Wt (%)	At (%)
CK	19.22	52.00
SiK	1.28	1.48
CrK	25.98	23.74
MnK	1.22	0.72
FeK	52.30	32.06
Matrix	Correction	ZAF

(b)

Figure 11. (**a**) SEM image of the nanostructural layer microstructure over the EDS analysis area—a precipitation of chromium carbide; (**b**) the EDS spectrum showing the above indicated spot-separation of chromium carbide and results of the quantitative elemental analysis.

The size of the crystallographic grains of the Fe-Cr-Nb-B nanocrystalline microstructure was measured using an Xpert PRO X-ray diffractometer by PANalytical and a computerized radiation recording system equipped with a cobalt lamp at 40 kV and 30 mA current with a strip detector, in the Bragg angle range of 30–120°. Based on calculations of crystalline sizes carried out using the Scherrer Equation (2), it was found that the average grain size of the layered microstructure measured in the direction perpendicular to the deposited substrate was approximately 20 nm.

$$D = \frac{K \cdot \lambda}{B_{struct} \cdot \cos\theta} \tag{2}$$

where: D—the average size of the crystallite in the direction perpendicular to the planes of deflection; K—Scherrer constant (0.98); λ—wavelength; B_{struct}—width of reflexes; θ—the angle of reflection.

Analysis of the diffraction pattern of the nanocrystalline layers of Fe-Cr-Nb-B showed the presence of reflections from the three types of carbides (Figure 12):

-   Cr_7C_3 from the lattice plane (002), (151), (321), (202), (222), (260), (081);
-   $Cr_{23}C_6$ from the lattice plane (400), (420), (422), (333), (440), (531), (620), (911);
-   NbC from the lattice plane (111), (200), (220).

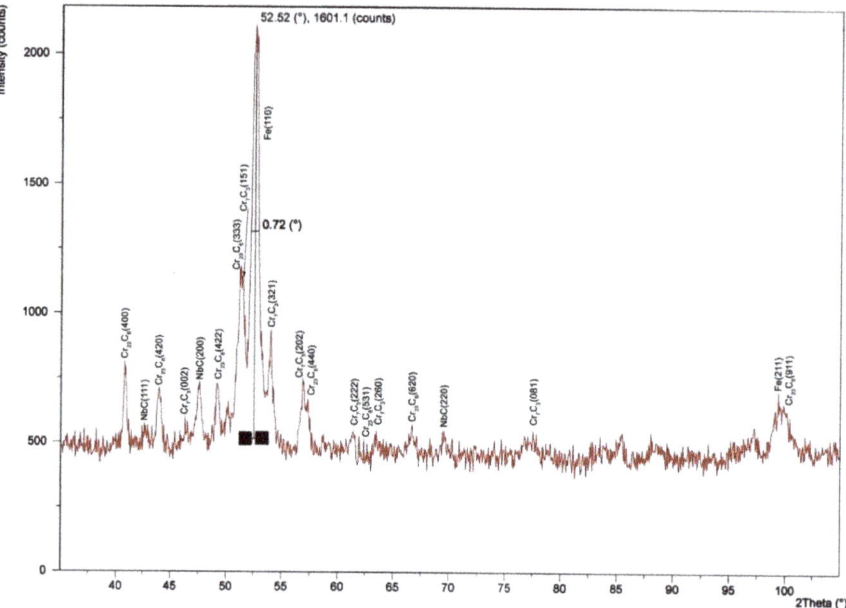

Figure 12. The diffraction pattern of the nanocrystalline Fe-Cr-Nb-B layer.

3.3. Hardness Testing

In order to determine the hardness of the tested materials, the Rockwell hardness measurement was carried out in five places on the weld face/sheet surface and in four locations on the deposit weld/sheet cross-section using the Vickers method at a load of 1000 g, shown in Figures 13 and 14. Hardness measurement results are shown in Table 8 and Figure 15.

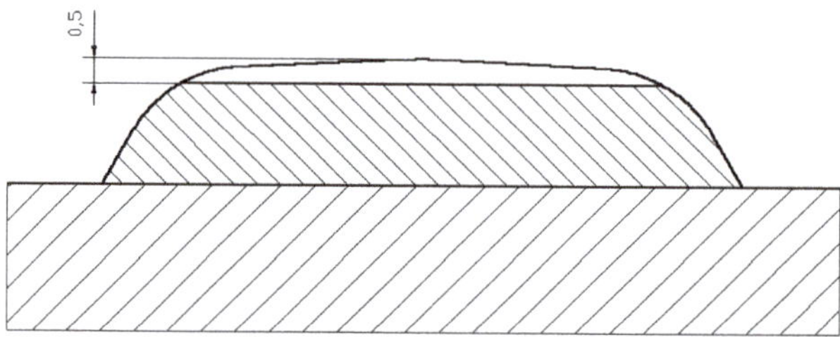

Figure 13. Preparation for hardness testing.

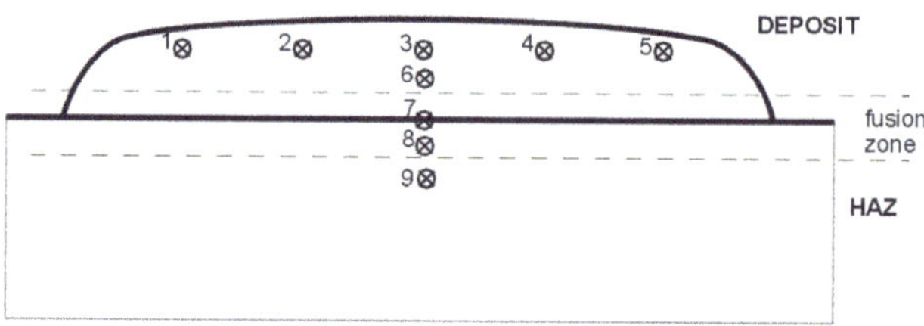

Figure 14. Hardness test area.

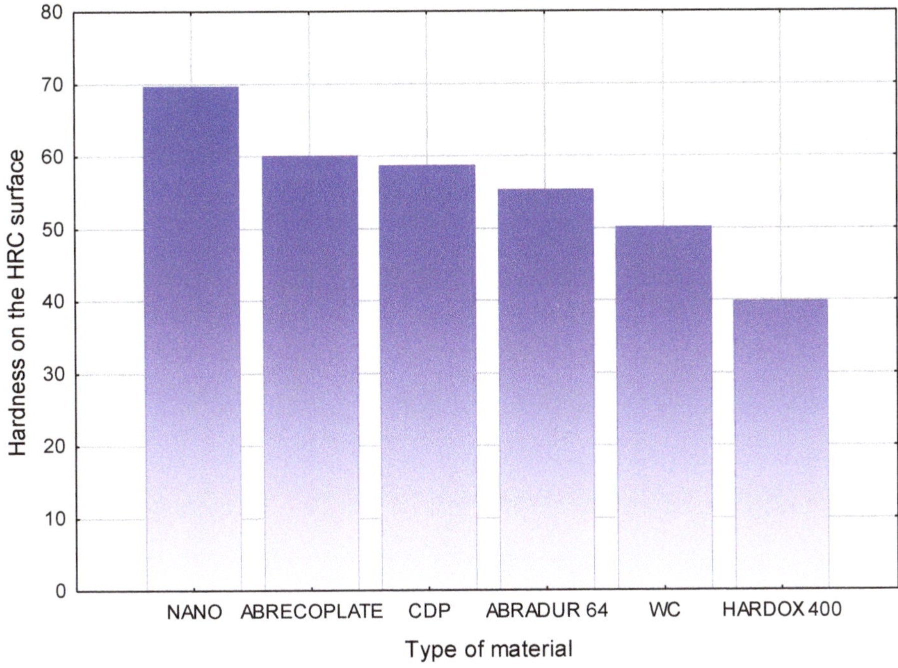

Figure 15. Average hardness of the surface of tested construction materials.

Table 8. Hardness testing results on the face and cross-section of deposit welds and sheets.

Material	HRC						HV 1			
	1	2	3	4	5	Average	6	7	8	9
NANO	69	70	69	71	69	69.60	679	289	179	188
ABRECOPLATE	60	58	61	61	62	60.04	663	154	148	106
CDP	62	59	58	61	58	58.70	643	299	179	182
ABRADUR 64	57	56	57	54	52	55.20	556	174	304	290
WC	54	50	48	51	49	50.04	563	486	180	167
HARDOX 400	41	40	39	40	39	39.80	380	378	377	378

4. Conclusions

The metallographic examinations of the materials selected for the tests did not show any internal or external defects in the layers formed by MMA welding with coated electrodes and GMA welding. Analysis of the test results of the obtained layers of arc-welded Fe-Cr-Nb-B-type alloy confirmed:

- The obtained layers are made of crystallites with a size of 20 nm, which classifies these layers as nanocrystalline.
- Numerous carbide precipitates with empirical formula Cr_7C_3, $Cr_{23}C_6$, and NbC were observed and detected in the samples.

The ABRECOPLATE plate has a structure of white cast iron with chromium-molybdenum carbide precipitates. To join this plate with a low-carbon steel support plate, a soft binder was used on the copper matrix, which perfectly transfers the stresses occurring between the layers. In the case of CDP, the surface layer structure is a chromium cast iron containing many primary carbides.

The layer created with ABDADUR 64 coated electrodes possesses a eutectic iron structure with numerous precipitates of niobium and chromium carbides. The use of an austenitic steel buffer layer, in this case, allowed for crack avoidance that could propagate through to the substrate material. The layer formed by GMA surfacing with a ceramo-metallic wire is characterized by a nickel matrix containing many tungsten carbides. As a result of the thermal cycle, it can be observed that the WC carbides are partially dissolved, which may reduce the abrasion resistance of such layers.

HARDOX 400 is characterized by a tempered martensite structure. The hardness measurements carried out on the ground face of abrasion-resistant layers showed that all materials have a hardness similar to the hardness quoted by the manufacturers. The highest hardness on the surface is characterized by the nanocrystalline layer, with a hardness of 70 HRC. The tests of resistance to abrasive wear of the metal-mineral type, according to ASTM G 65-00, have shown that the best usable properties are characterized by a layer made of Fe-Cr-Nb-B alloy. The metal-mineral abrasion resistance of this material is 11 times higher than a typical HARDOX 400 abrasion-resistant sheet.

Author Contributions: Conceptualization, J.G. and A.C.; Methodology, M.A.; Validation, A.K., M.Ż.; Formal Analysis, J.G. and A.C.; Investigation, M.A. and M.Ż.; Writing-Original Draft Preparation, J.G.; Writing-Review & Editing, A.K.

Funding: The publication was co-financed from the statutory subsidy of the Faculty of Mechanical Engineering of the Silesian University of Technology in 2018.

Conflicts of Interest: The authors declare no conflicts of interest.

References

1. Kato, K. Abrosive wear of metals. *Tribol. Int.* **1997**, *5*, 333–338. [CrossRef]
2. Grigoroudis, K.; Stephenson, D.J. Modelling low stress abrasive wear. *Wear* **1997**, *213*, 103–111. [CrossRef]
3. Zum Gahr, K.H. Wear by hard particles. *Tribol. Int.* **1998**, *10*, 587–596. [CrossRef]
4. Sundararajan, G.; Manisz, R. Solod particle erosion behaviour of metallic materials at room and elevated temperatures. *Tribol. Int.* **1997**, *30*, 339–359. [CrossRef]
5. Ma, X.; Liu, R.; Li, D.Y. Abrasive wear behavior of D2 tool steel with respect to load and sliding speed under dry sand/rubber wheel abrasion condition. *Wear* **2000**, *241*, 79–85. [CrossRef]
6. Masen, M.A.; de Rooij, M.B.; Schipper, D.J. Micro-contact based modelling of abrasive wear. *Wear* **2005**, *241*, 75–89. [CrossRef]
7. Stachowiak, G.W. Particle angularity and its relationship to abrasive and erosive wear. *Wear* **2000**, *241*, 214–219. [CrossRef]
8. Mbabazi, J.G.; Sheer, T.J.; Schandu, R. A model predict erozion on mild steel surfaces impacted by boiler fly ash particles. *Wear* **2005**, *257*, 612–624. [CrossRef]
9. Górka, J.; Czupryński, A. Testing of flame sprayed ZrO_2 matrix coatings containing CaO. *Appl. Mech. Mater.* **2015**, *809–810*, 501–506. [CrossRef]
10. Włosiński, W.; Chmielewski, T. Plasma–hardfaced chromium protective coatings-effect of ceramic reinforcement on their wettability by glass. In *Contributions of Surface Engineering to Modern Manufacturing and Remanufacturing 3rd International Conference on Surface Engineering*; Southwest Jiaotong University Press: Chengdu, China, 2002; pp. 48–53.
11. Czupryński, A.; Górka, J.; Adamiak, M.; Tomiczek, B. Testing of Flame Sprayed Al_2O_3 Matrix Coatings Containing TiO_2. *Arch. Metall. Mater.* **2016**, *61*, 1017–1023. [CrossRef]
12. Boncel, S.; Górka, J.; Shaffer, M.S.P.; Koziol, K. Shear-induced crystallisation of molten isotactic polypropylene within the intertube channels of aligned multi-wall carbon nanotube arrays towards structurally controlled composites. *Mater. Lett.* **2014**, *116*, 53–56. [CrossRef]
13. Chmielewski, T.; Golański, D.; Włosiński, W.; Zimmerman, J. Utilizing the energy of kinetic friction for the metallization of ceramics. *Bull. Pol. Acad. Sci. Tech. Sci.* **2015**, *63*, 201–207. [CrossRef]
14. Reisgen, U.; Stein, L.; Balashov, B.; Geffers, C. Nanophase hardfaced coatings. In *Materialwissenschaft und Werkstofftechnik*; Springer: Berlin, Germany, 2008; Volume 39, pp. 791–794.
15. Czupryński, A.; Górka, J.; Adamiak, M. Examining properties of arc sprayed nanostructured coatings. *Metalurgija* **2016**, *55*, 173–176.
16. Lisiecki, A. Titanium matrix composite Ti/TiN produced by diode laser gas nitriding. *Metals* **2015**, *5*, 54–69. [CrossRef]

17. Adamiak, M.; Tomiczek, B.; Górka, J.; Czupryński, A. Joining of the AMC composites reinforced with Ti$_3$Al intermetallic particles by resistance butt welding. *Arch. Metall. Mater.* **2016**, *61*, 847–851. [CrossRef]

18. Janicki, D. Disk laser welding of armor steel. *Arch. Metall. Mater.* **2014**, *59*, 1641–1646. [CrossRef]

19. Burdzik, R.; Konieczny, Ł.; Stanik, Z.; Folęga, P.; Smalcerz, A.; Lisiecki, A. Analysis of impact of chosen parameters on the wear of camshaft. *Arch. Metall. Mater.* **2014**, *59*, 957–963. [CrossRef]

20. Grajcar, A.; Różański, M.; Stano, S.; Kowalski, A. Microstructure characterization of laser-welded Nb-microalloyed silicon-aluminum TRIP steel. *J. Mater. Eng. Perform.* **2014**, *23*, 3400–3406. [CrossRef]

21. Chmielewski, T.; Golański, D.; Włosiński, W. Metallization of ceramic materials based on the kinetic energy of detonation waves. *Bull. Pol. Acad. Sci. Tech. Sci.* **2015**, *63*, 449–456. [CrossRef]

22. Janicki, D. High power diode laser cladding of wear resistant metal matrix composite coatings. In *Solid State Phenomena, Mechatronic Systems and Materials*; Trans Tech Publications: Zürich, Switzerland, 2013; Volume 199, pp. 587–592.

23. Chmielewski, T.; Golanski, A. New method of in-situ fabrication of protective coatings based on Fe-Al intermetallic compounds. *Proc. Inst. Mech. Eng. Part B J. Eng.* **2011**, *225*, 611–616. [CrossRef]

24. Golański, D.; Dymny, G.; Kujawińska, M.; Chmielewski, T. Experimental investigation of displacement/strain fields in metal coatings deposited on ceramic substrates by thermal spraying. In *Solid State Phenomena*; Trans Tech Publications: Zürich, Switzerland, 2015; Volume 240, pp. 174–182.

25. Poole, C.P., Jr.; Ownes, F.J. *Introduction to Nanotechnology*; Wiley: Hoboken, NJ, USA, 2003; pp. 1–8.

26. Boncel, S.; Górka, J.; Shaffer, M.S.P.; Koziol, K. 'Binary salt' of hexane-1,6-diaminium adipate and 'carbon nanotubate' as a synthetic precursor of carbon nanotube/Nylon-6,6 hybrid materials. *Polym. Compos.* **2014**, *35*, 523–529. [CrossRef]

27. Heath, G. Nanotechnology and welding–Actual and possible future applications. In Proceedings of the Castolin-Eutectic Seminar, Brussels, Belgium, 25 October 2006; pp. 15–19.

28. Sevostianov, I.; Kachanov, M. Effect of interphase layers on the overall elastic and conductive properties of matrix composites. Applications to Nanosize inclusion. *Int. J. Solids Struct.* **2007**, *44*, 1304–1315. [CrossRef]

29. Wagner, D.; Vaia, R. Nanocomposites: Issues at the Interface. *Mater. Today* **2004**, *7*, 38–42. [CrossRef]

30. Yao, J.; Zhang, Q.; Gao, M.; Zhang, M. Microstructure and wear property of carbon nanotube carburizing carbon steel by laser surface remelting. *Appl. Surf. Sci.* **2008**, *254*, 7092–7097. [CrossRef]

31. Laha, T.; Kuchibhatla, S.; Seal, S.; Li, W.; Agarwal, A. Interfacial phenomena in thermally sprayed multiwalled carbon nanotube reinforced aluminum nanocomposite. *Acta Mater.* **2007**, *55*, 1059–1066. [CrossRef]

32. Esawi, A.; Farag, M. Carbon nanotube reinforced composites: Potential and current challenges. *Mater. Des.* **2007**, *28*, 2394–2401. [CrossRef]

33. Chen, W.X.; Tu, J.P.; Wang, L.Y.; Gan, H.Y.; Xu, Z.D.; Zhang, X.B. Tribological application of carbon nanotubes in a metal-based composite coating and composites. *Carbon* **2003**, *41*, 215–222. [CrossRef]

 *materials*

Article

Deposition of Stainless Steel Thin Films: An Electron Beam Physical Vapour Deposition Approach

Naser Ali [1,2,*], **Joao A. Teixeira** [1], **Abdulmajid Addali** [1], **Maryam Saeed** [2], **Feras Al-Zubi** [2], **Ahmad Sedaghat** [3,4] **and Husain Bahzad** [5]

[1]   Cranfield University, School of Aerospace, Transport and Manufacturing (SATM), MK430AL Cranfield, England, UK; j.a.amaral.teixeira@cranfield.ac.uk (J.A.T.); a.addali@cranfield.ac.uk (A.A.)
[2]   Nanotechnology and Advanced Materials Program, Energy and Building Research Center, Kuwait Institute for Scientific Research, Safat 13109, Kuwait; msaeed@kisr.edu.kw (M.S.); fzubi@kisr.edu.kw (F.A.-Z.)
[3]   Department of Mechanical Engineering, Isfahan University of Technology, Isfahan 84156-83111, Iran; a.sedaghat@ack.edu.kw
[4]   Department of Mechanical Engineering, The Australian College of Kuwait, Safat 13015, Kuwait
[5]   Imperial College London, Department of Chemical Engineering, SW72AZ London, UK; hb3015@ic.ac.uk
[*]   Correspondence: naser.ali@cranfield.ac.uk or nmali@kisr.edu.kw

Received: 4 January 2019; Accepted: 4 February 2019; Published: 14 February 2019

Abstract: This study demonstrates an electron beam physical vapour deposition approach as an alternative stainless steel thin films fabrication method with controlled layer thickness and uniform particles distribution capability. The films were fabricated at a range of starting electron beam power percentages of 3–10%, and thickness of 50–150 nm. Surface topography and wettability analysis of the samples were investigated to observe the changes in surface microstructure and the contact angle behaviour of 20 °C to 60 °C deionised waters, of pH 4, pH 7, and pH 9, with the as-prepared surfaces. The results indicated that films fabricated at low controlled deposition rates provided uniform particles distribution and had the closest elemental percentages to stainless steel 316L and that increasing the deposition thickness caused the surface roughness to reduce by 38%. Surface wettability behaviour, in general, showed that the surface hydrophobic nature tends to weaken with the increase in temperature of the three examined fluids.

Keywords: coating; controlled deposition rate; EB-PVD; morphology; topography; wettability

---

1. Introduction

Stainless steels are passive alloys, which due to their chemical composition tend to form a thin oxide layer that inhibits the metal dissolution in corrosive environments [1]. Physical, mechanical, and anticorrosive properties of the alloy are highly related to its microstructure, where one or two phases (i.e., austenitic, ferritic, or both) may be formed [2]. Due to their unique properties, including adaptation to changes in solution salinity and pH level, these alloys are widely used in application areas such as construction and building [3], heat exchangers [4], and biomedicine [5]. In addition to these application areas, stainless steel (SS) in its powder form was reported to be used in fabricating nanofluids [6,7], which are heat transfer fluids; and surface coatings, via the cold spray deposition method [8]. Since the current trend in engineering industry, such as the automotive sector, is to rely on light construction materials (e.g., aluminium and its alloys) in order to reduce the overall weight of constructions and manufactured parts, hence SS surface coatings on metals are considered to be a promising solution for achieving this target while providing anticorrosion and wear resistance to the bulk material [9,10]. So far, all reported deposition procedures of SS films are seen as adaptation of cold gas dynamic spraying of premixed powders onto the surface [9,11], wire feedstock melt down on surfaces via electron beam solid freeforming (EB-SFF) [12,13], ionic sputtering of a target source [14–21], thermal evaporation of a source and ionic bombardment of the particles by ion beam assisted deposition (IBAD) approach [22], and pulsed laser evaporation technique [23,24]. However, some of these routes can be incompatible for industrial usage because of the lack in precision of controlling the deposited layer thickness, the thin film can be associated with contamination, and the cold spray deposited particles and/or its coated surface can suffer from intensive plastic deformation. Furthermore, the aforementioned methods raise processability and cost concerns due to the large number of parameters involved in the coating procedure. For example, when employing cold gas dynamic spraying approach, parameters such as the nozzle dimensions, jet velocity, particles size, and particles impact temperature need to be considered cautiously before starting the process.

On the other hand, electron beam physical vapour deposition (EB-PVD), which is a high vacuum thermal coating technology, is considered to be a simple and relatively cheap process in which a focused high energy electron beam is directed towards melting an evaporant material inside a vacuumed chamber. The evaporating material is then condensed on the surface of a substrate or component to form the film layer [25]. The distinct advantages of this approach are the high deposition purity, enlarged coating area, precise film thickness, in-situ growth monitoring, and smoothness control [26]. In addition to the associated benefits, the aforementioned technique has proven its capability of depositing alloys, as demonstrated by Almeida et al. [27] with their MCrAlY film fabrication study. On the industrial scale, EB-PVD has been widely employed for coating materials, including SS bulk materials, but to the authors of this article's knowledge, has never been reported as the means to deposit SS thin films [25,28].

Herein, we demonstrate the deposition of SS thin films on metallic substrates using an EB-PVD approach. The present study, based on the conducted literature review, is the first reported EB-PVD process for forming SS films and does not aim to challenge other film fabrication processes but rather unlocks opportunities for new ways of depositing SS thin films. Furthermore, to illustrate the crucial role of the controlled deposition rates on the uniformity and elemental distribution within the fabricated thin layer, a comparison between the morphologies of the SS films, of 150 nm, coated on copper (Cu) substrates with fixed deposition rates of 0.05 Å/s to 1.45 Å/s was performed. The main reasons behind selecting Cu as the hosting substrate is due to: (1) the fact that Cu is not part of the forming elements of the SS 316L alloy and can therefore be easily diverted from the deposited film when elemental characterisations is performed, and (2) most commercial heat pipes are made of Cu, because of the material high thermal conductivity, but usually faces erosion damages from water flows [29]; so providing an insight into SS coatings on Cu would be desirable for the industry since it can help reduce such common phenomena with minimum degradation effects on the heat transfer properties of the bulk material. Moreover, since there is still a need for further investigation and clarification of the wettability behaviour of SS 316L surfaces and the effect of different parameters on their wetting phenomena [30], the impact of the 0.05 Å/s as-prepared films on the surface topography and water wettability behaviour was explored for 50, 100, and 150 nm SS layers coated on SS 316L substrates. The expected applications that can benefit from this study include, but are not limited to, medical equipment, automotive parts, and heat transfer devices.

2. Experimental

2.1. Materials

All chemicals were used as-received from the manufacturer without further purification. Acetone ($CH_3COCH_3 \geq 99.5\%$) grade ACS reagent and hydrochloric acid (HCl ~37%) grade ACS reagent were purchased from SIGMA-ALDRICH, and sodium hydroxide pellets (NaOH ~98%) grade AR were purchased from LOBA Chemie (Mumbai, India). Stainless steel AISI 316L bearing balls, of grade 100 and 8.5 mm diameter, were purchased from Bearing Warehouse Limited (Sheffield, UK). Thirteen square shaped substrates, divided into: (1) four of 3 cm² surface area and 0.6 mm thickness stainless steel 316L (supplied by YC Inox Co., Chang-Hwa, Taiwan), (2) four of 0.5 cm² surface area and 0.6

mm thickness stainless steel 316L (supplied by YC Inox Co.), and (3) five of 1 cm^2 surface area and 0.127 mm thickness copper of 99.9% purity (supplied by Precision Brand Products, Downers Grove, IL, USA), were manufactured using a computer controlled machine. The substrates were then cleaned with acetone, using a bath type Soniclean limited 250TD ultrasonicator (Thebarton, Australia), for 15 min at room temperature then carefully wiped to remove any remaining residuals. Three litres of deionised water (DIW), of pH 6.11, produced by an Elga PR030BPM1-US Purelab Prima 30 water purification system (Buckinghamshire, UK) was used after being divided into 3 sets of 1 L's, then their pH value at 25 °C were adjusted, via a HACH HQ11D portable pH meter (Loveland, CO, USA) of 0.002 pH accuracy, to 4, 7, and 9, respectively.

2.2. X-ray Fluorescence and X-ray Diffraction Characterisation

Elemental analysis of the fabricated substrates was performed three times and averaged using a BRUKER TITAN S1 X-ray fluorescent (XRF) handheld analyser (Coventry, UK) to ensure that the bulk components in the manufactured substrates match the composition standards. This was done by placing the substrate on a working station then adjusting the lens of the XRF device vertically on the substrate before starting the measurements, which required 10 s to complete for each single measurement. Moreover, X-ray diffraction (XRD) analysis was performed using a 9 kW Rigaku SmartLab, Tokyo, Japan, XRD device that utilizes a CuKα X-ray source at a diffraction angle of 2θ and an incidence beam angle of 0.02°. This was done in order to identify the Bragg's peaks of each element contained within the substrates and to define the phase of the stainless steel alloy used. The diffraction scanning angle ranged from 20° to 80°, with a scanning rate of 2 °/min.

2.3. Stainless Steel Film Deposition

The thin film production method used consists of fabrication at constant deposition rates, as commonly seen in literature [31], and is demonstrated in Figure 1. Eleven of the thirteen substrates were individually placed inside the electron beam physical vapour deposition (EB-PVD) device chamber after being tightly adjusted on the sample holder and were screwed vertically above the evaporation source. The remaining two stainless steel 316L substrates were kept as references for characterisation purposes. Stainless steel AISI 316L bearing balls, which were used as the deposition source, were placed in an 8 cm^3 graphite crucible located at the bottom of the EB-PVD chamber, thus having a fixed target-to-sample distance of 26 cm. The EB-PVD device chamber (40 cm inner diameter × 50 cm inner height) was then vacuumed to a pressure of 6 × 10^{-6} Torr to ensure the removal of all particle contaminations within it, and the film thickness was controlled via an INFICON SQC-310 electronic thickness monitor system (Bad Ragaz, Switzerland) connected to a sensor located inside the chamber. It is worth noting that there was no external heating or cooling applied to the substrates temperature during the film fabrication process. In the case of the copper substrates, the deposition source was evaporated at a set of starting power percentages after which the deposition rates were maintained so that a film layer of 150 nm thick can be achieved. The starting power percentages employed were of 3, 4, 6, 8, and 10%, and the maintained deposition rates were of 0.05, 0.16, 0.82, 1.07, and 1.45 Å/s, respectively. It is worth noting that power percentages less than 3% had no trace of evaporation and that power percentages higher than 12% are restricted by the manufacturer of the EB-PVD device due to safety concerns. As such, the range of power percentage was selected to be from 3% to 10%, with a maximum deviation of ±1% to sustain the deposition rate. As for the stainless steel 316L substrates, based on the elemental characterisation of the film coated on the copper substrates, the evaporation was selected to be at the lowest deposition rate (i.e., 0.05 Å/s) for a set of film thickness of 50, 100, and 150 nm. After the completion of each of the aforementioned particle deposition processes, the substrate was kept in the chamber for 4 h to cool down before removal from the EB-PVD chamber for further analysis.

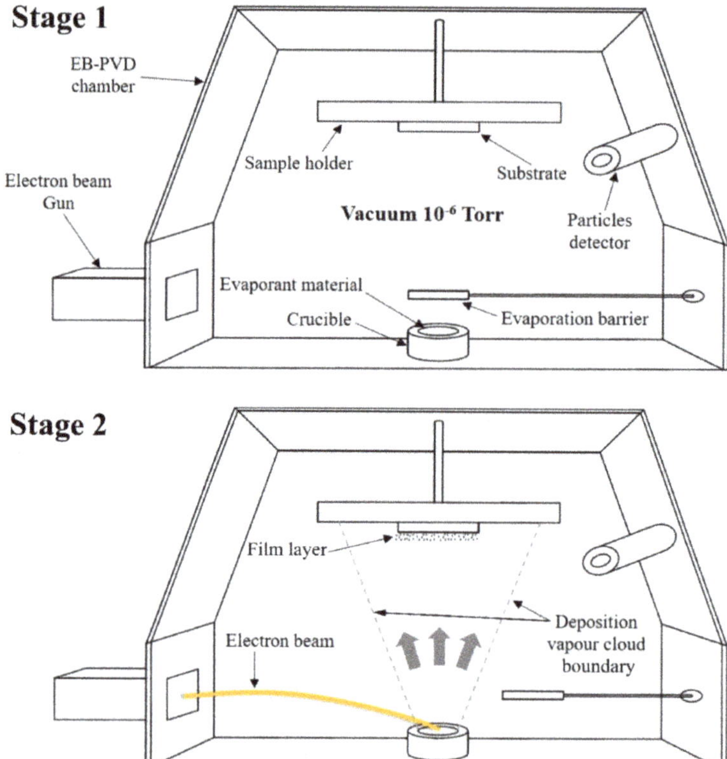

Figure 1. Electron beam physical vapour deposition process, where (Stage 1) shows the schematic illustration of the device configuration, and (Stage 2) demonstrates the source evaporation and film formation process.

2.4. Scanning Electron Microscopy and Elemental Mapping

The surface microstructure and the elemental mapping of the chemical composition of the 150 nm stainless steel film coated on the copper substrates at fixed deposition rates (i.e., from 0.05 Å/s to 1.45 Å/s); and the evaporated source before and after 150 nm film deposition at a 0.05 Å/s were studied using a JEOL JSM-6010LA InTouchScope™ scanning electron microscopy (SEM, Tokyo, Japan) device that is equipped with an integrated energy dispersive x-ray spectroscopy (EDS) analyser and operates via the InTouchScope 1.12 software. All SEM images were recorded by the secondary electron mode from the surface region of the samples and then recorded at different magnifications. Elemental distribution and percentages were obtained by the EDS analyser at a process real time of 100 s. Both SEM and EDS analysis were conducted at a working distance of 10 mm and an accelerating voltage of 20 kV, to reduce any possible damage to the tested samples. It is important to note that elements such as carbon, oxygen, and copper were excluded from the EDS elemental composition due to the presence of carbon in the adhesive tape used for mounting the samples into the device, traces of oxygen can remain in the chamber even at a high vacuum condition while copper was the tested substrate beneath the thin film, this being our area of interest.

2.5. Atomic Force Microscopy

Topography images of the coated and uncoated stainless steel 316L substrates were recorded at room temperature, using a PicoView 1.14.4 software (Woburn, MA, USA), at 10 μm², 1024 × 1024 pixels, and 0.84 line/s via an Agilent Technologies 5600LS atomic force microscopy (AFM, Santa Clara, CA, USA) instrument, equipped with a 90 μm N9521A multipurpose scanner, in tapping mode. NANOSENSORS^{TM} silicon tips (type: PPP-CONTPt-20; resonance frequency 6–21 kHz) were employed for the characterisation. Data analysis was performed, with a Pico Image Basic 6.2 software (Chandler, AZ, USA), via first enabling the gaussian filter with 0.25 μm^2 cut-off feature, for the background corrections, then having the software calculate the main height parameters and particles height distribution of the samples.

2.6. Deionised Water Properties Measurements and Theoretical Calculation

Deionised water, of pH 4, 7, and 9, kinematic viscosity and density changes with temperature were characterised at a temperature range from 20 °C to 60 °C, via an Anton Paar DMA 4500M density meter (Graz, Austria) of accuracy 5×10^{-5} g/cm^3 and a PAC Herzog HVM 472 multirange viscometer device (Houston, TX, USA), respectively. Both of the aforementioned devices have built-in calibration systems that are initiated before starting the measurements. The variation in pH value, of the DIW's, with temperature was obtained from previously published results [32,33], for pH 7 and 9, and theoretically calculated for pH 4, for the same range of temperatures, using the following equation [34,35]:

$$\mathrm{pH}_T = \mathrm{pH}_{25\,°C} + [(T - 25\,°C) \times \text{solution temperature coefficient}] \tag{1}$$

where pH_T, $\mathrm{pH}_{25\,°C}$ and T are the solution pH value for an examined temperature, the solution pH value at 25 °C, and the examined temperature in Celsius, respectively. The solution temperature coefficient of DIW of pH 4 was selected to be -0.004 pH/°C, based on the extrapolation of the available data of the two previous DIW's (i.e., pH 7 and 9).

2.7. Surface Wettability Characterisation

Three 1 mL Hamilton 1000 series syringes, containing DIW of pH 4, 7, and 9, were adjusted to a Dataphysics SHD syringe temperature controller (Reno, NV, USA), which is integrated with the Dataphysics OCA 100 automatic multi-liquid dispenser contact angle goniometer device (San Jose, CA, USA). This was done in order to increase/decrease the liquid temperature inside the syringes, while being monitored. The surface wettability of the coated and uncoated stainless steel 316L samples was measured using the Sessile drop method [36]. This was done by dispersing a 5 μL droplet (5 μL/s dosing rate) on the examined surface then capturing its image, at static condition, which is analysed afterwards through the device provided software SCA 20 to obtain the liquid – surface CA. The CA measurements were conducted three times, for the three types of liquids, at a fixed liquid temperature of 20 °C to 60 °C, with a $\pm0.1°$ contact angle precision.

3. Results and Discussion

3.1. Substrates Elemental Analysis

The average elemental content of the manufactured SS 316L and Cu substrates, which were each examined three times by the XRF, are given in Table 1. In addition, the XRD pattern corresponds well with the XRF analysis as suggested by the sharp diffraction Bragg's peaks shown in Figure 2a,b, where the SS substrate (Figure 2a) showed peaks at $2\theta = 43.6°$, $50.9°$, and $74.7°$ corresponding to the planes (111), (200), and (220) austenite gamma phase; and the Cu substrate (Figure 2b) illustrated diffraction peaks at $2\theta = 43.2°$, $50.4°$, and $74.1°$ corresponding to the planes (111), (200), and (220) of pattern (PDF Card No.: 03-065-9026). Hence, this confirms that the bulk formation of the as-prepared substrates used for the deposition experiments are of Cu and SS 316L in its austenite phase.

Table 1. Averaged XRF elemental analysis of SS 316L and Cu substrates.

	Stainless Steel 316L					Copper			
Element	**Content (%)**	**Max. (%)**	**Min. (%)**	**+/− Error (%)**	**Element**	**Content (%)**	**Max. (%)**	**Min. (%)**	**+/− Error (%)**
Iron	69.22	75	60	0.5	Copper	99.82	100	90	0.39
Chromium	16.78	18	16	0.18	Zirconium	0.02	-	-	0.01
Nickel	10.12	14	10	0.21	-	-	-	-	-
Manganese	2.02	3	2	0.04	-	-	-	-	-
Molybdenum	1.32	2	0	0.11	-	-	-	-	-
Silicon	0.27	1	0	0.05	-	-	-	-	-

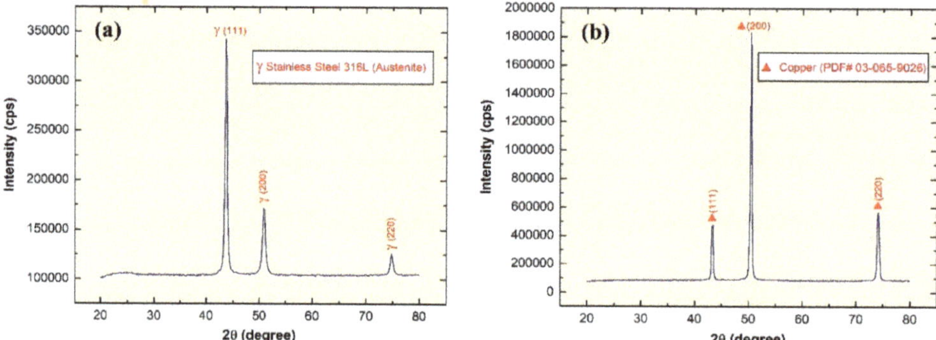

Figure 2. X-ray diffraction pattern of: (**a**) SS 316L substrate, and (**b**) Cu substrate.

3.2. Deposition Morphology

Fabricated samples of 150 nm SS deposited on Cu substrates, which were synthesised by the EB-PVD starting electron beam powers of 3% to 10%, are shown in Figure 3. The SS thin film layer produced can be visually observed (Figure 3(b2–b6)), except for the corners where the adhesive tapes were placed to adjust the samples on the sample holder. The time required for achieving the film thickness varied from 500 min, at a fixed deposition rate of 0.05 Å/s, down to 17.25 min, via a 1.45 Å/s controlled evaporation rate. Samples were then characterised to determine the effect of the controlled deposition rates on the film and evaporant material (Figure 3(c1–c6)) morphologies, via the SEM and EDS analysis. The SEM characterisation of the 0.05 Å/s and 0.16 Å/s fabricated samples has demonstrated a well-constructed film structure as shown in Figure 3(a1,a2) (alternatively, Supplementary Figure S1(a1,a2)), whereas higher deposition rates have resulted in the development of non-uniform films with partial detachments from the surface as seen in Figure 3 (alternatively, Supplementary Figure S1(a3–a5)). Both observations can be linked to the phase of the SS film formed due to the elemental ratio of the deposit, where a well-constructed film structure indicates a ferritic phase and a semi-detached structure illustrates an austenitic phase [37]. In addition, analysing the evaporant source, before and after 0.05 Å/s deposition, has shown a local melt down in the material, due to the electron beam being focused on a single location at a low power (i.e., the beam is fixed and does not follow a movement path across the evaporant source), thus resulting in surface microstructural changes as shown in Figure 3(d1–d4) (alternatively, Supplementary Figure S1(b1–b4)). Further inspection of the samples using the EDS device (Figure 4) has illustrated that, unlike higher controlled deposit rates, the film fabricated via 0.05 Å/s was within the SS 316L elemental composition acceptable ranges, except for the chromium, manganese, and nickel, which showed partial divergences of +3.33%, +1.15%, and −5.24%, respectively. This indicates that the film is indeed SS but of a different category [38–40], whereas to obtain SS 316L thin films the evaporant source would require some modification in its composition (i.e., fabricate our own EB-PVD evaporant source). In addition, the film elemental distribution of the fixed 0.05 Å/s deposition was seen to be uniformly distributed along the substrate surface as shown in Figure 5 (alternatively, Supplementary Figure S2a–b). It is important to note that due to the limitation of the EDS device used, elements of less than 0.5% of the composition mass could not be elementally mapped, as seen in the case of molybdenum. The EDS elemental composition percentages of the 150 nm SS deposited film at a fixed rate of 0.05 Å/s are tabulated in Table 2.

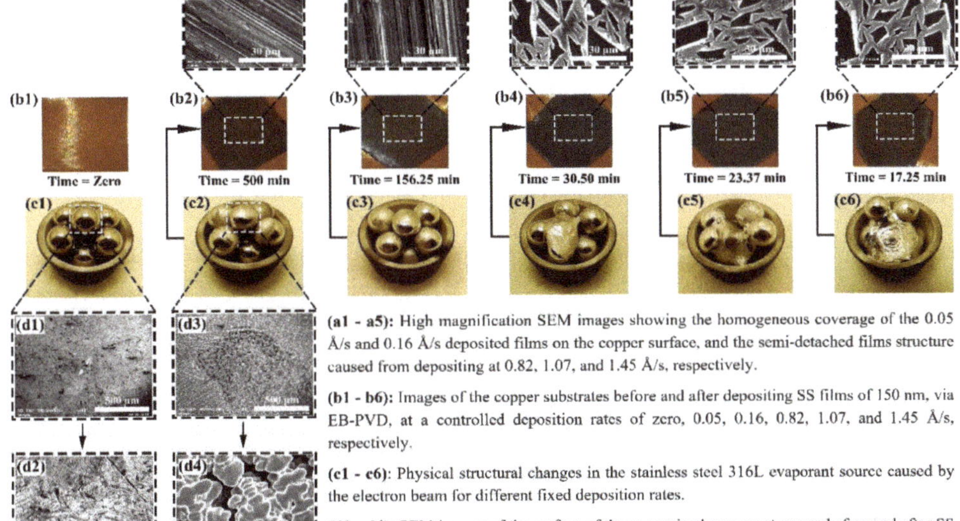

(a1 - a5): High magnification SEM images showing the homogeneous coverage of the 0.05 Å/s and 0.16 Å/s deposited films on the copper surface, and the semi-detached films structure caused from depositing at 0.82, 1.07, and 1.45 Å/s, respectively.

(b1 - b6): Images of the copper substrates before and after depositing SS films of 150 nm, via EB-PVD, at a controlled deposition rates of zero, 0.05, 0.16, 0.82, 1.07, and 1.45 Å/s, respectively.

(c1 - c6): Physical structural changes in the stainless steel 316L evaporant source caused by the electron beam for different fixed deposition rates.

(d1 - d4): SEM images of the surface of the as-received evaporant source before and after SS film fabrication at 0.05 Å/s.

Figure 3. Characterisation of the Stainless steel evaporant source and deposited thin films, where (**a1–a5**) shows the SEM images of the 0.05–1.45 Å/s as-deposited films structure, (**b1–b6**) illustrates the copper substrates before and after SS deposition, (**c1–c6**) demonstrates the physical changes in evaporant source caused by different deposition rates, and (**d1–d4**) shows the SEM images of the evaporant source before and after 0.05 Å/s film deposition.

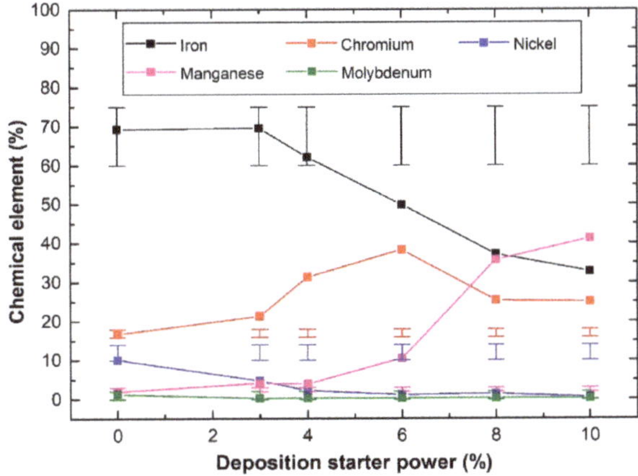

Figure 4. EDS characterisation of the chemical elemental percentages of the evaporant source (0% power), and the SS thin films deposited with a starting beam power of 3% up to 10%. The bars at the top and bottom of each data point indicate the maximum and minimum range of each element percentage of the stainless steel 316L composition and was attained from the XRF device installed database.

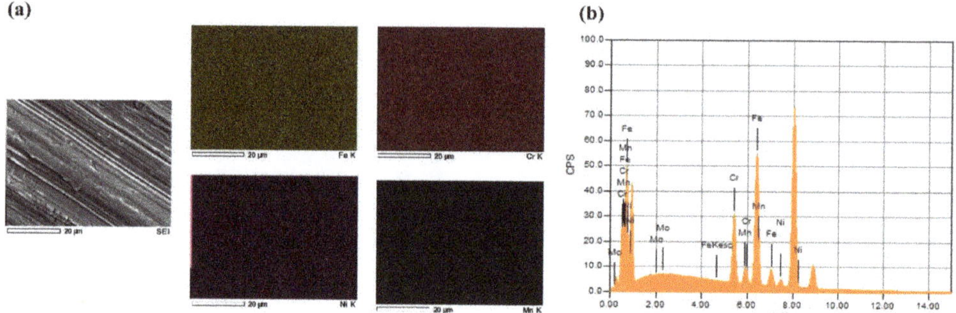

Figure 5. EDS elemental analysis, where (**a**) is the SEM image and its elemental maps of the characterised 150 nm deposited SS film at 0.05 Å/s on Cu substrate, and (**b**) demonstrates the EDS X-ray spectrum of the elements within the film shown in (**a**).

Table 2. EDS elemental composition percentage of the 150 nm SS fabricated film at 0.05 Å/s controlled deposition rate.

Element	Mass %	Atom %	Sigma	Net	K ratio
Iron	69.52	68.62	0.05	2879132	0.3919295
Chromium	21.33	22.61	0.03	1347161	0.1368063
Nickel	4.76	4.47	0.03	131065	0.0250911
Manganese	4.15	4.16	0.03	201285	0.0236898
Molybdenum	0.25	0.14	0.03	12922	0.0009314
Total	100	100	-	-	-

3.3. Surface Topography

Surface topography analysis of the uncoated and coated SS 316L samples, which were conducted using an AFM device, is shown in Figures 6 and 7 (alternatively, Figure 3a–d). The experimental results have revealed that the nanostructures on the uncoated surface have a range of height between 87.3 to 204 nm (Figure 7a), with almost 47.5% of the structure height being in the range of 116 to 130.5 nm. Moreover, the maximum height of surface (MHS), which was obtained by adding up the surface maximum peak height and maximum valley depth, and root mean square roughness (RMSR) of the uncoated sample, were shown to be 291 nm and 12 nm, correspondingly, as demonstrated in Figure 6. On the other hand, as the deposited layer thickness increased (Figure 7b–d), the structure height on the surface, RMSR, and MHS were seen to reduce, reaching values between 35.9 to 83.7 nm, 6.86 nm, and 120 nm, respectively. Furthermore, the degree of symmetry of the surface heights about the mean plane was also seen to improve with the deposited film thickness, as the obtained skewness (Ssk) values of the measured samples (Table 1) were shown to move closer to the mean plane (i.e., zero) with the increase in fabricated layer thickness. It is worth noting that the sign of Ssk represents the predominance of the comprising surface peaks (Ssk > 0) or valley structures (Ssk < 0). The aforementioned changes in surface conditions can be attributed to the deposited particles occupying the vacant spaces on the surface structure, which consist of valleys, hills, and micro gaps, leading to the height variation on the surface to narrow down [41]. Moreover, the presence of inordinately high peaks or deep valleys was also found on the examined substrates, as indicated by the surface kurtosis (Sku) values, where Sku > 3.00 suggests the existence of a peaks/valleys defect on the surface and Sku < 3.00 illustrates a lack thereof (i.e., insufficient surface information). Such an observation is not surprising as it is commonly present on most surfaces [42]. The average roughness values were found to be 7.87 nm, 7.48 nm, 6.00 nm, and 4.88 nm for the uncoated, 50 nm, 100 nm, and 150 nm coated substrates, correspondingly. These results confirmed the smoothening effect caused by the increase in EB-PVD deposited film thickness on the SS 316L substrate surface. The roughness results can also be used as a general indication of the corrosion behaviour, as it has been reported by other authors that decreasing the surface roughness of passive alloys tends to reduce the pitting susceptibility and corrosion rate [43,44]. The height parameters values obtained from the AFM analysis of the samples can be seen in Table 3.

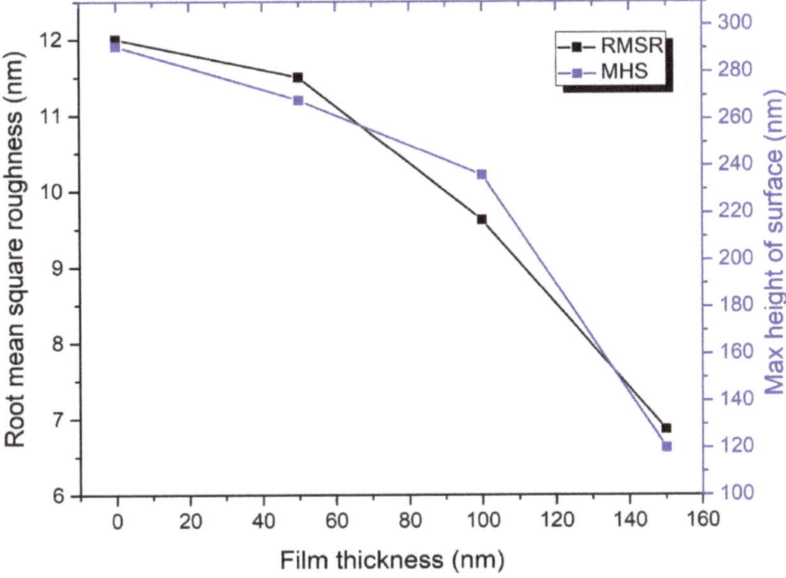

Figure 6. Root mean square roughness and maximum height of surface variation with deposition thickness on SS 316L substrates.

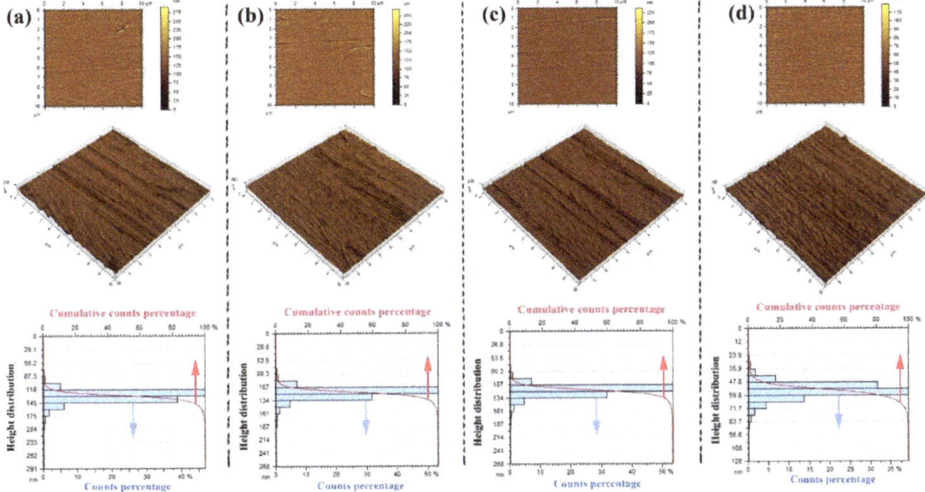

Figure 7. Surface topography analysis of SS films on SS 316L substrates, where (**a**) 2D and 3D rendered AFM topograph and height distribution of the surface of the uncoated SS 316L substrate, and (**b**–**d**) 2D and 3D rendered AFM topograph after 50, 100, and 150 nm SS deposition on substrates and their height distribution.

Table 3. Height parameters of the AFM analysis of the uncoated, 50 nm, 100 nm, and 150 nm coated SS substrates.

Height Parameters	Film Thickness (nm)			
	0	50	100	150
Root mean square roughness (nm)	12	11.5	9.62	6.86
Skewness	−1.08	−1.0	−0.654	−0.535
Kurtosis	16.8	18	19.6	6.39
Maximum peak height (nm)	131	119	119	55.4
Maximum valley depth (nm)	160	149	118	64.1
Maximum height of surface (nm)	291	268	236	120
Average roughness (nm)	7.87	7.48	6.0	4.88

3.4. Deionised Water Properties Variation with Temperature

The deionised waters used were selected to have a pH of 4, 7, and 9 to observe the acidity, neutrality, and alkalinity of the liquid effect on the surfaces wettability. Analyses results of the changes in properties, namely, kinematic viscosity (v), density (ϱ), and pH value of the three liquids, within our temperature range, are shown in Figure 8a. Comparing the ϱ and v characterisation outcomes, of the as-prepared DIW of pH 7, with the available data on pure water in literature [45] has shown a deviation of 0.015% and 3.67%, respectively, thus verifying the measurement approach conducted. Moreover, regardless of the examined DIW pH value, the as-fabricated liquids v and ϱ were seen to have a negligible difference in their values at each investigated point of temperature. For example, at 30 °C, the DIW ϱ_{pH4}, ϱ_{pH7} and ϱ_{pH9} ad an outcome of 0.99564, 0.99562, and 0.99563 g/cm^3, respectively. In contrast, manipulating the temperature was seen to have a notable influence on all three properties of the DIW's (i.e., v, ϱ, and pH value), as demonstrated in Figure 8a. This can be explained by the fact that v is inversely related to the ϱ, and that ϱ is a representation of substance mass to its volume, where at a constant volume, the mass is influenced by the bonds distance of the molecules and their forming atoms. Our as-prepared DIW's consist of four types of bonds: (1) Polar covalent bond between a single or a pair of hydrogen atoms and one atom of oxygen, (2) Dative covalent bond between a single atom of H$^+$ and a H$_2$O molecule, (3) hydrogen bond between the oxygen atom of a H$_2$O molecule and a hydrogen atom of a neighbouring H$_2$O molecule, and (4) Ion–dipole interaction between the H$_2$O molecules and a Cl$^-$ atom (e.g., DIW of pH 4) or Na$^+$ atom (e.g., DIW of pH 9). The four previous bounds are shown in Figure 8b–d. Based on the obtained data, it is believed that at a point of temperature, the bond distances of the newly introduced dative covalent bond (pH 4) and ion – dipole interaction (pH 9) are very close in distance to the other two initially existing bonds in neutral water, causing this neglectable changes in the liquid mass. On the other hand, raising the temperature weakens all four bonds, because of the increase in molecular vibrations, causing the bonds distance to widen; and hence the liquid mass reduces and becomes more acidic due to the release of H$^+$ and growth in its concentration [46]. The different formed reactions in our as-prepared DIW's, based on the Bronsted–Lowry theory of acids and bases [47], are demonstrated in Equations (2)–(4) as the following:

$$\text{pH 4: HCl} + \text{H}_2\text{O} \rightleftharpoons \text{H}_3\text{O}^+ + \text{Cl}^- \text{ (Reverse reaction)} \tag{2}$$

$$\text{pH 7: } 2\text{H}_2{}^+ + \text{O}_2{}^- \rightarrow 2\text{H}_2\text{O} \text{ (Chemical reaction)} \tag{3}$$

$$\text{pH 9: NaOH} + \text{H}_2\text{O} \rightarrow \text{Na}^+ + \text{OH}^- + \text{H}_2\text{O} + \text{Heat (Exothermic reaction)} \tag{4}$$

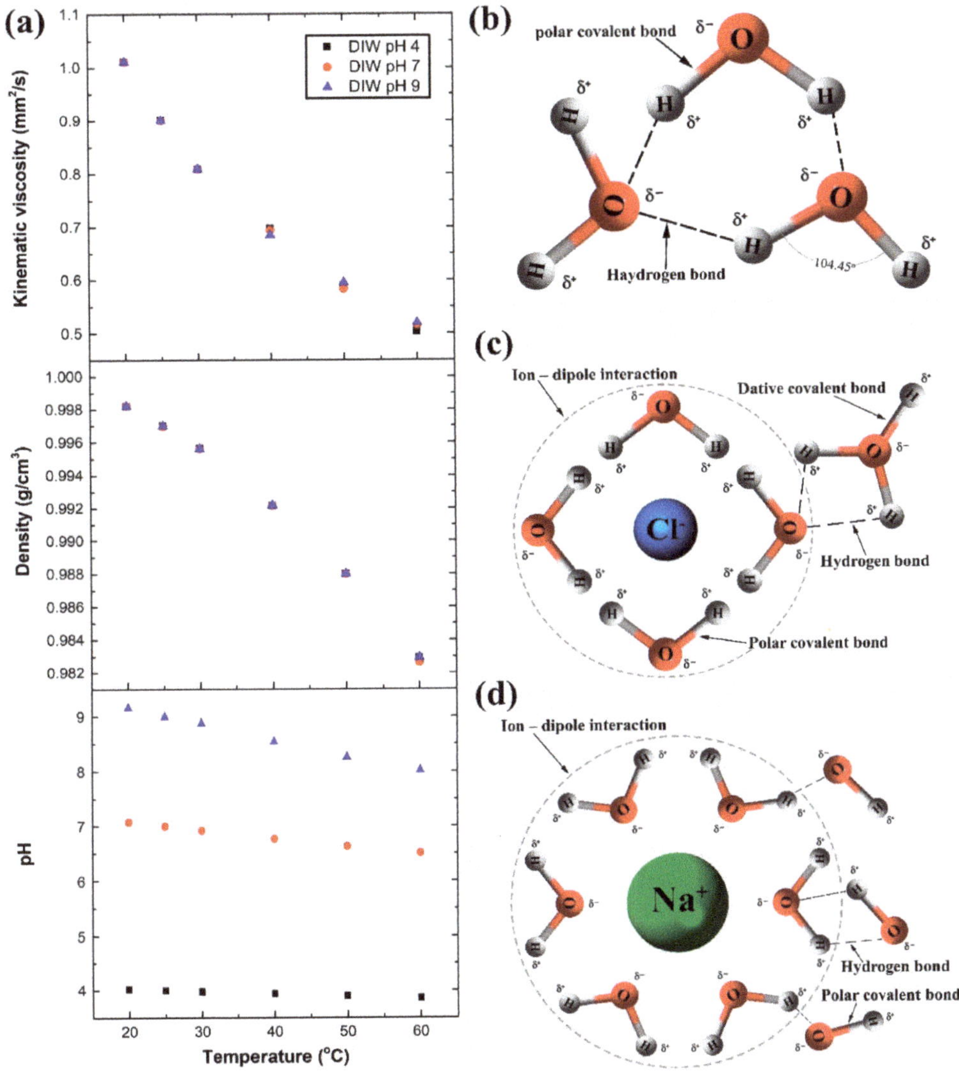

Figure 8. Water atoms and molecules bonds, and properties variation with temperature, where (**a**) shows the DIW's kinematic viscosity, density, and pH value changes with temperature, and (**b–d**) illustrates the bonds in water of pH 7, 4, and 9, respectively.

3.5. Contact Angle Measurement

Examining the wettability of the uncoated SS 316L sample (Figure 9a), with 20 °C of DIW of pH 7, showed that the surface had an average contact angle (ACA) of 131.7°, which illustrates a hydrophobic behaviour. The high ACA value is believed to be linked to the substrate Cassie–Baxter state via its surface roughness, as reported by other authors [30,48]. In general, there are two common texture states that explain the relationship between surface wettability and roughness, which are the Cassie-Baxter [49] and Wenzel [50] states. In the Cassie-Baxter state, the surface pores and valleys tend to trap the air, which leads to a reduction in the degree of liquid-surface interaction. On the other hand, in the Wenzel state, the liquid fully occupies the pores, thus improving the surface wettability. Applying DIW's of pH 4 and 9 have led the surface ACA to reduce to 124.9° and 117.4°, respectively, without additional surface modifications. Furthermore, it was found that raising the temperature of the as-prepared DIW's tends to weaken the hydrophobic nature of the uncoated surface, as demonstrated by the obtained data in Figure 9b–d. The grey dashed line in the plots (Figure 9b-d) illustrates the transition point between the surface hydrophobic (top) and hydrophilic (bottom) regions. Moreover, the deposition thickness was seen to be inversely related to the CA, where increasing the film thickness caused the ACA to reduce. For example, when examining DIW, of 20 °C and pH of 4, the ACA of the uncoated, 50 nm, 100 nm, and 150 nm film gave angles of 124.9°, 119.5°, 116.7°, and 110.9°, correspondingly. This can be attributed to the reduction in surface micro-roughness and air pockets formation at the interface between the substrate and liquid as a result of the deposited particles occupying the surface structure, thus enhancing the substrate surface energy to attract the liquid towards the surface (i.e., reducing the CA) [49,51]. In addition, the level of decrease in CA is seen to correspond to the liquid temperature, pH value, and fabricated film thickness due to their ability to modify the surface mode from a Cassie–Baxter state to a Wenzel state, and vice versa. For instance, the ACA of DIW of pH 4, 7, and 9, at 50 °C, showed a decrease from 110.1°, 114.8°, and 112.3° (uncoated) to 93.1°, 95.5°, and 97.0° (150 nm film), respectively. It was also possible to change the substrate surface wettability nature from hydrophobic to hydrophilic by manipulating the three aforementioned parameters, as shown in Figure 9b when investigating the 150 nm coated substrate with DIW of 60 °C and pH 4. Such findings are very attractive for heat transfer applications, as lowering the CA can enhance the heat transfer efficiency of SS by providing larger contact area between the liquid and the surface. On the other hand, the fluctuation in the data trend of DIW of pH 4 and 9 across the examined temperature range is believed to be caused by the free ions hosted by the liquid. Since both fluids are considered ionically unstable, attempting to change the surface functional group of the substrate from hydrophobic toward hydrophilic by inducing the transformation in surface charge, in what is known as the Hofmeister series reversal effect, could be the reason behind this kind of behaviour in the data trend [52]. In addition to the previously mentioned reason, there is a possibility that traces of hydrocarbon contamination from the surrounding atmosphere on the substrate surface are strongly interacting with the two aforementioned liquids free ions, since the DIW of pH 7 did not show such fluctuation in its data trend. Usually, such contamination is unavoidable in open atmospheric experiments, and would have some sort of influence on all the conducted measurements [53,54]. Supplementary Table S1 summarises the testing parameters and obtained contact angles of the characterised samples.

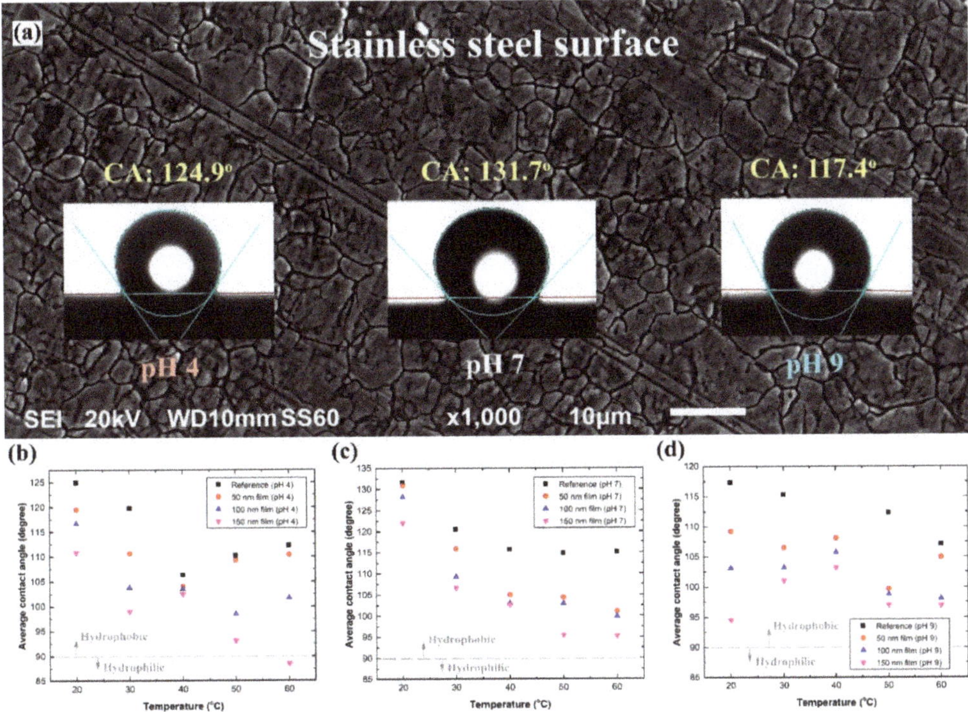

Figure 9. Effect of DIW temperature and pH value on the wettability behaviour of SS 316L surface, where (**a**) illustrates the contact angles between the 20 °C DIW's, of pH 4, 7, and 9, and the uncoated SS 316L substrate surface, and (**b–d**) demonstrates the average contact angle measurements of the uncoated and coated samples using the DIW's, at 20–60 °C, as the testing fluids.

4. Conclusions

Stainless steel films were fabricated via an electron beam physical vapour deposition method with starting electron beam power percentages of 3%–10%. The thin layers obtained with a controlled deposition rates of 0.05 Å/s and 0.16 Å/s have shown a uniform elemental distribution with a well-constructed film structure that covered the whole exposed area, while higher deposition rates have illustrated semi-detachments in the film structure. Furthermore, the closest film elemental content to SS 316L was achieved with a 0.05 Å/s, where higher deposition rates were seen to extend the maximum and minimum elemental limits of SS 316L. Surface topography of SS 316L before and after depositing 50 nm, 100 nm, and 150 nm films, using controlled 0.05 Å/s fabrication rate, was then examined. The results illustrated a reduction in structure height on the surface, MHS, RMSR, and average roughness from 87.3–204 nm, 291 nm, 12 nm, and 7.87 nm (uncoated substrate) to 35.9–83.7 nm, 120 nm, 6.86 nm, and 4.88 nm (150 nm coated substrate), respectively. It also showed, via the obtained Ssk values, that the degree of symmetry of the surface heights about the mean plane was improved by ~49.5% for the reference substrate after 150 nm film deposition. Surface wettability of the as-prepared samples were afterwards characterised with DIW's, of pH 4, 7, and 9, at a 20–60 °C liquid temperatures. The film thickness was seen to be inversely related to the liquid–surface CA, and hence the CA reduced with the increase in film thickness. Moreover, the rise in DIW's temperature has been shown to weaken the hydrophobic nature of the as-prepared substrates. It was also noticed that, unlike the DIW of pH 7, the liquids of pH 4 and 9 demonstrated some fluctuation in their CA data trend.

In summary, this article unlocks a new approach for depositing stainless steel thin films using an electron beam physical vapour deposition technique. The resulting film is ultrathin, uniform, conformal, and controllable. Moreover, an extension towards depositing different grades of stainless steel can be achieved by changing the composition of the evaporant source, based on exploratory experiments; and hence may facilitate a feasible route towards industrial usage of the process after further film properties investigation is provided (e.g., corrosivity, cohesion, hardness, and abrasiveness). Furthermore, as our approach is the first example of any stainless steel EB-PVD coating, the present work marks an important milestone in the future of stainless steel depositions on metallic surfaces and is expected to be beneficial to many applications such as medical equipment, automotive parts, and heat transfer devices.

Supplementary Materials: The following are available online at http://www.mdpi.com/1996-1944/12/4/571/s1, Figure S1: (a1) SEM image of the 0.05 Å/s deposited film; (a2) SEM image of the 0.16 Å/s deposited film; (a3) SEM image of the 0.82 Å/s deposited film; (a4) SEM image of the 1.07 Å/s deposited film; (a5) SEM image of the 1.45 Å/s deposited film; (b1) SEM image of the as-received evaporant source before film deposition; (b2) Higher resolution SEM image of the as-received evaporant source before film deposition; (b3) SEM image of the as-received evaporant source after 0.05 Å/s film deposition; (b4) Higher resolution SEM image of the as-received evaporant source after 0.05 Å/s film deposition; Figure S2: (a) SEM image and its elemental maps of the characterized 150 nm deposited SS film at 0.05 Å/s on Cu substrate; (b) EDS X-ray spectrum of the elements, Figure S3: (a) AFM images and analysis of the uncoated substrate; (b) AFM images and analysis of the 50 nm coated substrate; (c) AFM images and analysis of the 100 nm coated substrate; (d) AFM images and analysis of the 150 nm coated substrate, Table S1: Contact angle measurements data.

Author Contributions: N.A. conceived and designed all the experiments, and manufactured the substrates. N.A. and A.A. carried out all the EB-PVD film deposition. F.A.-Z. conducted the AFM characterisation experiments and their analysis. J.A.T. and H.B. conducted the XRD and XRF measurements. A.A. and A.S. prepared and characterised the density, viscosity, and change in pH value of the working fluids. M.S. conducted the SEM and EDS characterisations experiments. N.A. and F.A.-Z. conducted and analysed the contact angle measurements. The manuscript was primarily written by N.A. using the inputs provided from all the authors.

Funding: This work was financially supported by the Kuwait Institute for Scientific Research (KISR) and Cranfield University.

Acknowledgments: We acknowledge the help provided by M. Sherif El-Eskandarany, the program manager of Nanotechnology and Advanced Materials Program at KISR, for his help and support throughout the conducted work.

Conflicts of Interest: The authors declare no conflict of interest.

References

1. Landoulsi, J.; Genet, M.J.; Richard, C.; El Kirat, K.; Pulvin, S.; Rouxhet, P.G. Evolution of the passive film and organic constituents at the surface of stainless steel immersed in fresh water. *J. Colloid Interface Sci.* **2008**, *318*, 278–289. [CrossRef] [PubMed]
2. Tylek, I.; Kuchta, K. Mechanical properties of structural stainless steels. *Civil Eng.* **2014**, *4-B*, 59–80.
3. Corradi, M.; Di Schino, A.; Borri, A.; Rufini, R. A review of the use of stainless steel for masonry repair and reinforcement. *Constr. Build. Mater.* **2018**, *181*, 335–346. [CrossRef]
4. Bowden, D.; Krysiak, Y.; Palatinus, L.; Tsivoulas, D.; Plana-Ruiz, S.; Sarakinou, E.; Kolb, U.; Stewart, D.; Preuss, M. A high-strength silicide phase in a stainless steel alloy designed for wear-resistant applications. *Nat. Commun.* **2018**, *9*, 1–10. [CrossRef] [PubMed]
5. Trzaskowska, P.A.; Kuźmińska, A.; Butruk-Raszeja, B.; Rybak, E.; Ciach, T. Electropolymerized hydrophilic coating on stainless steel for biomedical applications. *Colloids Surf. B* **2018**, *167*, 499–508. [CrossRef] [PubMed]
6. Song, Y.Y.; Bhadeshia, H.K.D.H.; Suh, D.W. Stability of stainless-steel nanoparticle and water mixtures. *Powder Technol.* **2015**, *272*, 34–44. [CrossRef]
7. Ali, N.; Teixeira, J.A.; Addali, A. A Review on Nanofluids: Fabrication, Stability, and Thermophysical Properties. *J. Nanomater.* **2018**, *33*, 1–33. [CrossRef]
8. Spencer, K.; Zhang, M.X. Optimisation of stainless steel cold spray coatings using mixed particle size distributions. *Surf. Coat. Technol.* **2011**, *205*, 5135–5140. [CrossRef]

9. Sova, A.; Grigoriev, S.; Okunkova, A.; Smurov, I. Cold spray deposition of 316L stainless steel coatings on aluminium surface with following laser post-treatment. *Surf. Coat. Technol.* **2013**, *235*, 283–289. [CrossRef]

10. Vitos, L.; Korzhavyi, P.A.; Johansson, B. Stainless steel optimization from quantum mechanical calculations. *Nat. Mater.* **2002**, *2*, 25. [CrossRef]

11. Li, W.-Y.; Liao, H.; Douchy, G.; Coddet, C. Optimal design of a cold spray nozzle by numerical analysis of particle velocity and experimental validation with 316L stainless steel powder. *Mater. Des.* **2007**, *28*, 2129–2137. [CrossRef]

12. Wanjara, P.; Brochu, M.; Jahazi, M. Electron beam freeforming of stainless steel using solid wire feed. *Mater. Des.* **2007**, *28*, 2278–2286. [CrossRef]

13. Davé, V.R.; Matz, J.E.; Eagar, T.W. Electron beam solid freeform fabrication of metal parts. Available online: http://sffsymposium.engr.utexas.edu/Manuscripts/1995/1995-09-Dave.pdf (accessed on 12 February 2019).

14. Sahoo, B.; Schlage, K.; Major, J.; Von Hörsten, U.; Keune, W.; Wende, H.; Röhlsberger, R. Preparation and characterization of ultrathin stainless steel films. *AIP Conf. Proc.* **2011**, *1347*, 57–60.

15. Nomura, K.; Iio, S.; Ujihira, Y.; Terai, T. DCEMS study of thin stainless steel films deposited by RF sputtering of AISI316L. *AIP Conf. Proc.* **2005**, *765*, 108–113.

16. De Baerdemaeker, J.; Van Hoecke, T.; Van Petegem, S.; Segers, D.; Bauer-Kugelmann, W.; Sperr, P.; Terwagne, G. Investigation of stainless steel films sputtered on glass. *Mater. Sci. Forum* **2001**, *363*, 496–498. [CrossRef]

17. Kraack, M.; Boehni, H.; Muster, W.; Patscheider, J. Influence of molybdenum on the corrosion properties of stainless steel films. *Surf. Coat. Technol.* **1994**, *68–69*, 541–545. [CrossRef]

18. Eymer, J.P. On the hyperfine field of bcc 304 L stainless steel films. *J. Phys. IV France* **1992**, *2*, C3-211–C213-215.

19. Godbole, M.J.; Pedraza, A.J.; Allard, L.F.; Geesey, G. Characterization of sputter-deposited 316L stainless steel films. *J. Mater. Sci.* **1992**, *27*, 5585–5590. [CrossRef]

20. Fabis, P.M. Microporosity in 304 stainless steel films prepared by vapor quenching. *Thin Solid Films* **1985**, *128*, 57–66. [CrossRef]

21. Pedraza, A.J.; Godbole, M.J.; Bremer, P.J.; Avci, R.; Drake, B.; Geesey, G.G. Stability in aqueous media of 316L stainless steel films deposited on internal reflection elements. *Appl. Spectmsc.* **1993**, *47*, 161–166. [CrossRef]

22. Song, Y.S.; Lee, J.H.; Lee, K.H.; Lee, D.Y. Corrosion properties of N-doped austenitic stainless steel films prepared by IBAD. *Surf. Coat. Technol.* **2005**, *195*, 227–233. [CrossRef]

23. Nomura, K.; Yamada, Y.; Tomita, R.; Yajima, T.; Shimizu, K.; Mashlan, M. CEMS study of stainless steel films deposited by pulsed laser ablation of AISI316. *Czech J. Phys.* **2005**, *55*, 845–852. [CrossRef]

24. Koinkar, V.N.; Chaudhari, S.M.; Kanetkar, S.M.; Ogale, S.B. Deposition of stainless steel film using pulsed laser evaporation. *Thin Solid Films* **1989**, *171*, 335–342. [CrossRef]

25. Singh, J.; Wolfe, D.E. Review Nano and macro-structured component fabrication by electron beam-physical vapor deposition (EB-PVD). *J. Mater. Sci.* **2005**, *40*, 1–26. [CrossRef]

26. Arunkumar, P.; Aarthi, U.; Sribalaji, M.; Mukherjee, B.; Keshri, A.K.; Tanveer, W.H.; Cha, S.-W.; Babu, K.S. Deposition rate dependent phase/mechanical property evolution in zirconia and ceria-zirconia thin film by EB-PVD technique. *J. Alloys Compd.* **2018**, *765*, 418–427. [CrossRef]

27. De Almeida, D.S.; da Silva, C.R.M.; do Carmo, M. Ni Al alloy coating deposition by electron beam physical vapour deposition. In Proceedings of the 17th Brazilian Congress of Engineering and Materials Science, Paraná, Brazil, 15–19 November 2006.

28. Singh, J.; Wolfe, D.; Quli, F. Electron beam-physical vapor deposition technology: Present and future applications. In *IMAST Quarterly*; Pennsylvania State University: University Park, PA, USA, 2000; pp. 3–6.

29. Nam, Y.; Ju, Y.S. A comparative study of the morphology and wetting characteristics of micro/nanostructured Cu surfaces for phase change heat transfer applications. *J. Adhes. Sci. Technol.* **2013**, *27*, 2163–2176. [CrossRef]

30. Cai, Y.; Chang, W.; Luo, X.; Sousa, A.M.L.; Lau, K.H.A.; Qin, Y. Superhydrophobic structures on 316L stainless steel surfaces machined by nanosecond pulsed laser. *Precis. Eng.* **2018**, *52*, 266–275. [CrossRef]

31. Ali, N.; Teixeira, J.A.; Addali, A.; Al-Zubi, F.; Shaban, E.; Behbehani, I. The effect of aluminium nanocoating and water pH value on the wettability behavior of an aluminium surface. *Appl. Surf. Sci.* **2018**, *443*, 24–30. [CrossRef]

32. Light, T.S. Temperature dependence and measurement of resistivity of pure water. *Anal. Chem.* **1984**, *56*, 1138–1142. [CrossRef]

33. Down, R.D.; Lehr, J.H. *Environmental Instrumentation and Analysis Handbook*; John Wiley & Sons: Hoboken, NJ, USA, 2005.

34. Langelier, W.F. Effect of Temperature on the pH of Natural Waters. *J. Am. Water Works Assoc.* **1946**, *38*, 179–185. [CrossRef]

35. Yokogawa Electric Corporation, Technical Note TNA0924. Available online: https://web-material3.yokogawa.com/TNA0924.us.pdf (accessed on 11 February 2019).

36. Villa, F.; Marengo, M.; De Coninck, J. A new model to predict the influence of surface temperature on contact angle. *Sci. Rep.* **2018**, *8*, 6549. [CrossRef] [PubMed]

37. Tylek, I.; Kuchta, K. Physical and technological properties of structural stainless steel. *Civil Eng.* **2014**, *4-B*, 81–100.

38. Association, I.M. *Practical Guidelines for the Fabrication of Duplex Stainless Steels*; IMOA: London, UK, 2009; pp. 1–64.

39. Stainless Steel Grade Datasheets. Available online: http://www.worldstainless.org/Files/issf/non-image-files/PDF/Atlas_Grade_datasheet_-_all_datasheets_rev_Aug_2013.pdf (accessed on 12 February 2019).

40. Capus, J. 100 Years of Stainless Steel. *Met. Powder Rep.* **2013**, *68*, 12. [CrossRef]

41. Hariprasad, S.; Gowtham, S.; Arun, S.; Ashok, M.; Rameshbabu, N. Fabrication of duplex coatings on biodegradable AZ31 magnesium alloy by integrating cerium conversion (CC) and plasma electrolytic oxidation (PEO) processes. *J. Alloys Compd.* **2017**, *722*, 698–715.

42. King, T.G. rms skew and kurtosis of surface profile height distributions: Some aspects of sample variation. *Precis. Eng.* **1980**, *2*, 207–215. [CrossRef]

43. Zuo, Y.; Wang, H.; Xiong, J. The aspect ratio of surface grooves and metastable pitting of stainless steel. *Corros. Sci.* **2002**, *44*, 25–35. [CrossRef]

44. Hong, T.; Nagumo, M. Effect of surface roughness on early stages of pitting corrosion of type 301 stainless steel. *Corros. Sci.* **1997**, *39*, 1665–1672. [CrossRef]

45. Baboian, R. *Corrosion Tests and Standards: Application and Interpretation*; ASTM International: West Conshohocken, PA, USA, 2005.

46. Dougherty, R.C. Temperature and pressure dependence of hydrogen bond strength: A perturbation molecular orbital approach. *J. Chem. Phys.* **1998**, *109*, 7372–7378. [CrossRef]

47. Kauffman, G.B. The Bronsted-Lowry acid base concept. *J. Chem. Edu.* **1988**, *65*, 28. [CrossRef]

48. Gregorčič, P.; Šetina-Batič, B.; Hočevar, M. Controlling the stainless steel surface wettability by nanosecond direct laser texturing at high fluences. *Appl. Phys. A: Mater. Sci. Process.* **2017**, *123*, 1–8. [CrossRef]

49. Cassie, A.B.D.; Baxter, S. Wettability of porous surfaces, Trans. *Faraday Soc.* **1944**, *40*, 546–551. [CrossRef]

50. Wenzel, R.N. Resistance of solid surfaces to wetting by water. *Ind. Eng. Chem.* **1936**, *28*, 988–994. [CrossRef]

51. Steele, A.; Bayer, I.; Moran, S.; Cannon, A.; King, W.P.; Loth, E. Conformal ZnO nanocomposite coatings on micro-patterned surfaces for superhydrophobicity. *Thin Solid Films* **2010**, *518*, 5426–5431. [CrossRef]

52. Schwierz, N.; Horinek, D.; Sivan, U.; Netz, R.R. Reversed Hofmeister series—The rule rather than the exception. *Curr. Opin. Colloid Interface Sci.* **2016**, *23*, 10–18. [CrossRef]

53. Mantel, M.; Wightman, J.P. Influence of the surface chemistry on the wettability of stainless steel. *Surf. Interface Anal.* **1994**, *21*, 595–605. [CrossRef]

54. Rupp, F.; Gittens, R.A.; Scheideler, L.; Marmur, A.; Boyan, B.D.; Schwartz, Z.; Geis-Gerstorfer, J. A review on the wettability of dental implant surfaces I: Theoretical and experimental aspects. *Acta Biomater.* **2014**, *10*, 2894–2906. [CrossRef]

Article

Effect of Plasma Nitriding Pretreatment on the Mechanical Properties of AlCrSiN-Coated Tool Steels

Yin-Yu Chang * and Siddhant Amrutwar

Department of Mechanical and Computer-Aided Engineering, National Formosa University, Yunlin 632, Taiwan; siddhant.amrutwar@gmail.com
* Correspondence: yinyu@nfu.edu.tw; Tel.: +886-5-631-5332

Received: 19 February 2019; Accepted: 4 March 2019; Published: 7 March 2019

Abstract: Surface modification of steel has been reported to improve hardness and other mechanical properties, such as increase in resistance, for reducing plastic deformation, fatigue, and wear. Duplex surface treatment, such as a combination of plasma nitriding and physical vapor deposition, achieves superior mechanical properties and resistance to wear. In this study, the plasma nitriding process was conducted prior to the deposition of hard coatings on the SKH9 substrate. This process was done by a proper mixture of nitrogen/hydrogen gas at suitable duty cycle, pressure, and voltage with proper temperature. Later on, the deposition of gradient AlCrSiN coatings synthesized by a cathodic-arc deposition process was performed. During the deposition of AlCrSiN, CrN, AlCrN/CrN, and AlCrSiN/AlCrN were deposited as gradient interlayers to improve adhesion between the coatings and nitrided steels. A repetitive impact test (200k–400k times) was performed at room temperature and at high temperature (~500 °C) to assess impact resistance. The results showed that the tribological impact resistance for the synthesized AlCrSiN increased because of a progressive hardness support. The combination of plasma nitriding and AlCrSiN hard coatings is capable of increasing the life of molding dies and metal forging dies in mass production.

Keywords: duplex treatment; plasma nitriding; hard coating; tool steel

---

1. Introduction

For warm stamping of stainless steels and light alloys, the molds are operated at high temperature—thermal and load cycling—and often lose effectiveness early because of oxidation and wear. The plasma nitriding process (PN) is a well-known technique, which can be utilized to improve the surface properties of ferrous alloys [1,2]. Similar to PN, physical vapor deposition (PVD) techniques make coatings useful for tool steels with good tribological performance due to high hardness, anti-abrasion, and high-temperature oxidation resistance. However, their mechanical performance is restricted when they are deposited on a soft substrate due to plastic deformation under high loads without adequate support. Duplex treatments including PN and PVD are verified to be successful in creating wear and corrosion resistance with good mechanical support. The adhesion strength between the coating and the tool steel can be controlled by nitriding techniques and gradient coating designs to tailor the nitrided layer and by inserting a surface treatment method, such as polishing, before the deposition of hard coatings [3–6].

Currently, PVD hard coatings such as CrN, TiN, CrAlN, and TiAlN are used to improve the properties of tool steels and extend the lifetime of mechanical components and tools [7–11]. The introduction of silicon into CrAlN could decrease the grain size, change the coating structure, composition, and its mechanical properties [12–14]. In this CrAlSiN, silicon is apt to segregate into amorphous SiN_x along the grain boundaries. Many nanocomposite and nanolayered coatings have been studied due to their excellent mechanical and tribological properties. The coating structure and chemical content of the film can influence the mechanical and tribological properties of multicomponent coatings in the form of gradient and multilayered layers. Previous studies have shown that multilayered and graded coatings, such as AlCrSiN/TiVN and TiVN/TiSiN, are very effective in reducing wear in molding, forming, and machining applications [15–17]. High-speed steels and hot-working steels are the universal stamping, cutting, and machining tool steels because of their good strength at high temperature and are usually used for metal forming, cutting tools, and light alloy injection. However, these tool steels have poor resistance to tribological wear with a high coefficient of friction. The demand has been increased using surface modification and coating techniques to improve the mechanical and wear performances of the tool steel. Thermal-and-chemical surface modification such as PN is applied to these steels for prolonging the lifetime of tools [18]. This study tried to develop plasma nitriding before the AlCrSiN hard coating deposition to improve the mechanical properties of tool steel substrates and the adhesion strength. An impact fatigue test was conducted at room temperature and at high temperature. The purpose was to study the microstructure, mechanical properties, and periodic impact resistance of the high-speed steels with duplex treatment combing PVD AlCrSiN coatings and PN.

2. Materials and Methods

The heat-treated SKH9 tool steels (25 mm in diameter and 4 mm in thickness) having hardness of 60 HRC (Rockwell Hardness scale C) were used as substrates. A nominal chemical composition of the SKH9 steel in wt.% was as follows: 0.8–0.9 wt.% C, 3.8–4.6 wt.% Cr, 5.5–7 wt.% W, 0.45 wt.% Mn, 1.5–2.2 wt.% V, 4.5–5.5 wt.% Mo, 4.3–5.2 wt.% Co, and Fe as balance. The samples were metalographically prepared using a grinding machine (SiC emery paper up to 1500-grit size, Jet measurement Co., Taichung, Taiwan) and a polishing machine (suspension with Al_2O_3 particle size 0.05 μm, Jet measurement Co., Taichung, Taiwan) to have a surface roughness of Ra = 0.2 ± 0.05 μm. The tool steel samples were cleaned using ultrasonic agitation in an alcoholic solution for 15 min. After that, the samples were dried by hot air. The PN treatment was conducted using an industrial ion and plasma nitriding system, as shown in Figure 1a. The reactor was equipped with a rotary pump to obtain low vacuum before the PN process. Ar was introduced and the samples were sputtered prior to plasma nitriding to remove the native oxide layer and keep clean. Afterwards, nitrogen/hydrogen mixture was introduced to form a nitride layer on samples. The detailed processing parameters are shown in Table 1 [19,20]. After PN treatment, the tool steel samples were cooled down in the furnace to the room temperature under vacuum conditions. The nitriding process usually raises the surface roughness. In this study, the as-nitrided tool steels had a surface roughness of Ra = ~0.5 μm, and the subsequent polishing after PN decreased the surface roughness effectively. The nitrided samples were polished prior to PVD to have Ra = 0.2 ± 0.05 μm. The PVD technique followed by cathodic-arc evaporation used different types of targets: AlCrSi alloy, AlCr alloy, and pure chromium (Cr), as shown in Figure 1b. Table 2 shows the deposition parameters of AlCrSiN. The deposition process was initially Ar (99.999% pure) ion bombardment followed by the deposition of a Cr adhesion layer. The introduction of nitrogen into the chamber then led to the formation of CrN on top of the Cr layer. CrN was used as the bottom interlayer, and then CrN/AlCrSiN transition layers were deposited. The contents of Al and Si increased to form an AlCrSiN top layer. The temperature of the PN treatment and PVD was measured near the specimen holder using a thermocouple whose thermal junction was insulated from the holder with a quartz cup.

A field emission scanning electron microscope (FESEM, JSM IT-100, JEOL Ltd., Tokyo, Japan), which was equipped with an energy-dispersive X-ray spectroscopy (EDS) system, was used to investigate the coating morphology and microstructure. The texture and phase identification of the AlCrSiN with and without PN-treated samples were examined by X-ray diffraction MXP III MAC Science (Bruker's) with Cu radiation. The XRD system was operated at 40 kV, using a low glancing incidence angle of 4° to identify the coating structure. The X-ray beam incidence angle was fixed while the XRD detector was moved in the 2θ between 20° and 90° to obtain the diffraction pattern of the samples.

The adhesion strength of the PN and PN + AlCrSiN-coated samples was evaluated by Rockwell C indentation (HR-200, Mitutoyo Co., Japan) according to the ISO 26443 standard. The hardness of the films and PN-treated substrates was measured by the Vickers tester (MMT-X, Matsuzawa Co., Akita, Japan) with a load of 25 g and measured by an average value from seven measured points. In this study, a Calotest equipment was used to determine the thickness of the hard coatings. The coated samples were placed under an optical microscope to show a circular crater shape and then measure the coating thickness [21].

To study the impact fatigue resistance of the deposited AlCrSiN coatings with and without PN-pretreated samples, a cyclic impact test was conducted using a periodic loading device at room temperature and at a high temperature of 500 °C [22]. The distance between the sample and indenter and the impact frequency were set at 1 mm and 20 Hz, respectively. Continuous impacts of a reciprocating WC–Co ball indenter were applied onto the deposited AlCrSiN with and without PN. The test was performed without lubrication under a load of 9.8 N.

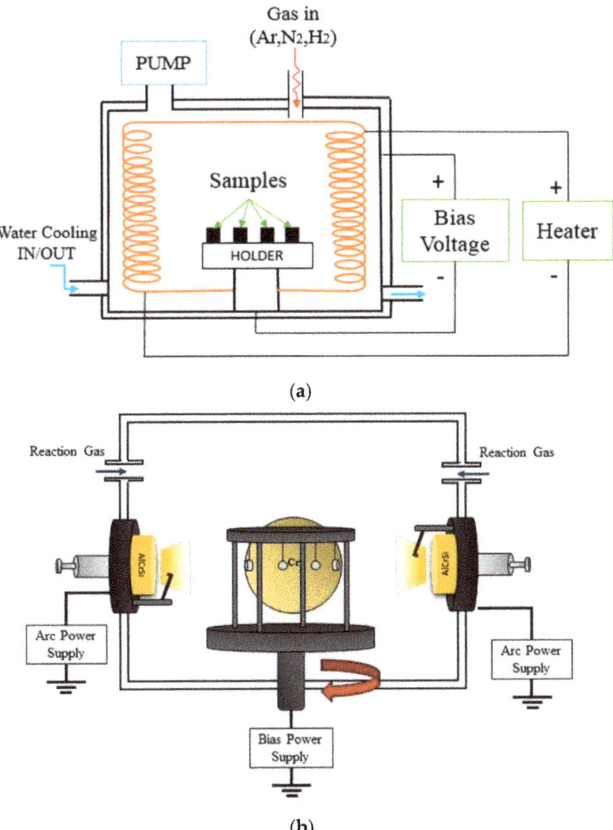

Figure 1. (**a**) Schematic diagram of plasma nitriding setup and (**b**) schematic diagram of physical vapor deposition (PVD) cathodic-arc evaporation system.

Table 1. Parameters for plasma nitriding.

Parameters	
Base pressure	<1 Pa
Ar ion cleaning	Bias voltage = −700 V, Pressure = 2 Pa, Time = 30 min
Nitriding	T = 450 °C, Time = 7 h, Bias voltage = −500 V, N_2/H_2 = 80/20 sccm
Duty Cycle	82%
Pressure	250 Pa

Table 2. Parameters for AlCrSiN thin-film deposition.

Parameters	
Target	Cr, $Al_{0.7}Cr_{0.3}$, $Al_{0.6}Cr_{0.3}Si_{0.1}$
Base pressure	<0.004 Pa
Ar ion cleaning	Bias voltage = −700 V, Pressure = 2 Pa, Time = 20 min
Reactive gas	Ar/N_2
Working pressure	3 Pa
Deposition temperature	250 °C
Target current	70 A
Bias voltage	−120 V
Rotation speed	2 rpm
Deposition time	90 min

3. Results and Discussion

3.1. Microstructure Characterization

The nitrided layer is usually composed of the compound layer and the diffusion layer. The compound zone mainly consists of porous ε-nitrides, whereas the diffusion layer is nitric martensite [23,24]. Duplex treatments combine PN with the subsequent deposition of hard coatings to create good tribological properties. In this study, samples of plasma nitrided SKH9 steel, AlCrSiN-coated sample, and PN + AlCrSiN (with PN pretreatment) coated SKH9 steel were tested for research. Figure 2 shows typical X-ray diffraction (XRD) patterns of plasma nitrided (PN), AlCrSiN-coated, and PN + AlCrSiN-coated samples. After plasma nitriding, it can be seen that the XRD patterns of the nitrided compound layer consists of diffraction peaks of CrN, Fe_xN, where x = 2–3, and Fe_4N, as shown in Figure 2a. The substrate α'-martensite peak at 82.4° was detected in the PN-treated layer, and it suggested that the nitrided layer was thin and less than the X-ray penetration depth, and resulted in observing the substrate α'-martensite phase. The main nitride precipitates that were identified were $Fe_{2-3}N$ and Fe_4N. These precipitates play the main role in increasing the surface hardness and improving the wear resistance, thereby, increasing the time of fatigue fracture [25]. For the AlCrSiN, when N_2 was introduced during the coating process, the Al, Cr, and Si plasma generated by arc evaporators induced the excitation of N_2 to form AlCrSiN on the substrate. Figure 2b shows the XRD pattern of the AlCrSiN without plasma nitriding treatment. Regardless of the low Si content (less than 10 at.%), the sequence of crystalline phases was observed, and no corresponding peaks of crystalline silicon nitride were found. The crystallography structure of single-phase face-centered cubic (FCC) AlCrN solid solution was detected. The four main peaks correspond to reflections (111), (200), (220), and (311) consisting of nanocrystalline FCC-AlCrN [26–28]. Figure 2c shows the diffraction pattern of PN + AlCrSiN consisting of AlCrN and Fe_4N phases. The Fe_4N compound layer was obtained during the nitriding process. The compound layer led to higher hardness and improved the adhesion quality of coatings. Similar to the previous AlCrSiN without plasma nitriding, the AlCr(Si)N phases exhibited cubic B1-NaCl structure. The lattice structure of the deposited coatings showed solid solution in these coatings. The lattice parameter of the AlCrSiN was 0.424 nm, which was between the values of cubic AlN (JCPDF file No.: #251495) and CrN (JCPDF file No.: #110065). The microstructure was obtained by the formation of an FCC phase, obtained by the substitution of Al and Si in the cubic

CrN phase. The Si-containing AlCrSiN coatings had better oxidation and tribological resistances than CrN and CrAlN and possessed good mechanical properties [14,29,30].

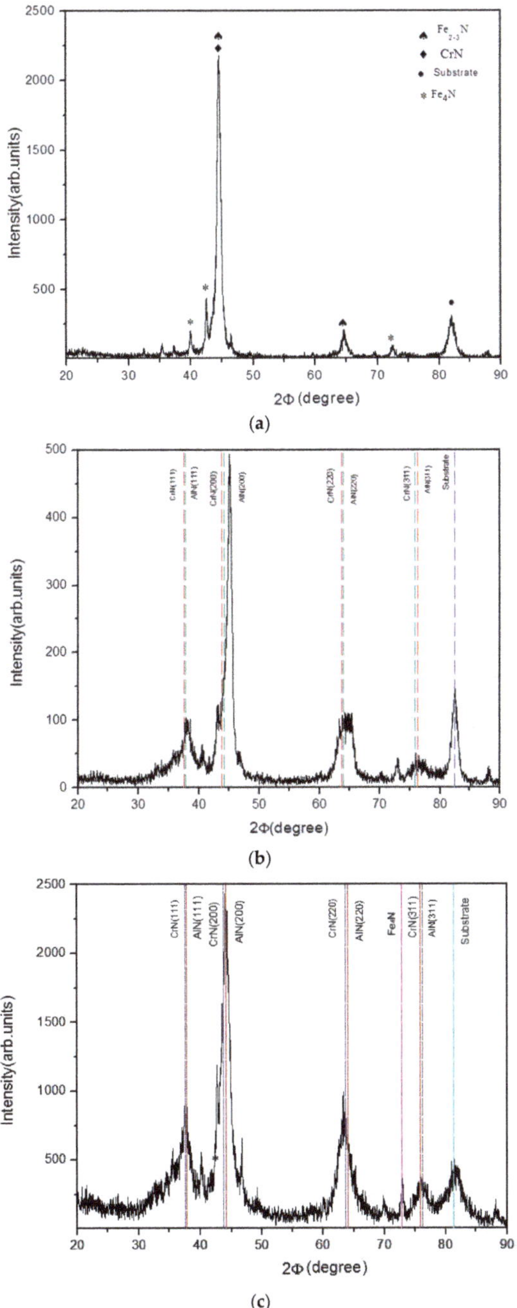

Figure 2. XRD pattern of (**a**) plasma nitriding (PN), (**b**) AlCrSiN, and (**c**) PN + AlCrSiN samples.

In order to study the microstructure of coating construction, Figure 3 shows the fractured cross-sectional SEM image of the AlCrSiN coating on an Si (100) wafer. Deposited at a high bias voltage of −120 V, the top AlCrSiN layer showed fine dense structures. CrN and AlCrN/CrN bottom interlayers showed obvious columnar structure. After AlCrSiN was co-deposited with AlCrN, the formation of the columnar structure was reduced. The deposited AlCrSiN with a transition layer of AlCrSiN/AlCrN possessed a much denser structure because of the presence of Al and Si to form compact nanocrystalline structures. Similar results by He et al. [31] also showed that the single-phase cubic CrAlN coatings exhibited a columnar structure, while dense and featureless characteristic morphologies for AlCrSiN coatings were observed. The total thickness of the graded AlCrSiN coating was 2.24 µm, which was similar to the result of Calotest (~2 µm), and consisted of the top AlCrSiN layer, transition layer of AlCrSiN/AlCrN, and columnar AlCrN/CrN interlayers. From the EDS measurement, the chemical composition of the deposited AlCrSiN top layer coating was 26.7 at.% Al, 17.6 at.% Cr, 4.2 at.% Si, and 51.5 at.% N. It showed atomic stoichiometries of $Al_{0.55}Cr_{0.37}Si_{0.08}N$ for the deposited AlCrSiN. Distinct interfaces were observed at the boundaries between the transition layer (AlCrSiN/AlCrN) and the columnar AlCrN/CrN interlayers. In this AlCrSiN, the addition of Si impeded grain growth and led to grain re-nucleation, and. Therefore. resulted in a compact and dense structure. The retardation of columnar growth by the incorporation of Si in the CrAlN coating was proved, and previous studies demonstrated that the overall oxidation resistance of the AlCrSiN coatings after Si doping was significantly improved at high temperature [32–34].

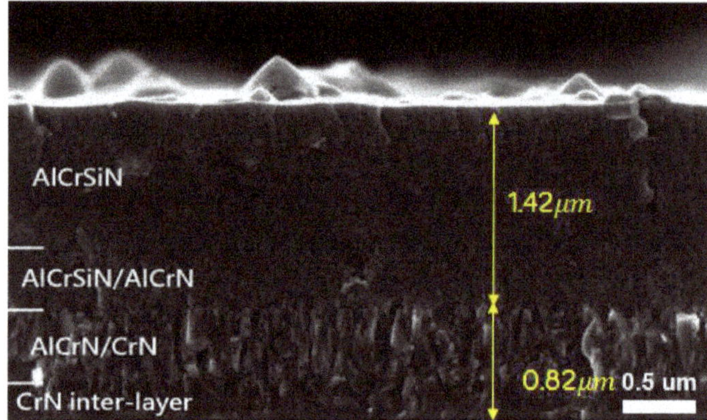

Figure 3. Cross-sectional SEM image of AlCrSiN coating.

3.2. Mechanical Properties

3.2.1. Hardness Test

High surface hardness of the tool steels can increase capacities of load support and high abrasion wear resistance. In this study, the measurement was performed using a Vickers microhardness tester. The applied load on the specimen was 25 gf. Figure 4 shows the hardness of SKH9 high-speed steel substrate, PN-treated, AlCrSiN-coated, and PN + AlCrSiN-coated samples under different processes. The hardness for PN samples was 1352 HV_{25g}, which was higher than that of SKH9 high-speed steel. From the results, the PN + AlCrSiN possessed hardness of about 3256 HV_{25g}, which was the highest hardness among all the samples, and the AlCrSiN-coated samples (without PN pretreatment) possessed lower hardness, i.e., 2285 HV_{25g}, than PN + AlCrSiN. The hardness of the PN + AlCrSiN-coated samples was 2.4 times higher than that of the PN-treated SKH9. In this case, it revealed that the AlCrSiN-coated sample with PN pretreatment could achieve the highest hardness among the samples. The duplex treatment combining PN and hard coating exhibited higher hardness as compared to non-duplex-treated samples. The nitrided layer was helpful for the progressive hardness distribution, inducing the gradual transition of mechanical properties improving the wear resistance [35,36].

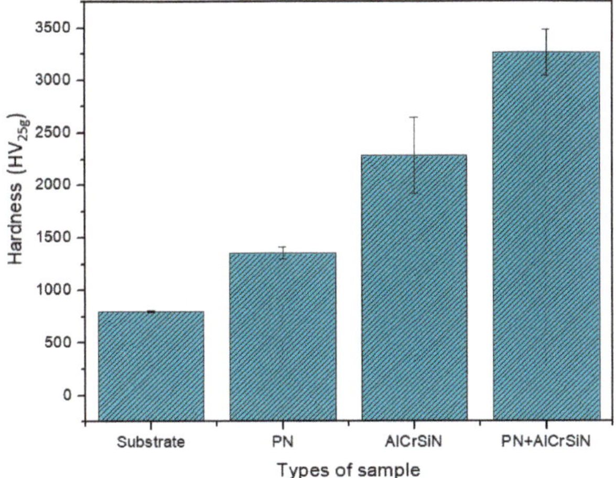

Figure 4. Vickers microhardnesses of the SKH9 high-speed steel substrate, PN-treated, AlCrSiN-coated, and PN + AlCrSiN-coated samples.

3.2.2. Adhesion Test

The adhesion strength of all the PN, AlCrSiN-, and PN + AlCrSiN-coated samples was investigated using the Rockwell indentation test. This destructive test, may exhibit two distinctive properties of the coated compound [37,38]. We used Rockwell indentation with a load of 150 kg based on the ISO 26443 standard to evaluate the adhesion strength of the coatings. According to the criteria, class 0 reveals acceptable adhesion. Class 1 shows no adhesive delamination; adhesion is acceptable. In the cases of class 2 and class 3, adhesion is unacceptable [39]. Adhesive delamination is defined as a removal of the coating, whereby the underlying substrate can be clearly seen, or a removal of one or more sublayers in a multilayered or graded coating, whereby the substrate or an underlying sublayer can be clearly distinguished. Figure 5 shows optical images to reveal the crack behavior of the AlCrSiN- and PN + AlCrSiN-coated samples. The optical image with high magnification shows the crack condition on the edge of the indented crater. The AlCrSiN coating (without PN) showed that, initially, circumferential cone cracks developed beneath the indenter, followed by radial and annular cracks as well as some coating spallation, which appeared at the edge of the indenter crater. The poor adhesion of class 2 is recognized on contrast. The coating was too stiff to buckle, which created the formation of compressive stress cracks at the wear interface of the coating and the specimen. These compressive stress-induced cracks propagated, and the formation of spallation in coatings occurred. This kind of adhesion failure was usually observed in brittle ceramic coatings with weak interfacial adhesion to the specimen. The plastic deformation easily took place due to its low hardness, resulting in the cracks formed in the indentation edge. If the interfacial adhesion of coatings was large in area, lesser wedge spallation occurred on the substrate coatings.

A significant change was observed in the PN + AlCrSiN-coated samples. In case of this study, PN + AlCrSiN coatings showed no cohesive failure and no coating delamination, which meant it had no cracks and good adhesion properties of class 0 due to a hard nitrided layer before the deposition of AlCrSiN coating on the substrate. The hardness was increased by the nitrided layer, which provided enough support for the hard coatings even under heavy loads. This fact was also confirmed by He and Deng et al. [35,40]. Surface hardening by plasma nitriding improved the heavier load bearing capacity for AlCrSiN-coated specimens. The nitrided layer provided good adhesion property and was useful in sustaining a high load. It may be helpful for the improvement of the adhesion on PN + AlCrSiN SKH9 substrate in this study.

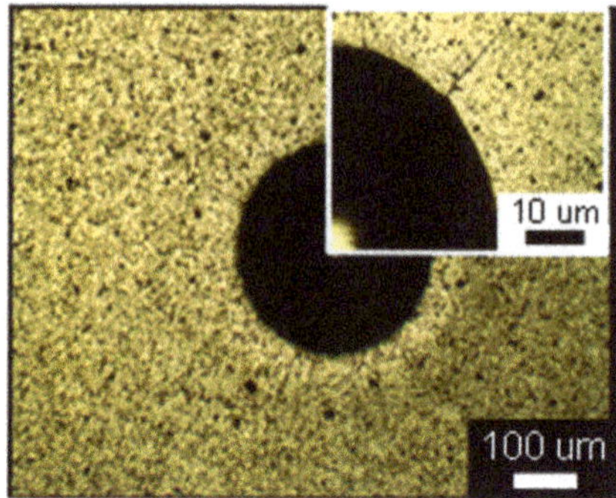

Figure 5. Optical images showing crack behavior evaluated by Rockwell indentation test applied on the AlCrSiN (**Left**) and PN + AlCrSiN (**Right**) coated samples.

3.2.3. Impact Fatigue Test

The impact tester was designed to simulate fatigue wear in tribological coating systems subjected to a dynamic stress on the coated samples [41]. In this study, we explored the anti-fatigue properties and impact resistance of the coated samples using a room temperature fatigue test and a high-temperature fatigue test (500 °C). The number of impacts was 200,000 (200k), 300,000 (300k), and 400,000 (400k). It was tested using a dynamic impact fatigue tester for the AlCrSiN- (without PN pretreatment) and the PN + AlCrSiN-coated samples. During the test, a tungsten carbide ball with a diameter of 2 mm was used as a punch. The frequency was controlled at 20 Hz and a load of 9.8 N was applied. The film was subjected to periodic reciprocating impact tests, and then the film was observed for damage, sticky and stacked conditions, and bare substrate. Referring to the previous study by Batista et al. [42], the damage of the film after the impact test was studied. During the process of the impact fatigue test, deformation of the surface takes place step by step depending upon the impacts and load. The failure zones can be divided into three types: Central zone with cohesive failure, intermediate zone with adhesion and cohesive failure, and lastly peripheral failure, which usually takes place at the boundary of the indentation area. Figure 6 shows the SEM image and EDS element mapping of the AlCrSiN film at room temperature after 200,000 (200k), 300,000 (300k), and 400,000 (400k) impacts. EDS mapping revealed that a composition signal with various compositions such as aluminum (Al), chromium (Cr), silicon (Si), tungsten (W), iron (Fe), and nitrogen (N) signals was obtained. After continuous cycles of the loading and unloading process, the surface morphology was observed through element mapping. The AlCrSiN-coated samples showed smooth morphology, where no significant transfer of the WC–Co ball material was observed after 200k impacts. Chipping outside of the impacted crater was usually the main failure for ceramic hard coatings in the impact fatigue test [22]. However, in this study there were no significant changes for the wear morphology with increasing impacts from 200k to 400k cycles. But in case of impact cavity size under a continuous process of loading–unloading at room temperature and the normal load of 9.8 N after 300k impacts, formation of the cohesive and adhesive failures of the AlCrSiN film in the intermediate and central zones of the cavity was observed. No iron (Fe) signal was observed on the surface. This indicated the substrate was not exposed at this stage. The results showed that the AlCrSiN-coated samples can sustain the load of around 400k impacts at room temperature. This phenomenon indicated that, after 400k continuous impacts at room temperature, the AlCrSiN-coated sample had possible formation of cohesive failure, which might later effect the hardness and cause failures in the mechanical performance of dies.

Figure 7 reveals the results of the PN + AlCrSiN-coated sample at the same impacts of 200k, 300k, and 400k at room temperature showing the surface is not yet exposed at this stage. No obvious substrate iron (Fe) signal can be found. The benefit of the duplex treatment in terms of improving the load bearing capacity of hard coatings originated from its higher critical load than those from non-duplex-treated parts. The change in crater volume with an increase in the number of impact cycles for the PN + AlCrSiN duplex-treated sample had no significant change observed. As the impact cycles increased from 200k to 400k there was an absence of rapid increases in wear as compared to the non-duplex AlCrSiN sample. This might be the reason for the increase in load of coatings and minimizing substrate deformation, which reduced the bending and stretching of the coatings. A similar result of impact wear resistance of duplex PN-treated/PVD-coated Ti–6Al–4V alloy was revealed by Cassar et al. [43].

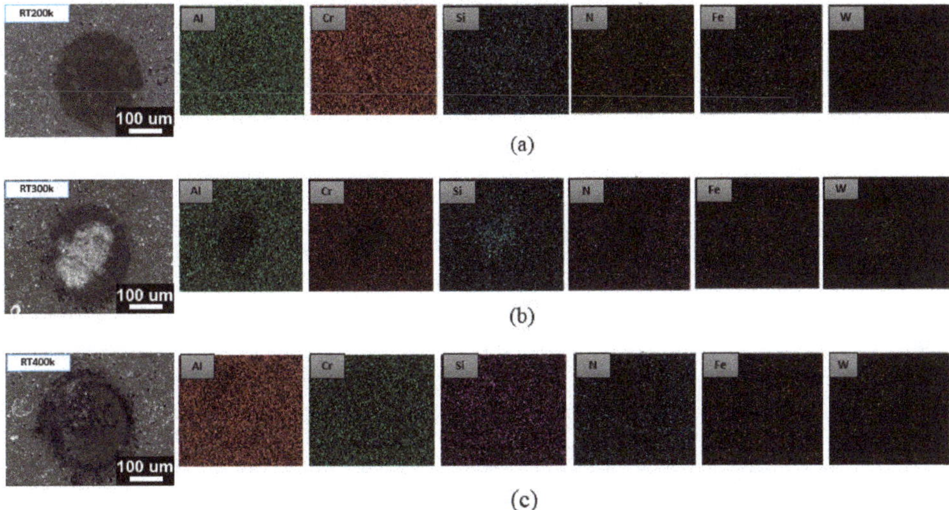

Figure 6. SEM and energy-dispersive X-ray spectroscopy (EDS) element mapping of AlCrSiN-coated samples at room temperature after (**a**) 200k, (**b**) 300k, and (**c**) 400k impacts.

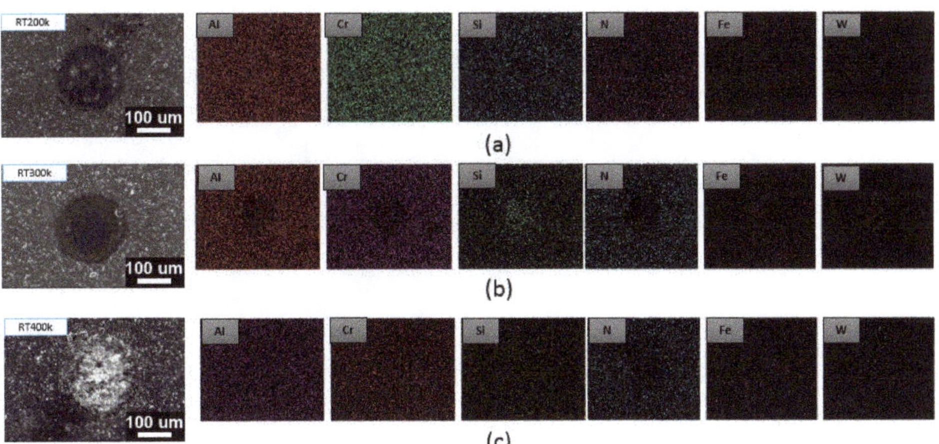

Figure 7. SEM and EDS element mapping of PN + AlCrSiN at room temperature after (**a**) 200k, (**b**) 300k, and (**c**) 400k impacts.

The high-temperature fatigue impact test at 500 °C was performed on the AlCrSiN-coated high-speed SKH9 steel. The SEM image with EDS element mapping image is shown in Figure 8. The elemental mappings of aluminum (Al), chromium (Cr), silicon (Si), tungsten (W), oxygen (O), iron (Fe), and nitrogen (N) signals are shown. It showed that the surface was exposed of iron signal from the substrate of the AlCrSiN-coated SKH9 steel. Therefore the substrate was exposed to 200k impacts at the high temperature of 500 °C. On the other hand, a high-temperature impact of PN + AlCrSiN after 200k impacts was conducted. As shown in Figure 9, no iron (Fe) signal was present on the impacted surface. The duplex-treated sample (PN + AlCrSiN) had the presence of an Fe–N compound and a diffusion layer beneath the hard AlCrSiN coating. These two layers supported the top AlCrSiN layer from heavy mechanical loads at high temperature, which also improved the hardness and adhesion strength to the tool steel. The surface modification of SKH9 steels by a combination of plasma nitriding and coatings resulted in excellent mechanical load bearing capacity and strong adhesion at high temperature. Compared to the non-duplex-treated AlCrSiN sample that only resists 200k impacts at high temperature, duplex-treated AlCrSiN coatings had better performance of periodic impact fatigue resistance at high temperature.

Previous studies showed that AlCrSiN possessed good thermal stability and oxidation resistance because of the formation of the epitaxial growth structure, which inhibited atomic diffusion. The nanocomposite coatings synthesized by amorphous silicon nitride could provide a stable phase and limited grain growth in nanocrystal AlCrN at high temperature [44–46]. However, in this study, the non-duplex-treated AlCrSiN sample only resisted 200k impacts at 500 °C. The high-temperature fatigue impact test for the PN + AlCrSiN further was carried out by continuous impacts from 300k to 400k. No obvious iron signal was detected after 300k impacts. As shown in Figure 10, after 400k impacts, it was found that the substrate failed and the iron substrate was exposed. It is concluded that the dense PN + AlCrSiN coating with good thermal stability can resist around 400k impacts at a high temperature of 500 °C. The results showed the PN + AlCrSiN-coated tool steels exhibit an excellent result in both room and high-temperature fatigue impacts.

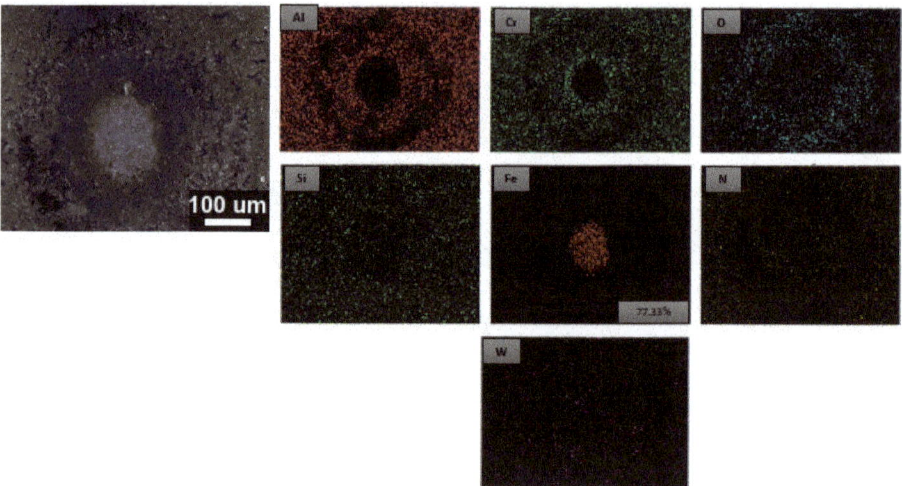

Figure 8. SEM and EDS element mapping of AlCrSiN at high temperature (500 °C) after 200k impacts.

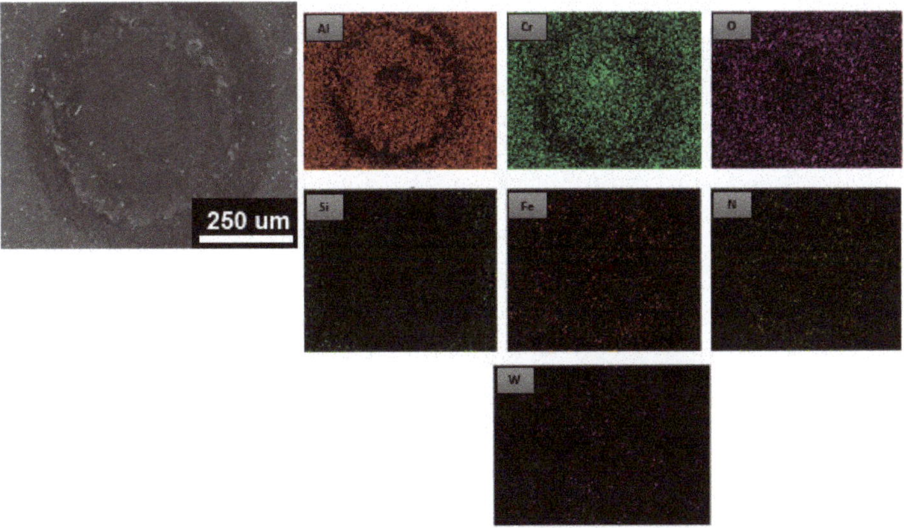

Figure 9. SEM and EDS element mapping of PN + AlCrSiN at high temperature (500 °C) after 200k impacts.

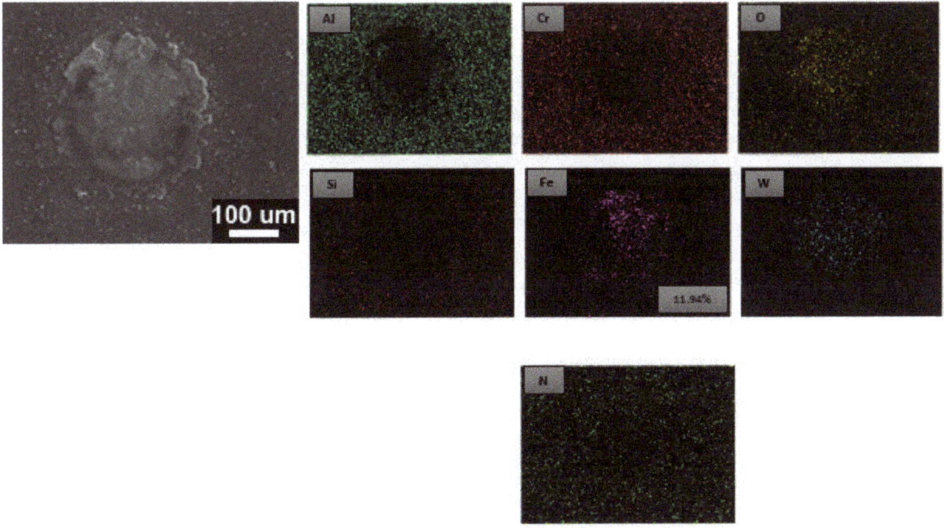

Figure 10. SEM and EDS element mapping of PN + AlCrSiN at high temperature (500 °C) after 400k impacts.

4. Conclusions

In this study, the AlCrSiN hard coatings were deposited on the plasma nitrided SKH9 steel using a duplex treatment combing plasma nitriding and cathodic-arc evaporation PVD technology. The deposited AlCrSiN had bottom interlayers of CrN and AlCrN and a transition layer (AlCrSiN/AlCrN). From SEM and XRD evaluation, it was observed that the AlCrSiN coating with gradient layer structure was a dense structure of nanocrystalline B1-NaCl, and it showed that the coating had obvious columnar bottom layers of AlCrN and CrN structure. AlCrSiN-coated samples (without PN pretreatment) possessed hardness of 2285 HV_{25g}. The PN + AlCrSiN had the highest hardness of 3256 HV_{25g} among all the samples. This significant change occurred due to the support of the modified layer formed as a result of the nitriding process and the reduction of plastic deformation under the applied load.

Impact fatigue tests for AlCrSiN- and PN + AlCrSiN-coated samples at room temperature and at high temperature (500 °C) were conducted. Both the AlCrSiN and duplex-treated samples possessed good impact fatigue performance at room temperature. The PN + AlCrSiN duplex-treated sample could resist 400k impacts at a high temperature of 500 °C, where only small area of iron signal from the substrate was exposed and no cohesive failure was observed. The PN + AlCrSiN duplex-treated tool steel had high hardness, and it possessed good impact fatigue performance at room temperature and at high temperature.

Author Contributions: Conceptualization: Y.-Y.C., S.A.; Methodology: Y.-Y.C., S.A.; Validation: Y.-Y.C.; Formal Analysis: Y.-Y.C., S.A.; Investigation: Y.-Y.C., S.A.; Resources: Y.-Y.C.; Data Curation: Y.-Y.C., S.A.; Writing—Original Draft Preparation: Y.-Y.C., S.A.; Writing—Review and Editing: Y.-Y.C.; Visualization: Y.-Y.C.; Supervision: Y.-Y.C.; Project Administration: Y.-Y.C.; Funding Acquisition: Y.-Y.C.

Funding: This research was supported by the Ministry of Science and Technology (MOST 107-2218-E-131 -001 and MOST 107-2221-E-150-003) of Taiwan.

Acknowledgments: The instrumental assistance from the Common Lab. for Micro/Nano Sci. and Tech. of National Formosa University is sincerely appreciated.

Conflicts of Interest: The authors declare no conflict of interest. The founding sponsors had no role in the design of the study; in the collection, analyses, or interpretation of data; in the writing of the manuscript; and in the decision to publish the results.

References

1. Höck, K.; Spies, H.J.; Larisch, B.; Leonhardt, G.; Buecken, B. Wear resistance of prenitrided hardcoated steels for tools and machine components. *Surf. Coat. Technol.* **1997**, *88*, 44–49. [CrossRef]

2. Paosawatyanyong, B.; Pongsopa, J.; Visuttipitukul, P.; Bhanthumnavin, W. Nitriding of tool steel using dual DC/RFICP plasma process. *Surf. Coat. Technol.* **2016**, *306*, 351–357. [CrossRef]

3. Das, K.; Alphonsa, J.; Ghosh, M.; Ghanshyam, J.; Rane, R.; Mukherjee, S. Influence of pretreatment on surface behavior of duplex plasma treated AISI H13 tool steel. *Surf. Interfaces* **2017**, *8*, 206–213. [CrossRef]

4. Škorić, B.; Kakaš, D.; Rakita, M.; Bibić, N.; Peruškob, D. Structure, hardness and adhesion of thin coatings deposited by PVD, IBAD on nitrided steels. *Vacuum* **2004**, *76*, 169–172. [CrossRef]

5. Hawryluk, M.; Gronostajski, Z.; Widomski, P.; Kaszuba, M.; Ziemba, J.; Smolik, J. Influence of the application of a PN+Cr/CrN hybrid layer on the improvement of the lifetime of hot forging tools. *J. Mater. Process. Technol.* **2018**, *258*, 226–238. [CrossRef]

6. Fontes, M.A.; Pereira, R.G.; Fernandes, F.A.P.; Casteletti, L.C.; de Paula Nascente, P.A. Characterization of plasma nitrided layers produced on sintered iron. *J. Mater. Res. Technol.* **2014**, *3*, 210–216. [CrossRef]

7. Navinšek, B.; Panjan, P.; Milošev, I. Industrial applications of CrN (PVD) coatings, deposited at high and low temperatures. *Surf. Coat. Technol.* **1997**, *97*, 182–191. [CrossRef]

8. Lee, D.B.; Kim, M.H.; Lee, Y.C.; Kwon, S.C. High temperature oxidation of TiCrN coatings deposited on a steel substrate by ion plating. *Surf. Coat. Technol.* **2001**, *141*, 232–239. [CrossRef]

9. Falsafein, M.; Ashrafizadeh, F.; Kheirandish, A. Influence of thickness on adhesion of nanostructured multilayer CrN/CrAlN coatings to stainless steel substrate. *Surf. Interfaces* **2018**, *13*, 178–185. [CrossRef]

10. Liu, W.; Chu, Q.; Zeng, J.; He, R.; Wu, H.; Wu, Z.; Wu, S. PVD-CrAlN and TiAlN coated Si3N4 ceramic cutting tools—1. Microstructure, turning performance and wear mechanism. *Ceram. Int.* **2017**, *43*, 8999–9004. [CrossRef]

11. Zhang, K.; Deng, J.; Guo, X.; Sun, L.; Lei, S. Study on the adhesion and tribological behavior of PVD TiAlN coatings with a multi-scale textured substrate surface. *Int. J. Refract. Met. Hard Mater.* **2018**, *72*, 292–305. [CrossRef]

12. Haršáni, M.; Sahul, M.; Zacková, P.; Čaplovič, Ľ. Study of cathode current effect on the properties of CrAlSiN coatings prepared by LARC. *Vacuum* **2017**, *139*, 1–8. [CrossRef]

13. Chang, C.-C.; Duh, J.-G. Duplex coating technique to improve the adhesion and tribological properties of CrAlSiN nanocomposite coating. *Surf. Coat. Technol.* **2017**, *326*, 375–381. [CrossRef]

14. Chang, Y.-Y.; Chang, C.-P.; Wang, D.-Y.; Yang, S.-M.; Wu, W. High temperature oxidation resistance of CrAlSiN coatings synthesized by a cathodic arc deposition process. *J. Alloys Compd.* **2008**, *461*, 336–341. [CrossRef]

15. Leyland, A.; Matthews, A. On the significance of the H/E ratio in wear control: A nanocomposite coating approach to optimised tribological behaviour. *Wear* **2000**, *246*, 1–11. [CrossRef]

16. Chang, Y.-Y.; Chiu, W.-T.; Hung, J.-P. Mechanical properties and high temperature oxidation of CrAlSiN/TiVN hard coatings synthesized by cathodic arc evaporation. *Surf. Coat. Technol.* **2016**, *303*, 18–24. [CrossRef]

17. Chang, Y.-Y.; Chang, H.; Jhao, L.-J.; Chuang, C.-C. Tribological and mechanical properties of multilayered TiVN/TiSiN coatings synthesized by cathodic arc evaporation. *Surf. Coat. Technol.* **2018**, *350*, 1071–1079. [CrossRef]

18. Basso, R.L.O.; Pastore, H.O.; Schmidt, V.; Baumvol, I.J.R.; Abarca, S.A.C.; de Souza, F.S.; Spinelli, A.; Figueroa, C.A.; Giacomelli, C. Microstructure and corrosion behaviour of pulsed plasma-nitrided AISI H13 tool steel. *Corros. Sci.* **2010**, *52*, 3133–3139. [CrossRef]

19. Naeem, M.; Waqas, M.; Jan, I.; Zaka-ul-Islam, M.; Díaz-Guillén, J.C.; Rehman, N.U.; Shafiq, M.; Zakaullah, M. Influence of pulsed power supply parameters on active screen plasma nitriding. *Surf. Coat. Technol.* **2016**, *300*, 67–77. [CrossRef]

20. Naeem, M.; Iqbal, J.; Abrar, M.; Khan, K.H.; Díaz-Guillén, J.C.; Lopez-Badillo, C.M.; Shafiq, M.; Zaka-ul-Islam, M.; Zakaullah, M. The effect of argon admixing on nitriding of plain carbon steel in N2 and N2-H2 plasma. *Surf. Coat. Technol.* **2018**, *350*, 48–56. [CrossRef]

21. Rother, B.; Jehn, H.A.; Gabriel, H.M. Multilayer hard coatings by coordinated substrate rotation modes in industrial PVD deposition systems. *Surf. Coat. Technol.* **1996**, *86–87*, 207–211. [CrossRef]

22. Beake, B.D.; Isern, L.; Endrino, J.L.; Fox-Rabinovich, G.S. Micro-impact testing of AlTiN and TiAlCrN coatings. *Wear* **2019**, *418–419*, 102–110. [CrossRef]

23. Yan, M.F.; Chen, B.F.; Li, B. Microstructure and mechanical properties from an attractive combination of plasma nitriding and secondary hardening of M50 steel. *Appl. Surf. Sci.* **2018**, *455*, 1–7. [CrossRef]

24. Tschiptschin, A.P.; Varela, L.B.; Pinedo, C.E.; Li, X.Y.; Dong, H. Development and microstructure characterization of single and duplex nitriding of UNS S31803 duplex stainless steel. *Surf. Coat. Technol.* **2017**, *327*, 83–92. [CrossRef]

25. Menthe, E.; Bulak, A.; Olfe, J.; Zimmermann, A.; Rie, K.T. Improvement of the mechanical properties of austenitic stainless steel after plasma nitriding. *Surf. Coat. Technol.* **2000**, *133–134*, 259–263. [CrossRef]

26. Tien, S.-K.; Lin, C.-H.; Tsai, Y.-Z.; Duh, J.-G. Effect of nitrogen flow on the properties of quaternary CrAlSiN coatings at elevated temperatures. *Surf. Coat. Technol.* **2007**, *202*, 735–739. [CrossRef]

27. Kuo, Y.-C.; Wang, C.-J.; Lee, J.-W. The microstructure and mechanical properties evaluation of CrTiAlSiN coatings: Effects of silicon content. *Thin Solid Films* **2017**, *638*, 220–229. [CrossRef]

28. Soldán, J.; Neidhardt, J.; Sartory, B.; Kaindl, R.; Čerstvý, R.; Mayrhofer, P.H.; Tessadri, R.; Polcik, P.; Lechthaler, M.; Mitterer, C. Structure–property relations of arc-evaporated Al–Cr–Si–N coatings. *Surf. Coat. Technol.* **2008**, *202*, 3555–3562. [CrossRef]

29. Chang, Y.-Y.; Lai, H.-M. Wear behavior and cutting performance of CrAlSiN and TiAlSiN hard coatings on cemented carbide cutting tools for Ti alloys. *Surf. Coat. Technol.* **2014**, *259*, 152–158. [CrossRef]

30. Sun, S.Q.; Ye, Y.W.; Wang, Y.X.; Liu, M.Q.; Liu, X.; Li, J.L.; Wang, L.P. Structure and tribological performances of CrAlSiN coatings with different Si percentages in seawater. *Tribol. Int.* **2017**, *115*, 591–599. [CrossRef]

31. He, L.; Chen, L.; Xu, Y. Interfacial structure, mechanical properties and thermal stability of CrAlSiN/CrAlN multilayer coatings. *Mater. Charact.* **2017**, *125*, 1–6. [CrossRef]

32. Wang, Y.X.; Zhang, S.; Lee, J.-W.; Lew, W.S.; Sun, D.; Li, B. Toward hard yet tough CrAlSiN coatings via compositional grading. *Surf. Coat. Technol.* **2013**, *231*, 346–352. [CrossRef]

33. Chang, C.-C.; Chen, H.-W.; Lee, J.-W.; Duh, J.-G. Influence of Si contents on tribological characteristics of CrAlSiN nanocomposite coatings. *Thin Solid Films* **2015**, *584*, 46–51. [CrossRef]

34. Chen, H.-W.; Chan, Y.-C.; Lee, J.-W.; Duh, J.-G. Oxidation behavior of Si-doped nanocomposite CrAlSiN coatings. *Surf. Coat. Technol.* **2010**, *205*, 1189–1194. [CrossRef]

35. Deng, Y.; Tan, C.; Wang, Y.; Chen, L.; Cai, P.; Kuang, T.; Lei, S.; Zhou, K. Effects of tailored nitriding layers on comprehensive properties of duplex plasma-treated AlTiN coatings. *Ceram. Int.* **2017**, *43*, 8721–8729. [CrossRef]

36. Tan, C.; Kuang, T.; Zhou, K.; Zhu, H.; Deng, Y.; Li, X.; Cai, P.; Liu, Z. Fabrication and characterization of in-situ duplex plasma-treated nanocrystalline Ti/AlTiN coatings. *Ceram. Int.* **2016**, *42*, 10793–10800. [CrossRef]
37. Tillmann, W.; Grisales, D.; Stangier, D. Effects of AISI H11 surface integrity on the residual stresses and adhesion of TiAlN/substrate compounds. *Surf. Coat. Technol.* **2019**, *357*, 466–472. [CrossRef]
38. Vidakis, N.; Antoniadis, A.; Bilalis, N. The VDI 3198 indentation test evaluation of a reliable qualitative control for layered compounds. *J. Mater. Process. Technol.* **2003**, *143–144*, 481–485. [CrossRef]
39. Haršáni, M.; Ghafoor, N.; Calamba, K.; Zacková, P.; Sahul, M.; Vopát, T.; Satrapinskyy, L.; Čaplovičová, M.; Čaplovič, Ľ. Adhesive-deformation relationships and mechanical properties of nc-AlCrN/a-SiNx hard coatings deposited at different bias voltages. *Thin Solid Films* **2018**, *650*, 11–19. [CrossRef]
40. He, Y.; Apachitei, I.; Zhou, J.; Walstock, T.; Duszczyk, J. Effect of prior plasma nitriding applied to a hot-work tool steel on the scratch-resistant properties of PACVD TiBN and TiCN coatings. *Surf. Coat. Technol.* **2006**, *201*, 2534–2539. [CrossRef]
41. Borodich, F.M. Chapter Three—The Hertz-Type and Adhesive Contact Problems for Depth-Sensing Indentation. In *Advances in Applied Mechanics*; Bordas, S.P.A., Ed.; Elsevier: New York, NY, USA, 2014; Volume 47, pp. 225–366.
42. Batista, J.C.A.; Godoy, C.; Matthews, A. Impact testing of duplex and non-duplex (Ti,Al)N and Cr–N PVD coatings. *Surf. Coat. Technol.* **2003**, *163–164*, 353–361. [CrossRef]
43. Cassar, G.; Banfield, S.; Avelar-Batista Wilson, J.C.; Housden, J.; Matthews, A.; Leyland, A. Impact wear resistance of plasma diffusion treated and duplex treated/PVD-coated Ti–6Al–4V alloy. *Surf. Coat. Technol.* **2012**, *206*, 2645–2654. [CrossRef]
44. Zhang, S.; Wang, L.; Wang, Q.; Li, M. A superhard CrAlSiN superlattice coating deposited by a multi-arc ion plating: II. Thermal stability and oxidation resistance. *Surf. Coat. Technol.* **2013**, *214*, 153–159. [CrossRef]
45. Chen, H.-W.; Chan, Y.-C.; Lee, J.-W.; Duh, J.-G. Oxidation resistance of nanocomposite CrAlSiN under long-time heat treatment. *Surf. Coat. Technol.* **2011**, *206*, 1571–1576. [CrossRef]
46. Chang, Y.-Y.; Cheng, C.-M.; Liou, Y.-Y.; Tillmann, W.; Hoffmann, F.; Sprute, T. High temperature wettability of multicomponent CrAlSiN and TiAlSiN coatings by molten glass. *Surf. Coat. Technol.* **2013**, *231*, 24–28. [CrossRef]

MDPI

St. Alban-Anlage 66

4052 Basel

Switzerland

Tel. +41 61 683 77 34

Fax +41 61 302 89 18

www.mdpi.com

Materials Editorial Office

E-mail: materials@mdpi.com

www.mdpi.com/journal/materials